TERRITORIES OF DIFFERENCE

NEW ECOLOGIES FOR THE TWENTY-FIRST CENTURY

Series Editors: Arturo Escobar, *University of North Carolina, Chapel Hill*

Dianne Rocheleau, *Clark University*

A John Hope Franklin Center Book

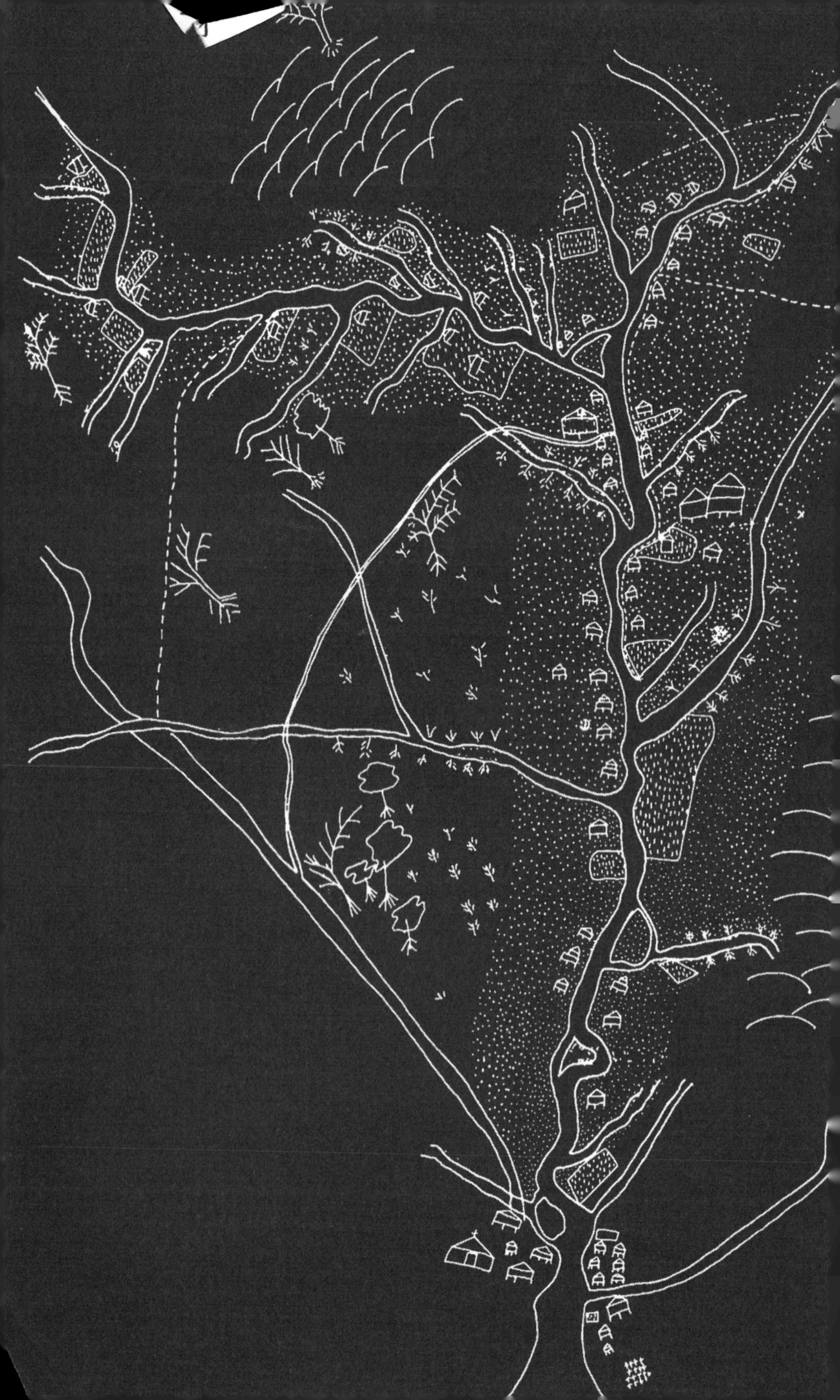

Arturo Escobar

TERRITORIES OF DIFFERENCE

place, movements, life, *redes*

DUKE UNIVERSITY PRESS Durham and London 2008

Printed in the United States of America
on acid-free paper ∞
Designed by C. H. Westmoreland
Typeset in Quadraat with Magma Compact display
by Achorn International, Inc.
Library of Congress Cataloging-in-Publication Data
appear on the last printed page of this book.

Some material in chapter 5 previously appeared
in David Nugent and Joan Vincent, eds., *A Companion to the Anthropology of Politics* (Oxford: Blackwell Publishing, 2004), 248–66, and is reprinted here
with permission of Blackwell.

Frontispiece and title page art:
page ii: based on an illustration from *Los sistemas productivos de la comunidad negra del río Valle, Bahía Solano, Chocó*, by Carlos Tapia, Rocío Polanco, and Claudia Leal, 1997.
page iii: based on an engraving produced by the Gente Entintada y Parlante project, Tumaco, early 1990s.

contents

About the Series

This series addresses two trends: critical conversations in academic fields about nature, sustainability, globalization, and culture, including constructive engagements between the natural, social, and human sciences; and intellectual and political conversations among social movements and other non-academic knowledge producers about alternative practices and socio-natural worlds. Its objective is to establish a synergy between these theoretical and political developments in both academic and non-academic arenas. This synergy is a sine qua non for new thinking about the real promise of emergent ecologies. The series includes works that envision more lasting and just ways of being-in-place and being-in-networks with a diversity of humans and other living and non-living beings.

New Ecologies for the Twenty-First Century aims to promote a dialogue between those who are transforming the understanding of the relationship between nature and culture. The series revisits existing fields such as environmental history, historical ecology, environmental anthropology, ecological economics, and cultural and political ecology. It addresses emerging tendencies, such as the use of complexity theory to rethink a range of questions on the nature–culture axis. It also deals with epistemological and ontological concerns, building bridges between the various forms of knowing and ways of being embedded in the multiplicity of practices of social actors worldwide. This series hopes to foster convergences among differently located actors and to provide a forum for authors and readers to widen the fields of theoretical inquiry, professional practice, and social struggles that characterize the current environmental arena.

preface

This book has been twelve years in the making. It has grown and stalled over this period in tandem with the demands and vicissitudes of my intellectual, personal, and professional life. I started on the journey that resulted in this book in 1991–92, when I first developed the proposal which took me to Colombia in January of 1993 for a year of field research, then simply entitled "Afro-Colombian Responses to Modernization and Development." During that initial year, I assembled a small research team to work in the southern Pacific region, at that point still customarily described as a poor, forgotten, hot, humid forest crisscrossed by innumerable rivers and inhabited by black and indigenous groups—a *litoral recóndito*, as Sofonía Yacup, a local author and politician, put it in the 1930s. By 1993, the region was fully immersed in an ambitious strategy of development that had started in the mid-1980s; armed with the tools given to researchers by the discursive critique of development of the 1980s, I set out to investigate ethnographically both the cultural and ecological impact of the various projects and the forms of resistance they faced from the black groups of the river communities. Or so I thought. What I discovered soon after my arrival was that the situation was far more complex than I had realized from a distance.

Indeed, it has not ceased to grow in complexity, posing unprecedented challenges to research method, politics, and understanding. First, two or three months into the project, we recognized that besides state-sponsored development and nascent capitalist enterprises (chiefly African oil palm plantations and industrial shrimp farming), albeit closely linked to them, there were two crucial factors in the struggle over the representation and fate of the region. The first was the concern with the region's biodiversity; the region was identified as one of the most important "biodiversity hot spots" in the world, and our arrival there coincided with the beginning of a novel, internationally funded conservation strategy of ambitious scope. As in other hot spots of this kind, *la conservación de la biodiversidad* had become the battle cry of the state, nongovernmental organizations (NGOs), academics, and local leaders alike. Closely related to conservation was a small but highly committed and articulate social movement of black communities. Our initial conversations with activists in this movement, while not immediately trusting, were nevertheless auspicious. In mid-June of that year (1993), our small research team held the first daylong workshop with a group of these activists and a handful of black intellectuals from the Pacific who were close to the movement. Our explicit intent was to discuss concrete ways of articulating our project with the interests and agenda of the movement. From this meeting on, the group of activists known as Proceso de Comunidades Negras (Process of Black Communities, PCN) became our closest interlocutor. On January 3, 1994, we concluded our first year of research with an intense, productive daylong interview with about ten PCN leaders on issues ranging from ethnicity, environment, and cultural difference to gender, strategy, and movement heterogeneity. The result of this first year of work, plus five additional months in 1994, was a collective volume, *Pacífico, desarrollo o diversidad? Estado, capital y movimientos sociales en el Pacífico colombiano*, published in Bogotá in 1996, with contributions from academics, intellectuals, and activists from the region.

My relation to PCN remains close to this day. Indeed, this book owes as much to this group's knowledge and acute political sensibility as to scholarly fields. The book can be seen, in fact, as an ethnography of the practices, strategies, and visions of this particular group of activists, including their own knowledge production. While the book is largely conceived from this perspective, however, it is more than that. Infused with PCN conceptualizations and engaging various strands in critical scholarship, the book proposes a way of analyzing some of the most salient social, cultural, and ecological issues of the present day. How does one examine and understand the momentous, complex processes engulfing regions such as the Pacific today? I should say at this point that finding biodiversity, the social movement, and activist knowledge was not the end of the complexification of the investigation. Further layers of complexity came from both social ("real world") and scholarly sources. Setting the social and the scholarly into productive dialogue is rarely an easy task, unless one decides to adopt a relatively straightforward framework of interpretation. But here the sources of complication were multiple. For instance, how does one go behind and beyond the forceful emergence of black identities of the 1990s to illuminate both historical processes of subalternity and possible ways forward in the struggles over difference? Second, particularly given the spread of armed conflict into the southern Pacific after 2000 and the ensuing massive displacement of local peoples from the region, how do events in the Pacific reflect and enact in distinct ways forces and conditions that go well beyond the region? To cite another example, the transnationalization of the region's social movements after 1995 certainly raised issues that could not be accommodated within the original research design and called for a more complex approach to both the movements and the region.

What resulted from this process was a series of nested frameworks that are articulated in the book in various ways and at several levels—not so much as in the proverbial layered onion, although the frameworks did grow somewhat organically, but in the fashion of a hypertextual formation. These include frameworks concerning political ecology, social movements, development, political economy, modernity and coloniality, science and technology, cultural politics, space and place, identity, networks, globalization, and complexity. This means the book is profoundly interdisciplinary or, as I explain in one of the chapters, it follows the trend toward "undisciplinarity" proposed by the Latin American group on modernity/coloniality/decoloniality. One might argue that this multiframing is the kind of approach that anthropology or cultural studies allows one to develop today; this is true up to a point but with clear limits since, after all, anthropology, particularly in the United States, wants to remain a discipline. Over the past ten years, I have written a series of largely theoretical papers that constituted bits and pieces of the framework presented here in a more comprehensive and integrated manner. In 1999, I started thinking about this book in terms of six key concepts: place, capital, nature, development, identity, and networks. This representation—the one in this book—interweaves both ethnographic research and theory around each notion. At the time, my decision to structure the text around these six concepts suggested to me that the book could be seen as a treatise in political ecology; today, I find it difficult to reduce it to this one field—or to any other, for that matter.

Another aspect of the book's character is that many of the frameworks summoned here have been developed collectively. These include the frameworks of "politics of place" (developed with the "Women and the Politics of Place," WPP, project since the late 1990s, which I have coordinated with Wendy Harcourt, of the Society for International Development in Rome); "modernity/coloniality/decoloniality," or MCD (developed by the Latin American MCD group); "diverse economies" (originally proposed by J. K. Gibson-Graham and elaborated in specific ways within the WPP project and the Cultures of Economies group at Chapel Hill); world anthropologies (out of the World Anthropologies Network, WAN, project); and a particular analysis of social movements (which I owe to my work with both the loose research group on Latin American social movements, maintained throughout the years with Sonia E. Alvarez and Evelina Dagnino, and the vibrant interdisciplinary Social Movements

Working Group, SMWG, at Chapel Hill since 2003). Even my political ecology approach has important collective dimensions in terms of groups in both Latin America and the United States. One thing these projects have in common is that they take the production of knowledge itself as a problematic; in a way, they all represent a social movement within the academy for a different kind of knowledge production. My own processing of these frameworks for this book, again, owes a lot to my ongoing engagement with the collectivity called PCN. This engagement has been an important aspect of the interface between theory and ethnographic research, as I have tried to articulate the production of knowledge by PCN with those arising more explicitly from academic sites, although in the process the boundary between academic and activist worlds and knowledges is blurred.

I am convinced that this collective dimension of framework building in the social sciences pays off in terms of theoretical grounding, interpretive power, social relevance, and sense of politics. It is a far cry from the persona—particularly accentuated at the level of the doctoral experience—of the lonely academic laboriously building a framework from literatures he or she has mastered by him- or herself. The collective approach might not follow the rule of one book every three years imposed on its practitioners by the maddening pace of the neoliberal academy in the United States, but it allows for plenty of creativity and, if one wishes, output, without being driven exclusively by the latter. The practice of working groups bringing together faculty and graduate students has become more common in recent years (certainly at Chapel Hill), and it is a way to bring this collective dimension of framework building to the fore.

acknowledgments

An explicit, built-in collective dimension to a book means that the influences and hence the debts of intellectual and personal gratitude multiply in ways one can never account for to complete satisfaction. I would like to highlight, however, the main ones, and I ask for the indulgence of the many more I should have mentioned and do not in order to make this task manageable. My deepest gratitude goes to the Proceso de Comunidades Negras, PCN; without their support and engagement this book would have been drastically different and poorer, in that the intellectual input of PCN thought has been definitive to the work. My thanks particularly to Libia Grueso and Carlos Rosero, my main PCN collaborators, with whom I have shared many experiences and conversations in activist, institutional, and academic milieus in Colombia and elsewhere over the years. Thanks also to other members of PCN (most of whom remain with the organization to this day), including Hernán Cortés, Leyla Arroyo, Victor Guevara, Julia Cogollo, Alfonso Cassiani, Félix Banguero, Edelmira Mina Roses, José Absalom Suárez, Konti Bikila Cifuentes, Mario Angulo, Yellen Aguilar, and, in Washington, D.C., Marino Córdoba and Charo Mina. A number of black intellectuals from the region were important sources of insight, knowledge, and information in the early stages, especially the scholar and activist of popular communications Jaime Rivas, the Guapi poet and expert in oral traditions Alfredo Vanín, the educator from Timbiquí (located on one of the legendary gold mining rivers of the Pacific since colonial times) Mary Lucía Hurtado, the cultural activist from Tumaco Dayra Quiñones, the anthropologist from Tumaco Ofir Hurtado, and the cooperative leaders Arismendi Aristizábal, and Rubén Caicedo. Our 1993–94 research team included my good friend Alvaro Pedrosa, whose remarkable work on popular education and communications at the Universidad del Valle in Cali was essential to understanding the early development initiatives in the Pacific; Jesús Alberto Grueso, the anthropologist from Timbiquí; and Betty Ruth Lozano, the sociologist from Cali, who started to think about black perspectives on gender and ethnicity in the early 1990s, when it was still a taboo subject. Also in Colombia, I am pleased to thank the staff of the biodiversity conservation project,particularly Enrique Sánchez and Juan Manuel Navarette, for sharing their grounded yet critical understanding of the project; and Margarita Flórez and Germán Vélez, of ILSA and Grupo Semillas in Bogotá, whose critical work on biodiversity and intellectual property rights provided a much-needed counterpoint to the early economistic approaches to the "benefits" of biodiversity. My understanding of alternative perspectives on biodiversity has also been enriched by exchanges with Timmi Tillman and Maruja Salas (Indigenous Knowledge and People's Network for Capacity Building, Chiang Mai, Thailand), Víctor Manuel Toledo in Mexico, María Fernanda Espinosa of IUCN in Quito, Nelson Alvarez and Henk Hobbelink (GRAIN, Barcelona), and the postings by Silvia Ribeiro of ETC Group in Mexico. More recently, I have benefited from my participation in the project to discuss science–policy interfaces in biodiversity governance, particularly the workshop on the subject in Leipzig in October 2006, and I thank the conveners and participants.

Among the *pacificólogos* none has contributed more substantially to this study than Eduardo Restrepo, my good friend and fellow anthropologist. Eduardo's prolific writings are the single most important oeuvre on the Pacific at present; besides, his solidarity knows no limit. Much of what I know about the Pacific I have learned from those anthropologists, ecologists, geographers, historians, communicators, and so on who entered a field of Pacific studies that by the mid- to late 1990s was burgeoning with activity. I can mention here only the few scholars and intellectuals with whom I have interacted most directly: Juana Camacho, Manuela Alvarez, Mauricio Pardo, Claudia Leal, Peter Wade, Kiran Asher, Michael Taussig, Jeannetee Rojas Silva, Jaime Arocha, Astrid Ulloa, Carlos Tapia, Alberto Gaona, Jesús Alberto Valdés, Aurora Sabogal, Sonia del Mar Gonzalez, Claudia Mosquera, Oscar Almario, Luis Carlos Castillo, and Ulrich Oslender. Many others in Colombian anthropology, ecology, and other social and human sciences, including some working in NGOs, have also been important to this book.

The debts incurred beyond Colombia are also immense—as large as the networks that have produced the frameworks that inform the book. Over the years, I learned much about ecology from Enrique Leff, Dianne Rocheleau, Joan Martínez Alier, and James O'Connor. I would also like to mention those with whom I worked most closely in the collective projects mentioned above: Wendy Harcourt, J. K. Gibson-Graham, and Smitu Kothari (Women and the Politics of Place); Walter Mignolo, Santiago Castro-Gómez, Edgardo Lander, Aníbal Quijano, Agustin Laó, and Catherine Walsh (Modernity/Coloniality/Decoloniality); Marisol de la Cadena, Eduardo Restrepo, Gustavo Lins Ribeiro, and Susana Narotzky (WAN); and, for the social movements field, Sonia E. Alvarez, Evelina Dagnino, and a number of doctoral students with whom I have worked closely on social movements, including Mary King and Chaia Heller (University of Massachusetts, Amherst), Juliana Flórez (Universitat Autónoma de Barcelona), and, at Chapel Hill, Michal Osterweil, Maribel Casas-Cortés, Elena Yehia, Juan Ricardo Aparicio, Dana Powell, Catalina Cortés Severino, and the one-time postdoctoral fellows Mario Blaser and Xochitl Leyva. These students' acute sense of the need to combine intellectual and political life places them in an outstanding group of intellectuals who are transforming the social movements research field. I want to mention also Jeff Juris and Harry Halpin, who provided feedback on the chapter on networks. To my former colleagues in anthropology at the University of Massachusetts and my current colleagues at the University of North Carolina, my deepest thanks for creating a nourishing climate for academic work. My life at UNC has been enriched by many people, of whom I would like to highlight colleagues from the various working groups, including Dorothy Holland, Charles Price, Don Nonini, Peter Redfield, Karla Slocum, Charles Kurzman, and Wendy Wolford; Larry Grossberg and John Pickles, with whom I have shared the exciting and trying aspects of transforming cultural studies at UNC at a time when the public university is under strong normalizing and privatizing pressures; and Joseph Jordan, for his sense of vision about the meaning of Afro–Latin American struggles and identities. Thanks to Natasha McCurley, Juan R. Aparicio, and Catalina Cortés for their careful assistance in proofreading and preparing the manuscript for submission.

Throughout the years, I have presented bits and pieces of the chapters that make up this book in many places and in many countries. I thank all of those who invited me to share with them the fragments of a work in progress. Besides those in Colombia, which include universities and research centers in several cities, such as the Instituto Colombiano de Antropología e Historia, and NGOs such as Fundación Habla/Scribe and the World Wildlife Fund, both in Cali, I would like to highlight groups in Helsinki, Barcelona, and Quito and in Mexico, England, and Brazil for repeated invitations. The intellectual and often personal input of many friends and colleagues has been very important on a number of subjects over the years, and I can mention only a few here: Cristina Rojas, Orlando Fals Borda, Claudia Steiner, Ana María Ochoa, Ingrid Bolívar, Maria Fernanda Espinosa, Anthony Bebbington, Lynn Stephen, Jacqueline Urla, Brooke Thomas, Orin Starn, David Slater, George Yúdice, David Hess, Daniel Mato, Boaventura de Sousa Santos, Jesús Martín Barbero, Eduardo Gudynas, Martín Hopenhayn, Shiv Visvanathan, Jordi Pigem, Pieter de Vries, Monique Nuijten, Manuel de Landa, Brian Goodwin, Peter

Waterman, Jai Sen, Jeremy Gould, Elina Voula, Andreu Viola, Subir Sinha, and Søren Hvalkof; thanks to Søren also for helping us to obtain Danish funding for projects in the Pacific proposed by PCN, and to Denise Bebbington and Beto Borges in the United States for the same reason. I would like to make very special mention of those friends who supported us repeatedly in times of personal hardship over the past few years in Colombia and Chapel Hill/Carrboro, but the list would be too long, and they know, I am sure, of our deep gratitude.

The 1993–94 research period was funded by grants from the Social Science Research Council, the Heinz Endowment, and the Arts and Humanities Division of the Rockefeller Foundation (now the Culture and Creativity Program). I also received funding from the John Simon Guggenheim Memorial Foundation (1997), the MacArthur Foundation's Global Security and Sustainability Program (2000), and the University of North Carolina. I am grateful to these funding sources for enabling various phases of research and writing.

My thanks, finally, to Valerie Millholland, senior editor at Duke University Press, with whom I have been working for over five years on our series New Ecologies for the Twenty-First Century. This has been a most rewarding endeavor and working relationship, and I thank her for her advice and decided support of the present book ever since I talked to her about it for the first time many years ago. Thanks also to the copyeditor Lawrence Kenney and Mark Mastromarino, Assistant Managing Editor at the Press, for his care in shepherding the manuscript through the production process. This book is dedicated to the memory of my mother, Yadira, and of my younger brother, Chepe, both of whom left this world hardly within a year of each other, in Cali, but whose presences continue to be with me; to my *compañera*, Magda, *con todo el amor del mundo*, and my sister Maria Victoria, for their steady support; and to the black, indigenous, and otherwise environmental and cultural activists in the Pacific, who day in and day out and against all odds continue their struggle to make of the Pacific a livable and joyful socionatural world—in PCN activists' conceptualization, a *Territorio de Vida, Alegría y Libertad* (a territory of life, happiness, and freedom).

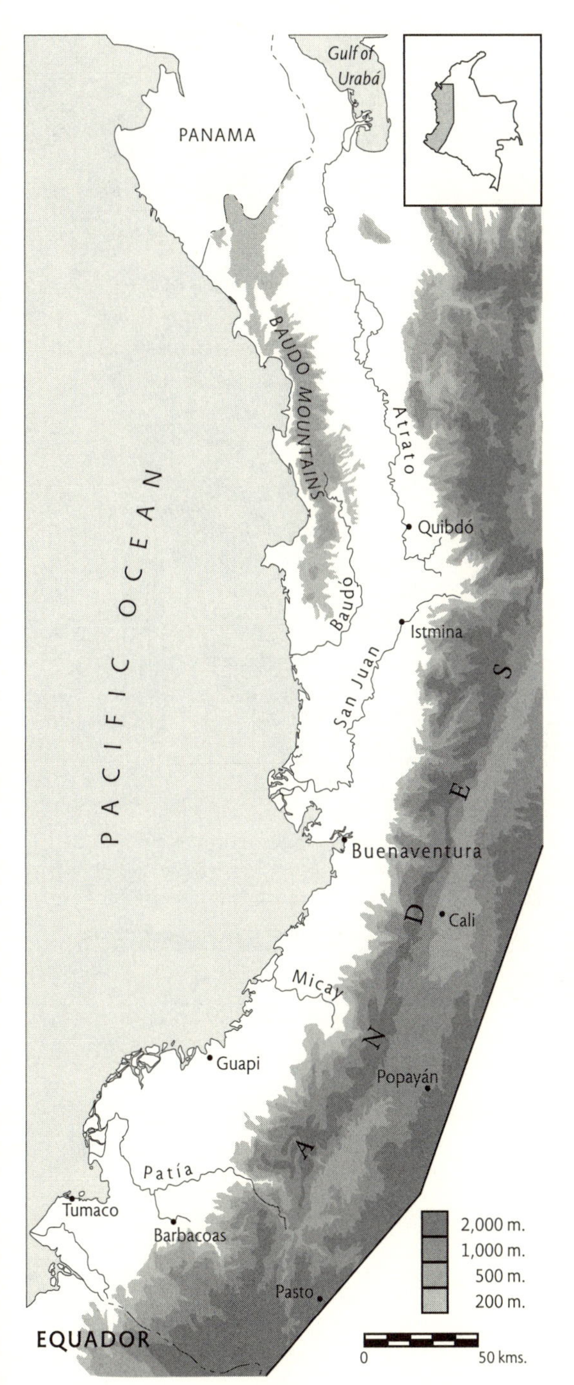

Map 1. The Pacific Region. Elaborated by Claudia Leal and Santiago Muñoz, Department of History, Universidad de los Andes, Bogotá

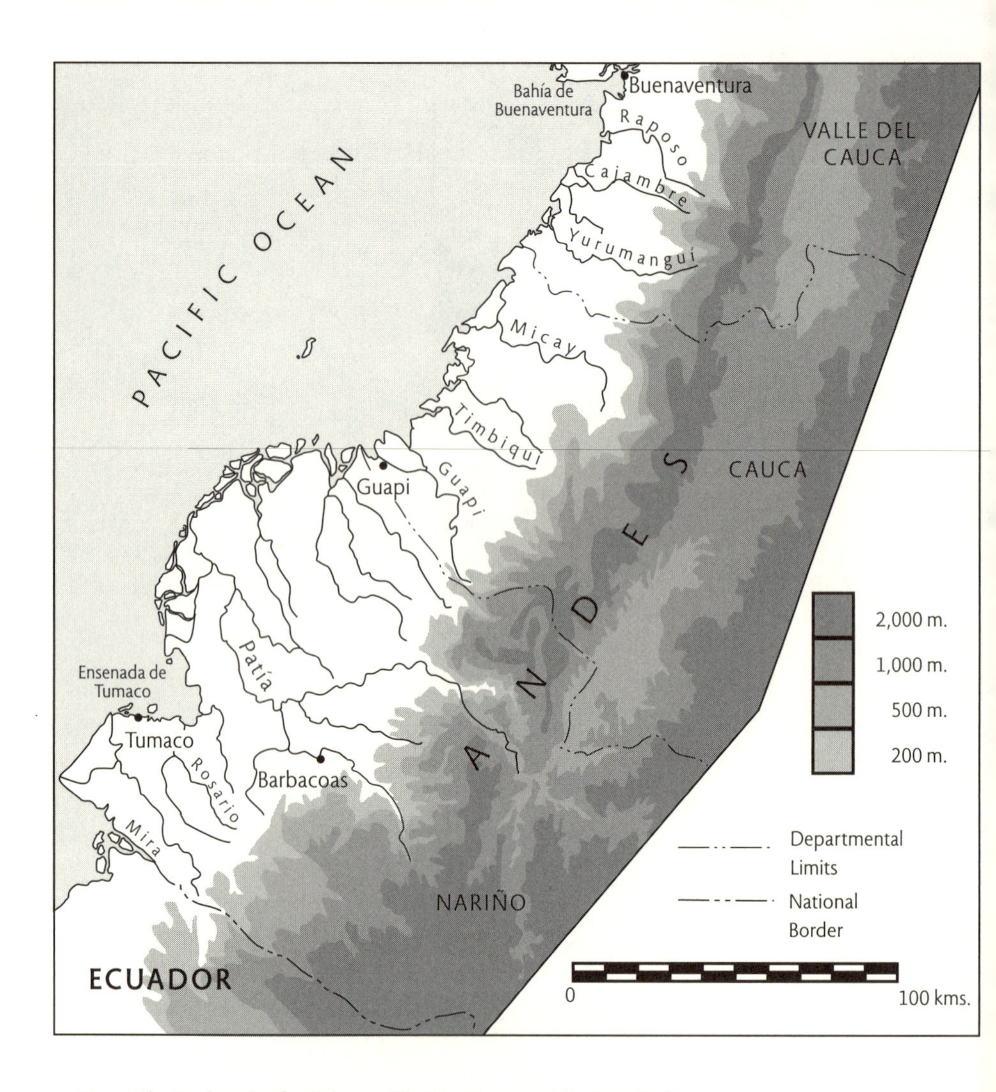

Map 2. The Southern Pacific. Elaborated by Claudia Leal and Santiago Muñoz, Department of History, Universidad de los Andes, Bogotá

introduction

Regions and Places in the Global Age

Sea mentira o sea verdad
se abra la tierra
y se vuelva a cerrar
que el que lo está oyendo
lo vuelva a contar

[Be it a lie or be it the truth
Let the earth open up
and close again
whoever's listening
will tell it again]

—Popular refrain often said at the beginning of a round of storytelling in the riverine regions of the southern Pacific

This book is about many diverse yet closely interrelated aspects of social, cultural, and biological life at present. It is, above all, about place-based and regional expressions or articulations of difference in contexts of globalization; this implies setting place-based and regional processes into conversation with the ever-changing dynamics of capital and culture at many levels. This conversation, however, is neither about the imprint left on a particular world region by an allegedly unstoppable process of globalization nor solely about how this region responds to it. Instead, it is about a complex, historically and spatially grounded experience that is negotiated and enacted at every site and region of the world, posing tremendous challenges to theory and politics alike.

The flesh, blood, and thoughts of the story come from a particular region in Latin America, the Colombian Pacific. Two contrasting positions arising from this region exemplify the range of responses to globality. The first comes from a meeting of about forty-five leaders and activists representing the most important indigenous and black social movements of the region held on June 18–22, 1995, in the predominantly black town of Puerto Tejada, an hour south of Cali. The goals of the meeting were to examine the social and environmental situation of the Pacific, discuss interethnic relations, and come up with joint strategies of negotiation with the state on various plans and policies. Under the rubric *Territorio, etnia, cultura e investigación en el Pacífico colombiano* (Territory, ethnicity, culture, and research in the Colombian Pacific), the activists made it clear that what was at stake went well beyond the situation of the moment to involve the definition of life itself, in particular the defense of what they called, in the heady climate of the moment, the *cosmovisiones* (worldviews) of the black and indigenous groups. Four principles for interethnic relations and relations with the state were identified: the fact that the Pacific is "an ancestral territory of ethnic groups"; that these groups are culturally diverse and seek to respect differences both among themselves and in relation to Colombian society; that from this position of mutual respect and difference they assume the coordination of the defense of their territories; and that their traditional knowledges are fundamental to their relation to nature and to their identity and should be recognized as such. Analyses and conclusions followed from there, referring to the defense of territory, culture, and identity. The notion that the Pacific is a territory of "ethnic settlements," first formulated at this meeting, was to result a couple of years later in a sophisticated conception of the Pacific as a "region-territory of ethnic groups," a notion that will occupy a salient place in this book.[1]

At about the same time, the recently established Gerencia de Proyectos para el Litoral Pacífico Vallecaucano (Office of Development Projects for the Pacific Region of the Valle del Cauca Province) held a daylong meeting at a luxury hotel in the also predominantly black city of Buenaventura, which is the most important port in the nation and is located approximately two hours from Cali. The gerencia presented a range of social and economic projects for the subregion that called for a dramatic expansion of investment with the help of international capital and included projects focusing on sewage and water supply, education, and health and a panoply of infrastructural and industrial schemes dealing with electricity, port facilities, roads, a local airport, tourism, large-scale fishing facilities, timber industries, and so forth. The aim was to create the right investment climate in order to turn Buenaventura, including the multiple rivers of its vast rural areas, into a development pole for the nation. The meeting, attended by over two hundred people representing the government and private sectors, was held in response to rising rhetoric about the "age of the Pacific" in which this vast, rich rain-forested territory was seen as the platform for an aggressive neoliberal strategy of integrating the country with the Pacific basin economies.

Also telling was the subtitle of the activists' meeting: *Conceptos de los Pueblos Indígenas y Negros del Pacífico Colombiano*. The idea that black and indigenous peoples could have knowledge, let alone concepts, was new, although it was becoming more common thanks to discussions about local knowledge in debates over the conservation of biological diversity, particularly after the Earth Summit (the UN Conference on Environment and Development) held in Rio de Janeiro in 1992. Needless to say, no aspect of this knowledge was contemplated at the Buenaventura meeting. But as this book will show in detail, the fact is that groups such as the black and indigenous activists of the Colombian Pacific do indeed produce their own knowledge about the situations they face, and furthermore this knowledge often constitutes sophisticated frameworks that can no longer be overlooked in any discussion of globalization, whether from an economic, cultural, or ecological perspective. Moreover, these frameworks are integral to the struggles mounted by subordinate groups over the terms of globality and also to the effectiveness of such struggles.

Places and Regions in the Age of Globality

The examination of place making and region making from multilevel economic, ecological, and cultural perspectives affords novel opportunities for understanding the politics of difference and sameness that accompanies enactments of globality. According to some arguments, today's politics of difference and sameness is still deeply shaped by the myths of universality and cultural superiority that from the dawn of modernity—the conquest of America by Spain in 1492—have allowed the West to define the identity of others. Ever since, an ensemble of Western, modern cultural forces (including particular views of the economy) has unceasingly exerted its influence—often its dominance—over most world regions. These forces continue to operate through the ever-changing interaction of forms of European thought and culture, taken to be universally valid, with the frequently subordinated knowledges and cultural practices of many non-European groups throughout the world. Eurocentric globality thus has an obligatory counterpart in the systematic act of *encubrimiento del otro* (the covering up of the other), to use the expression of the Latin American philosopher Enrique Dussel

(1992)—that is, in a kind of global coloniality. This book is, in a very abstract but real sense, about the dynamic of an imperial globality and its regime of coloniality as one of the most salient features of the modern colonial world system in the early twenty-first century. It is thus also about the geopolitics of knowledge: Whose knowledge counts? And what does this have to do with place, culture, and power?[2]

Described as a poor and forgotten hot and humid forest crisscrossed by innumerable rivers—a *litoral recóndito* (Yacup 1934)—the Colombian Pacific had been integrated into the world economy from the early post-conquest period through exploration, slavery, gold mining, and the subjection or elimination of indigenous inhabitants. Boom and bust cycles tied to the extraction of raw materials such as gold, platinum, precious woods, timber, rubber, and, recently, biodiversity have succeeded each other over the past two centuries, each leaving its indelible imprint on the social, economic, ecological, and cultural makeup of the place.

Only in the early 1980s was the region subjected to an explicit strategy of incorporation into the national and transnational spheres in the name of development. As a result, by the early 1990s the region had become a stage for an intense cultural politics that brought together development experts, black and indigenous activists, biodiversity conservation advocates, capitalists, fortune seekers, government officials, and academics into a tight space of dialogue, negotiation, and confrontation that, albeit for a brief moment, seemed to have an unclear resolution, with local movements and their allies making a valiant and brilliant attempt at providing a workable alternative. Two other factors were crucial in creating the context for this complex encounter: first, the decisive opening of the national economy to world markets after 1990 under neoliberal dictates; and second, the reform of the national constitution in 1991, which, among other things, resulted in a law that granted cultural and territorial rights to the black communities (Ley [Law] 70 of 1993). By the late 1990s, however, the regime of imperial globality had forcefully reasserted itself, and the region became submerged in a quagmire of violence, merciless capitalist expansion, and massive displacement that has affected black and indigenous communities and the environment with particular virulence—a reassertion of the coloniality of knowledge, power, and nature.

Such are the intent and material of the book in their broadest strokes. Emerging from this historical materiality, the book is about the incredibly complex intersections of nature and culture, space and place, landscape and human action, culture and identity, knowledge and power, economy and politics, modernity and globalization, and difference and sameness associated with imperial globality and global coloniality in a particular corner of the world; it is also about what has been called uneven geographies of poverty and livelihoods, and how they are related to historical political economies and culturally inflected patterns of development intervention (Bebbington 2004). As noted in the preface, I render these geographies manageable by a particular design in terms of six basic concepts: place, capital, nature, development, identity, and networks. These concepts are both chapter titles and notions that articulate my argument throughout the book; thus, while each concept is developed primarily in its respective chapter, most concepts are dealt with in several chapters. To give an example: while place is the central subject of chapter 1, it makes significant appearances in chapters on capital and nature. Another example: biodiversity is discussed at length in the chapter on nature but also figures prominently in that on development and is also treated in chapters on place, capital, and networks. This means the book has a networked or recursive logic in that a number of central subjects are treated in somewhat different ways in various chapters, as partial displacements of the same topic. This also means that while the chapters can to some extent be read independently, only by reading the entire book can the reader develop a comprehensive sense of the work. One further detail: while each chapter interweaves theory and ethnographic research, in most cases the more lengthy theoretical debates are, with some exceptions, relegated to the notes.[3]

I mentioned above that the book is about many subjects. Among them are a set of geobiological and cultural conditions making the Pacific into a distinct socionatural world; the state policies of development and pluriculturality that created conditions for the emergence of black and indigenous social movements, and these movements' efforts at steering the region in specific directions; the attempts by capitalists to appropriate the rain forest for extractive activities, by developers to set the region onto the path of modernist progress, by biologists and conservationists to defend this incredibly rich biodiversity hot spot from the most predatory activities of capitalists and developers, and by academics, activists, and intellectuals to understand the whole thing, this complicated process that took them by surprise and found them largely unprepared in terms of having solid studies and theoretical and political approaches. In addressing these questions, the book highlights the tremendous value of activists' knowledge for both understanding and action. For this very reason, the book is above all about difference and its politics and the difference this politics makes in places such as the Pacific. It is, by the same token, about what theorists call modernity—is it still a viable project in regions like the Pacific? or, on the contrary, do the events happening there suggest that the project of modernity, whatever it means, has to be abandoned once and for all? Finally, it is about ethnography and social theory and their efforts to respond more effectively to the dynamics of today's world: are there novel approaches in social theory that provide better accounts in this regard, perhaps because they are based not only on more inclusive epistemologies but on more diverse ontologies? If this is the case, scholars would be facing a significant reorientation of theory. As I shall discuss in the chapter on networks, some are making this bold claim.

I want now to provide a more explicit account of the book's content and structure, although this will still be barely a sketch in relation to the chapters that follow.

A Political Ecology of Difference

Joan Martínez Alier (2002) defines *political ecology* as the study of ecological distribution conflicts. By this he means conflicts over access to and control of natural resources, particularly as a source of livelihood, as well as the costs of environmental destruction. In many places, local groups engage in struggles against translocal forces of many types to defend their place. It is not easy to conceptualize this defense in all of its dimensions, and this is an important part of the story. In a nutshell, I argue that people mobilize against the destructive aspects of globalization from the perspective of what they have been and what they are at present: historical subjects of particular cultures, economies, and ecologies; particular knowledge producers; individuals and collectivities engaged in the play of living in landscapes and with each other in distinctive ways. I shall say that in regions like the Pacific people engage in the defense of place from the perspective of the economic, ecological, and cultural difference that their landscapes, cultures, and economies embody in relation to those of more dominant sectors of society. What follows is a brief description of the book's chapters and concepts. This constellation of concepts provides a basis for a political ecology framework focused on difference. Some important concepts are missing (e.g., state, gender, culture, science, and knowledge itself), although they will be treated to some extent in various chapters.

Place ◈ Why start with place? For three reasons. First and most immediately because the mobilizations of the past two decades in the Pacific are seen locally as struggles over culture, territory, and place. Black and indigenous movements see the aim of their struggle as one of retaining control of their territory; it is not far-fetched to see these movements as expressions of ecological and cultural attachment to place. In fact, in the mid-1990s indigenous and black activists together came up with a conceptualization of the Pacific as a "region-territory of ethnic groups," as noted above, that became a gravitating principle of political strategies and conservation policies alike. Place-based struggles more generally link body, environment, culture, and economy in all of their diversity (Harcourt and Escobar, eds. 2005). Second, in a philosophical vein, because place continues to be an important source of culture and identity; despite the pervasive delocalization of social life, there is an embodiment and emplacement to human life that cannot be denied. This is readily acknowledged by people such as the black and indigenous groups of the Pacific, who maintain more embodied and embedded practices of social and ecological existence. Third, because scholarship of the past two decades in many fields (geography, anthropology, political economy, communications, and so on) has tended to deemphasize place and to highlight, on the contrary, movement, displacement, traveling, diaspora, migration, and so forth. Thus, there is a need for a corrective theory that neutralizes this erasure of place, the asymmetry that arises from giving far too much importance to "the global" and far too little value to "place." To this end, I shall review the deeply historical and always changing character of this region, aiming to understand how, against this bioregional background—that is, the long history of geobiological life, landscape, and human settlement—today's cultural, economic, and ecological struggles make full sense.

Capital ◈ One of the main ways in which places have been transformed in the past centuries the world over is through capitalism. No account of place making can overlook the production of place by capital, and there are few examples of this as vivid as the transformation of a complex, self-organizing humid forest ecosystem into a monocultural landscape, as it continues to happen in many parts of the Pacific with the spread of African oil palm plantations, or the replacement of the meandering and rooted mangroves by a monotonous succession of rectangular pools for industrial shrimp farming. Marxist political economy has been the main corpus of theory enlightening scholars on these processes, yet Marxism was not very good at dealing with nature. The engagement of Marxism with culture and nature in recent decades has been very productive; applying these new frameworks to questions of place allows one to see the actions of capital in the Pacific in a new light. However, this is only half of the picture; for the other half, one has to look at the plethora of economic practices that local groups have either maintained over the long haul or created in recent times. Could some of these actually be seen as noncapitalist practices? How does one decide? As we shall see, mainstream political economy has been unable to see noncapitalist economies in their own right. Besides economic practices oriented toward self-subsistence, some collective shrimp farming practices created by local groups in the southern Pacific, in the very encounter with industrial shrimp farming, could be seen in terms of noncapitalist economies. Is this reinterpretation a naïve conceit in the minds of the social groups engaged in them or, worse still, wishful thinking on the part of the analyst? Or could there be something different in these practices that capitalocentric frameworks have previously rendered invisible? Finally, could theorists and activists plausibly entertain the project of cultivating subjects of economic difference, particularly subjects who desire noncapitalist economies? A positive answer to the question may shift academic and activist perspectives onto a new plane, as we shall see.

Nature ◈ Many environmentalists argue that there is a generalized ecological crisis today. Humans are destroying their biophysical environments at record speed and in unprecedented magnitude. Capitalist modernity, it would seem, has declared war on every ecosystem on the planet, and few places exemplify the scale of this destruction as patently as the Pacific. Philosophically minded ecologists argue that the ecological crisis is a crisis of modern systems of thought. Modern science and technology not only contribute to rampant destruction, but no longer seem able to devise workable solutions to it. This is why epistemological questions are fundamental in discussing questions about nature, and as such they will be given due importance in this chapter; there is, in short, a coloniality of nature in modernity that needs to be unveiled. Again, this is only half of the picture, and, as in the case of the economy, one needs to search for the other half in the place-based ecological practices existing in the Pacific. In the river settlements, black groups have historically enacted a grammar of the environment—a local model of nature—that exhibits a striking disparity in relation to modern understandings of nature. This grammar, embedded in rituals, languages, and forms of classification of natural beings that might look strange to moderns, constitutes the cultural-ecological basis of how they farm and utilize the forests. These traditional production systems, as biodiversity experts and activists came to call them in the mid-1990s, have had a built-in notion of sustainability, one that, however, has become impracticable in recent decades owing to a variety of pressures. Here lies one of the most difficult predicaments for conservation advocates and activists: pushed to rationalize ecological and environmental practices to ensure "conservation," they are aware that in doing so they are moving away from the long-standing, place-based notions and practices which ensured a reasonable level of sustainability until recent decades. Is it still possible to argue in favor of ecological difference for the Pacific this late in the game? Or are activists and conservationists forever doomed to bring nature into the realm of modern planning to ensure conservation? And if so, how can this be done without reinforcing coloniality (that is,, the subalternization or even elimination of local grammars and knowledge of the environment) at both cultural and ecological levels?

Development ◈ Development, in conjunction with capitalism, has been the single most important transformative force in the Pacific. In the early 1980s, the first plan for the putative comprehensive and integral development of the Pacific created the region as an entity susceptible of development. Few before the 1970s would spend a dime on developing this insalubrious, backward region. By the early 1990s, however, speaking of the development of the Pacific had become de rigueur, and institutions like the World Bank were quick to jump on the bandwagon. Not everything went according to the developmentalist script, however. While much conventional development did indeed take place, often with negative consequences for local peoples and ecosystems, many projects, especially those under participatory development schemes, enabled a certain degree of creative appropriation, even subversion in terms of intended goals, by local groups. This was the case with a number of projects that became linked to the social movements for cultural and territorial rights that had swept over the southern Pacific since the very early 1990s, for example, the establishment of cooperatives for the commercialization of cocoa and coconut; women's associations for shellfish marketing; innovative popular art-cum-literacy projects that brought low-technology radio and printing materials to local groups, through which they creatively linked literacy, history, and identity; and the biodiversity conservation project for the region, which black and indigenous activists profoundly transformed. Drawing on debates on postdevelopment and coloniality, this chapter suggests that these acts of counterwork by locals can reasonably be seen as producing alternative modernities—modern yet different ecological, economic, and cultural configurations—but also an inkling of alternatives to modernity, what could be termed decolonial configurations of nature, culture, and economy.

Identity ◈ The constitutional reform of 1991 created the conditions for an intense period of cultural and political activism by local black and indigenous groups. One of the most defining aspects of the period was the indisputable emergence of the category of *comunidades negras* (black communities) as a central cultural and political fact—so central that collectivities such as the network of ethnoterritorial organizations known as Proceso de Comunidades Negras (PCN) adopted the category for itself, the state issued development plans "for the black communities," and so forth. The category took on local force, at least at the level of movements, NGOs, and church organizations, and of course the development apparatus. Most analysts concur that there was indeed a veritable "relocation of 'blackness' in structures of alterity," as Peter Wade (1997: 36) descriptively put it. Were these identities the product of the neoliberal state (e.g., Ley 70 of territorial and cultural rights)? Were they the result of the decided action by social movement activists? Or something in between? An adequate answer to these questions can be attempted only by a careful tacking between theory and ethnography. Contemporary theory, including poststructuralism, feminist and critical race theories, and cultural and psychological frameworks in various fields, has given great salience to questions of identity as an expression of the politics of difference; activists unambiguously described their actions in terms of the right to cultural difference and to a black or indigenous identity. Generally speaking, what relations between the individual and the collective, between culture and politics, between state and social movement action, between activist and expert knowledge account for the making of particular identities in place-based yet translocal situations? Moreover, even if there were no "ethnic identities" or "black communities" before 1990, Pacific local peoples did not have any trouble knowing who they were and how to talk about themselves and others, including whites—it's just that they did it according to a very different regime of representation of difference and sameness, of belonging. What happened with these "traditional" identities once the post-1990s regime erupted onto the picture? As we shall see, there are no easy answers to these questions, and, again, I will highlight the knowledge about identities produced by the social movements.

Networks ◈ Biodiversity, social movements, capital, knowledge, and so on are decentralized, dispersed, and transnationalized ensembles of processes that operate at many levels and through multiple sites. No current image captures this state of affairs at present more auspiciously than that of the network. The salience of the network concept has to do with cultural and technological processes fueled by digital information and communication technologies (ICTS). A lot of hype in network talk arises from many quarters (from physics and mathematics to systems science, sociology, anthropology, geography, and cultural studies), but interesting ideas also emanate from network approaches. For example, the concept of biodiversity was barely known in the late 1980s; by the early 1990s it had become a transnational assemblage bringing together all kinds of organizations, actors, knowledges, endangered species, and genes. A movement organization such as PCN, which started as a regional force, embarked on a strategy of transnationalization by 1995–96. By the late 1990s there were networks of Afro–Latin American movements and Afro–Latin American women activists, where only sporadic contacts had existed a few years before. But there are networks of all kinds. Is it possible to differentiate between dominant and oppositional networks, for instance? Or are they all so inextricably tied that even an analytical separation of them becomes useless? Or between local and regional and transnational networks? Or between the hierarchical and centralized networks that have characterized most modern organizations, on the one hand, and the more self-organizing, decentralized, and nonhierarchical "meshworks" that characterize many contemporary movements, on the other? Or how does one reconcile being-in-place and being-in-networks? Finally, what are the implications of network thinking for social theory, including concepts of scale, space, ecosystem, and the real itself? If what some theorists are arguing is correct, the network concept would be a reflection of a more substantial reinterpretation of how social reality comes into being; the notions of actor network, assemblages, flat ontology, and flat sociality push one to think about the real in relational and contingent, not structural and law-driven, terms.

Thinking from the Colonial Difference

Coloniality, according to Walter Mignolo, is, on the one hand, "what the project of modernity needs to rule out and roll over in order to implant itself as modernity and—on the other hand—the site of enunciation where the blindness of the modern project is revealed, and concomitantly also the site where new projects begin to unfold. Coloniality is [. . .] the platform of pluri-versality, of diverse projects coming from the experience of local histories touched by western expansion; thus coloniality is not an abstract universal, but the place where diversality as a universal project can be thought out, where the question of languages and knowledges become[s] crucial" (cited in Escobar 2004a: 218; see also Mignolo 2000; Walsh 2007). The notion of coloniality thus signals two parallel processes: the systematic suppression of subordinated cultures and knowledges (*el encubrimiento del otro*) by dominant modernity; and the necessary emergence, in the very encounter, of particular knowledges shaped by this experience that have at least the potential to become the sites of articulation of alternative projects and of enabling a pluriverse of socionatural configurations. The modernity/coloniality/decoloniality perspective (MCD), to be discussed at length in the chapter on development, is interested in alternatives which, arising from the epistemic borders of the modern-colonial world system, might pose a challenge to Eurocentric forms of modernity. Succinctly put, this perspective is interested not only in alternative worlds and knowledges, but also in worlds and knowledges otherwise.

To give a more intuitive entry into this notion: The fact that the Pacific has always been connected with a dominant national Euro-Andean modernity has entailed the persistent suppression (often violent exclusion) of black and indigenous knowledges and cultures. This very situation, nevertheless, has been accompanied by an ongoing production by these groups of diverse knowledges about nature, economy, person, and the world in general. These knowledges are generated in the ceaseless process of living at the epistemic borders of the modern colonial world system, as happens in so many instances of border thinking by black and indigenous inhabitants. Literally speaking, black and indigenous groups of the Pacific—like, surely, many other groups in the world—have always lived in a pluriverse of culture and knowledge. But they have done so as *dominated* groups, which makes all the difference. Activists of local movements, as I suggest in this book, emerge from this border and produce knowledge that shuttles back and forth alongside the modernity/coloniality, universality/pluriversality interface. This border, furthermore, constitutes an exteriority of sorts (not an ontological outside) to modernity. I shall say that these activists conduct their struggle from the colonial difference—in this case, a colonial difference that has to do with blackness or indigeneity and with living in particular landscapes and ecosystems.[4]

A number of notions enable the construction of a framework for thinking theoretically and ethnographically about and from the colonial difference. The framework presented below incorporates elements from a variety of proposals, chiefly those of political ecology, modernity/coloniality/decoloniality, politics of place, and diverse economies, all of which I will present at some length in subsequent chapters. For now, I need only point out the rudiments of the framework.

For several reasons I have found it useful to think about the colonial difference under three interrelated rubrics: economic, ecological, and cultural difference (Escobar 2006 [1999]). First, the transformation of regions such as the Pacific by imperial globality is indeed a triple transformation, or conquest; it entails the transformation of local diverse economies, partly oriented to self-reproduction and subsistence, into a monetized, market-driven economy; of particular ecosystems into modern forms of nature; and of place-based local cultures into cultures that increasingly resemble Euro-Andean modernity. Dussel has similarly suggested that the political field is traversed by the three domains I have just described: the ecological, the economic, and the cultural. For him, the primary end of politics today is the perpetuation of life on the planet (Dussel 2006: 55–61, 131–40). Second, even if the transformation of regions such as the Pacific never stops, it is never complete. Academics have thought about these processes in terms of resistance, hybridization, accommodation, and the like. These have been useful notions, yet they have tended to obliterate the potential of difference for worlds and knowledges otherwise. I shall see if it is possible to arrive at an alternative formulation.

I already mentioned the definition of political ecology as the study of ecological distribution conflicts, meaning by this, conflicts over access to and control of natural resources. In providing this definition, Martínez Alier (2002) was making an extension from political economy as the study of economic distribution conflicts—class distribution of wealth, income, assets, and so forth—to the field of ecology. This two-pronged, political ecology perspective is missing an important dimension of conflict, namely, the cultural. It is necessary, in other words, to consider those conflicts that arise from the relative power, or powerlessness, accorded to various knowledges and cultural practices. To continue with the example above: by culturally privileging the capitalist (e.g., plantation) model of nature over the local diverse agroforest, ecosystem model, not geared to a single product and to accumulating capital, a cultural distribution conflict has been created. This conflict has ecological and economic consequences, so that economic, ecological, and cultural distribution conflicts are intimately intertwined.

There is added value in including the cultural, and this is to neutralize the tendency to ascribe determining importance to the economic or to the ecological, depending on the taste of the researcher. In other words, economic crises are ecological crises are cultural crises. It is important not to separate these three domains but to let them interpenetrate each other. When considered together, the domains of subjectivity and culture, economy, and ecology provide the basis for theoretical insights about how to reorient societies away from the nightmarish arrangements of the present and toward cultural, ecological, and cultural practices and singularities that could constitute tangible alternatives to capitalist significations and realizations, fostering the construction of new existential territories.[5]

Two further points about the cultural dimension. First, cultural distribution conflicts arise from the difference in effective power associated with particular cultural meanings and practices. They do not emerge out of cultural difference per se, but out of the difference that this difference makes in the definition of social life: whose norms and meaning-making practices define the terms and values that regulate social life concerning economy, ecology, personhood, body, knowledge, property, and so forth. Power inhabits meaning, and meanings are a main source of social power; struggles over meaning are thus central to the structuring of the social and of the physical world itself. This concept shifts the study of cultural difference from the modernist concern with multiculturalism to the distributive effects of cultural dominance (coloniality) and struggles around it; more than cultural justice movements at present emphasize interculturality. I define *interculturality* as a project, that of bringing about effective dialogue of cultures in contexts of power (Escobar 2006 [1999]). On the movement side, these dialogues are often enacted from the colonial difference. This is clearly the case with groups such as PCN, as we shall see in abundant detail.

Second, cultural conflicts are often the reflection of underlying ontological differences, that is, different ways of understanding the world and, in the last instance, different worlds. These differences are more patently clear in the case of, say, indigenous peoples and ethnic minorities. While they are increasingly recognized, for instance, in conservation programs (e.g., indigenous knowledge) they are rarely incorporated into program and project design, for to do so would mean very significant transformations in the existing frameworks and ultimately a radical questioning of foundational modern assumptions, such as the divide between nature and culture. The fact that dominant modern ontologies are connected to these other ontologies asymmetrically through the very same projects means that the latter are almost inevitably refunctionalized at the service of the former. This is why it can be said, with Blaser (forthcoming), that political ecology implies a political ontology in many cases. The political ontology framework thus constitutes a further elaboration of coloniality and of the coloniality of nature in particular.

Akin to the "women and politics of place" conception (Harcourt and Escobar, eds. 2005), the above argument brings together into one framework discourses and struggles around culture, often the focus of ethnic, gender, and other movements for identity; environment, the interest of ecology and environmental justice movements; and diverse economies, the concern of social and economic justice movements.[6] This conceptual framework aims to analyze the interrelations created within subaltern struggles (black people's, in the case of the Pacific) around identity, environment, and economies, in all of their diversities. In doing so, I aim to demystify theory that ignores subaltern experiences and knowledge of the local economy, environment, and culture *in order to relocate their politics of place as key to our understanding of globalization*. As we shall see in the last chapter, on networks, many subaltern struggles can be seen today in terms of place-based yet transnationalized strategies—or, more succinctly, as forms of place-based globalism (Osterweil 2005a). At the theoretical-political level, the focus on difference can also be interpreted in terms of the logic of articulation outlined by Laclau and Mouffe (1985); emerging out of the antagonisms that necessarily pervade social life, the logic of difference is a means to widen the political space and increase its complexity. The articulation of struggles across differences may lead to the deepening of democracy—indeed, to questioning the very principles of liberal democracy, if conceived from the perspective of the colonial difference. The following table summarizes the framework (see Escobar 2006 [1999] for further detail).

A final word about why I place so much emphasis on difference and conflict. First, as the Brazilian liberation theologian Leonardo Boff says, the valuation of difference needs to be accompanied by the acceptance of complementarities and by convergences constructed out of the

Table 1. A Political Ecology of Difference: Economic, Ecological, and Cultural Distribution Conflicts

Context/ Historical process	Concept/Problem	Theoretical/ Academic response	Intellectual/ Political project	Social/ Political responses
Global capitalism	Economic distribution (negation of economic difference)	Internalization of externalities	"Sustainable" capitalist development	Environmental governmentality
Reductionist science and technology	Ecological distribution (negation of ecological processes)	Highlighting of incommensurability of (modern) economy and ecology. Ecological economics and political ecology	Need to re-embed economy in society and ecosystems	Struggle over the environment as a source of livelihood; environmentalism of the poor
Modernity/ Coloniality (modern colonial world system)	Cultural distribution (negation of cultural difference)	Highlighting of Incommensurability of (Modern) Economy and Pluri-Culturality; Politics of Place Frameworks; Articulatory Politics; De-Coloniality	Need to re-embed economy in society, ecosystems, and culture	Place-based struggles for economic, ecological, and cultural difference; social movement networks; autonomy, counterhegemony; decolonial projects

diversity of worldviews and practices (2002: 26; see also Maturana and Varela 1987). Second, while highlighting power, conflict should not be seen as reducing everything to power or to quantitative assessments of inequalities. The emphasis on conflict and difference is not about exclusion or segregation, as some might fear. To continue with Boff, if talk of conflict and interculturality is about justice, it is also about forgiveness; if it commands, for instance, reparations, it does so in the sense of the acknowledgment of historical injustice rather than revenge. In the best of cases, the language of distribution conflicts entails serious individual and collective confrontations with difference but without (having to) fear; it entails bridge building and technologies of crossing across difference (Anzaldúa and Keatin, eds. 2002). As the biologists Maturana and Varela (1987: 246) put it, "A conflict can go away only if we move to another domain where coexistence takes place. The knowledge of this knowledge constitutes the social imperative for a human-centered ethics. . . . As human beings we have only the world which we create with others—whether we like them or not." This is, in fact, the deepest lesson of biology in the opinion of these two thinkers: "Without love, without acceptance of others living besides us, there is no social process and, therefore, no humanness" (246).

This emphasis of the framework also signals the widespread desire for peace that exists in many places like Colombia. To an order of imperial globality enforced through violence, the ecology of difference answers with a debate on distribution understood as the search for a shared sense of peace and justice. As a value, peace-with-justice does not belong completely to the domain of rationality but to that of ethics; it requires an attitude of transformation, caring, and solicitude in the face of difference and injustice. Peace-with-justice should be seen as always in process, something that can be approached only asymptotically but can never really be reached. To the declaration of war on nature and humanity by neoliberal globalization, there can only be a declaration of peace in which peace is both the means and an end. It is in the light of a planetary sense of ethics and spirituality like that found in the best of ecology and pluralist religious thought and in the best humanist traditions of secular modernity that one may find elements for a workable strategy of peace out of the recognition of conflict. Peace—understood as a set of economic, cultural, and ecological processes that bring about a measure of justice and balance to the natural and social orders—is the deepest meaning of the ecology of difference that aims toward worlds and knowledges otherwise.[7]

As a PCN activist put it, "Las diferencias son para enriquecer la acción y el pensamiento" (Differences are meant to enrich action and thought). For these activists, difference is the very source of a *pensamiento propio* (a thought of one's own), of differentiation in thought. One often finds among movement intellectuals the notion that difference is the very core of existence, that what persists is difference itself, not any unchanging essence. Difference is what defines being, and as difference is always in the process of being transformed, so is being. The oppressor, the colonizer, the dominant seek to occupy the time and energy of the subaltern to preclude difference from becoming an active social force. In places such as the Pacific today, this occupation of the time and space of difference is effected through brutal acts of repression and imperial models of war, economy, and development. Confronted with this situation and with the ideology of a *pensamiento único* (single thought) that pervades much of the world, activists attempt to create a breathing space for difference.[8]

The framework of the political ecology of difference (the integrated framework of diverse economies, environments, and cultures) is offered here as a contribution to a "global outline of a practical politics" that works by reading world events for difference, rather than just for dominance, and by weaving connections among languages and practices of economic, ecological, and cultural difference (Gibson-Graham 2006: 30). The framework is not offered as a universal approach; on the contrary, *it is a theory of difference that is historically specific and contingent*; it is a response to the present moment that builds on intellectual and political developments in many places, particularly some parts of Latin America. It is also partly a response to Eurocentric teleological arguments about the alleged universality of modernity and globalization. Above all, it is an attempt to think with intellectual activists who aim to go beyond the limits of Eurocentric models as they confront the ravages of neoliberal globalization and seek to defend their place-based cultures and territories; it is, finally, about projects of decoloniality in and for the present.

Colombia as a Theater of Imperial Globality

Talk of peace seems paradoxical when the world is increasingly nasty and in chaos. It is not well known that there are three million internally displaced people in Colombia, a disproportionate number of them black and indigenous. So much violence, often fueled by hatred and racism, one might think, surely needs to be confronted on its own terms. Indeed, to give just the roughest sketch of Colombia—one of five or six regions in the world where the struggle for the imposition of the terms of imperial globality is most fierce—let us say that in this country the "cruel little wars" (Joxe 2002) of imperial globality have been ongoing for several decades, as every year its fatidic embrace of territories and domains of social life broadens. Colombia represents patterns of historical exclusion found in many parts of Latin America but rarely in such depth. While inequality has been aggravated over the past twenty years by successive neoliberal regimes, it has a long historical basis in the structure of land tenure and industrial and financial capital. Today, 1.1 percent of landowners control over 55 percent of all arable land (and as much as one-third of this may be linked to drug money). Over 60 percent of the Colombian population live on incomes below the UN-established poverty line of two dollars a day. The armed conflict that affects the country is well known. It brings together a disparate set of actors—chiefly left-wing guerrillas, the army, and right-wing paramilitary groups—into a complex military, territorial, and political conflict, often intertwined with and aggravated by wealthy drug mafias.[9] From the perspective of imperial globality, all of these armed groups can be seen as war machines more interested in their survival than in peaceful solutions to the conflict. Massacres and human rights abuses are the order of the day, inflicted primarily by paramilitaries but also by guerrillas, and the civilian population is most often brought into the conflict as unwilling participants or sacrificial victims.

The subnational dynamics of imperial globality is pathetically illustrated by the experience of the Pacific region. This rain forest area has been home to about one million people, 95 percent of them Afro-Colombian, with about fifty thousand indigenous peoples belonging to various ethnic groups. Since the late 1990s, guerrillas and paramilitaries have been steadily moving into the region in order to gain control of territories that are either rich in natural resources or the site of planned large-scale development projects. In many river communities, guerrillas and paramilitaries both have pushed people to either plant coca or move out. Displacement has reached staggering levels, with several hundred thousand people displaced from this region alone. Displacement in some areas has been caused by paramilitaries, and this has often resulted in the expansion of African palm plantations by rich growers. The expansion of the palm frontier is supported by the government as a development strategy, including funds from Plan Colombia; this promotion has been linked in recent years to the international demand for biofuels.[10] In many world regions similar to the Colombian Pacific, ethnic minorities inhabit territories rich in natural resources that are now coveted by national and transnational capital (Mander and Tauli-Corpuz, eds. 2006; Blaser, Feit, and McRae, eds. 2004). Beyond this empirical observation, however, lies the fact that imperial globality is also about the defense of white privilege worldwide. By *white privilege* I mean not so much phenotypically white, but the defense of a Eurocentric way of life that worldwide has historically privileged white peoples (and, particularly since the 1950s, those elites and middle classes around the world who abide by this outlook) at the expense of non-European and colored peoples. This is global coloniality at its most material.

The case of Colombia and its Pacific region thus reflects key tendencies of imperial globality and global coloniality. The first tendency is the link between the economy and armed violence, particularly the still-prominent role of national and subnational wars over territory, peoples, and resources. These wars contribute to the spread of social fascism, defined as a combination of social and political exclusion whereby increasingly large segments of the population live under terrible material conditions and often under the threat of displacement and even death (Santos 2002; Escobar 2003a). In Colombia, the government's response has been to step up military repression and surveillance within a conception of "democratic security" that mirrors the U.S. global strategy as seen in the Iraqi case: democracy by force and without the right to dissent—a deterrence against common people.

Second, Colombia also shows that despite what could be seen as excellent conditions for achieving a peaceful society and democracy (e.g., very rich natural endowments, a large and highly trained professional class of both sexes, and determined cadres of activists that continue their labor of love against all odds), the opposite has happened. Why? Because the local war is in part a surrogate for global interests; because of intransigent national elites who refuse to entertain needed social reforms; and because the war logics (including drug mafias) have taken on a self-perpetuating dynamic. Finally, the Colombian case makes patently clear the exhaustion of modern models. Development and modernity, to be sure, were always inherently processes that created displacement. Yet what has become evident in the face of the excesses of imperial globality in places like the Pacific (but one can think also of the Sudan, the Middle East, and others) is that the gap between modernity's displacement-producing tendencies and displacement-averting mechanisms is not only growing but becoming untenable—that is, unmanageable within a modern framework (Escobar 2003a).

Which brings me back to the question with which I started this section: Does it make sense to talk about peace in this context, and if so, how is one to have a reasonable expectation that this talk will not play into the designs of the powerful? I shall take up this question again in the concluding chapter when I discuss the problematic of transition based on the idea that modernity's ability to provide solutions to modern problems has been increasingly compromised, making discussion of a transition beyond modernity feasible again. The intuitive question for now is this: Is globalization the last stage of capitalist modernity or the beginning of something new? I will address this question from the perspective of a politics of peace, place, and difference in which it finds its raison d'être.

Some Scholarly Contexts

The reader will find many references to specialized debates in the chapters that follow. I want to make only some very general remarks about literatures here. To start with political ecology: emerging in the 1970s out of the marriage of several ecological-oriented frameworks and political economy, political ecology has been an established field since the 1980s; today it is an interdisciplinary field drawing on many disciplines (geography, anthropology, ecology, ecological economics, environmental history, historical ecology, development studies, science and technology studies) and bodies of theory (liberal theory, Marxism, poststructuralism, feminist theory, phenomenology, postcolonial theory, complexity, and natural science approaches such as landscape ecology and conservation biology). More important is the range of questions with which it deals: the relation between environment, development, and social movements; between capital, nature, and culture; gender, race, and nature; space, place, and landscape; knowledge and conservation; economic valuation and externalities; population, and land and resource use; and so forth. This range of questions, conversely, refers to problems the very salience of which lends relevance to the field; these include, among others, destruction of biodiversity, deforestation, resource depletion, unsustainability, development, environmental racism, control of genetic resources and intellectual property rights, biotechnology, and global problems such as climate change, transboundary pollution, loss of carbon sinks, and the like.[11]

This work is situated within two domains of recent anthropological inquiry. The first is the trend started in the 1980s with the study of modernity and continued today, in a fruitful manner, with theoretical and methodological proposals focused explicitly on the ethnography of expert (Western) knowledge practices. This approach, pioneered by such scholars as Marilyn Strathern (e.g., 1991, 1992) and Paul Rabinow (e.g., 2003), is seeing a sophisticated development particularly in the field of the anthropology of science and technology and science and technology studies (STS), including informatics and cyberspace (e.g., Hess 2001, Hakken 2003). A crucial issue here is how to do the ethnography of situations that are fundamentally shaped by the same knowledge formations of which the ethnographer's knowledge is itself a product. This has led to ideas about critical anthropology (Marcus, ed. 1999), emergent forms of life (Fischer 2003), anthropology of the contemporary (Rabinow 2003), network and distributed studies (Riles 2000; Fortun 2003; Osterweil 2005b), and reconstructivist agendas in science studies (Woodhouse et al. 2002; this latter proposal seeks to bring together academic and nonacademic knowledge production spheres). The recasting of critical modernist anthropology is important to my study because it centers largely on activist knowledge practices—in many ways, as we shall see, a modernist enterprise. The second trend in which this book is situated is that of world anthropologies, an approach intended to de-essentialize anthropology and to pluralize anthropological inquiry by building on nonhegemonic anthropological practices. As in "worlds and knowledges otherwise," the world anthropologies project aims to foster "other anthropologies and anthropology otherwise." My book can be read in this light.[12]

With its acute reflexivity, the anthropological study of modernity pushes the boundaries of inquiry toward a renewed critical modernism; however, as I see it, it does not aim to question the project of modernity in the way that, say, Santos, Boff, or Mignolo do, nor do they call for a change of paradigm beyond modernity. This is why my book also adopts a framework that has been variously called "geopolitics of knowledge" (e.g., Mignolo 2000) in the humanities and "critical geopolitics" in geography (Slater 2004). These proposals, though connected to poststructuralist and postcolonial theory, are based on more than that; in particular, they bring fully into the picture the contributions from outside Eurocentric theory in order to put these theories' categories in question; these tendencies also engage with attempts to reimagine the world's geographies of power and knowledge in conjunction with social movements and experiences like the World Social Forum process. Besides questioning Western discourses, these trends pay attention to the epistemic potential of local histories embedded in or arising from the colonial difference, locating there some of the most meaningful sources for political action and for alternative world constructions. These local histories have remained largely invisible in Eurocentric theory precisely because they have been actively produced as nonexistent—as noncredible alternatives to what exists—calling for what Santos (2004) labels a sociology of absences that brings them into visibility, and a sociology of emergences that enables the enlargement of the range of knowledges that could be considered credible alternatives. This book is devoted to this sociology of emergences by foregrounding the contributions of a particular social movement.

Cultural studies is another important scholarly context for the present work. More than any other field, and after a period of apparent complacency, cultural studies today maintains a built-in reflection on its own contextuality. As Grossberg sees it, "Cultural studies is a project not only to construct a political history of the present, but to do so in a particular way, a radical contextualist way, to avoid reproducing the very sort of universalisms (and essentialisms) that all too often characterize the dominant practices of knowledge production. . . . Cultural studies seeks to embrace complexity and contingency, and to avoid the many faces and forms of reductionism" (2006: 2). Besides being contextualist and relational, cultural studies is centered on the study of conjunctures, understood in terms of articulations or condensations of contradictions within a given social formation that need to be examined along multiple axes, planes, and scales. This conception fits well with my focus on regions and places in the age of globality and on the particular conjuncture of the Colombian Pacific. The goal is not only to ascertain where and how the Pacific is at present, but also to illuminate how it can move from one conjuncture to another; these tasks, again, need to be developed in tandem with local social movements.[13] By focusing on the cultural potential of the knowledges produced by social movements and the epistemic force of local histories such as those of the black communities of the Pacific, the MCD perspective seeks to articulate cultural studies as a decolonial project.[14]

The last and perhaps most relevant body of work within which I would like to situate my book is the study of social movements. This field has been largely cultivated by sociologists and, to a lesser extent, political scientists and historians. Anthropologists are a late arrival to the field (although not completely; see Nonini, Price and Fox-Tree forthcoming), but there are reasons to believe that the interdisciplinary approaches arising from anthropology will have a noticeable influence on the field as a whole. For one thing, contemporary social movement theories are inadequate to explain the complexity of current forms of collective action—from place-based ecological, women's, and ethnic movements to antiglobalization protests (Leyva Solano 2003; Osterweil 2004, 2005b; Escobar 2004b). For another, a number of emphases are emerging from anthropological approaches, including the focus on activists as knowledge producers (and hence, the ethnography of knowledge production practices in this context); the blurring of the boundary between academic and activist worlds and knowledges, which a growing number of anthropologists are promoting for both theoretical interest and political disposition; and a series of concepts and domains of inquiry that arise readily from anthropological situations and reflections or in particular ways, such as, among others, network ethnography, ethnocartography, mapping of knowledges, ethnography of identities and activist figured worlds, cultural politics, and activist, partisan, or militant anthropology. Some of these notions are derived from encounters with movements themselves and with, so to speak, the activist within, in the sense that many of those wishing to understand today's movements also want to act and think with them and from their social and epistemic locations.[15] This means that the point of departure for working with activists is the political position of the movement, not academic interests; this creates a different basis for arriving at an enriched knowledge that, while allowing for disagreements, is arrived at from the perspective of the activists' reading and valuation of their own collective knowledge.[16]

The idea that social movements should be seen as knowledge producers is one of the main insights of this trend. This insight has many dimensions, beginning with an emphasis on the articulation between knowledge and resistance established by movements themselves; the identification of knowledge as a tool for struggle; the fact that activists more than ever engage in research on their own experiences—sometimes even drawing on critical academic theories; the relation between activist knowledge production and critical genealogies of thought; and the challenges all of the above pose for more conventional understandings and institutions of knowledge production. This trend is leading to engagements focused on knowledge production practices with particular movements, in the belief that knowledge is embedded in local contentious practice and in larger historical struggles (Holland and Lave 2001). The aim is to study the embeddedness of knowledge in social relations, that is, knowledge being produced in dialogue, tension, and interaction with other groups, and how this knowledge is enacted and networked.[17]

My attempt in this book is to build on ethnographic research in order to identify the knowledge produced by activists and to use this knowledge and these analyses to conduct my own analyses about related topics—or, as I like to put it, to build bridges between political-intellectual conversations in social movements about environment, development, and so on and conversations in the academy about corresponding issues. The book is only partly an ethnography of knowledge practices per se, although it is largely based on ethnographic research focused on activist knowledge, showing the tremendous complexity of its production and its embeddedness in social, political, and cultural processes. To say it metaphorically, as the Afro-Colombian historian Oscar Almario put it in his keynote address at the conference "Afro-Reparations: Memories of Slavery and Contemporary Social Justice," which included an opening act by an accomplished drummer from the city where it was held (Cartagena, October 19–21, 2005), my book is "un esfuerzo de la academia para estar más cerca de los tambores" (an effort by the academy to be closer to the drumming).[18]

A word about the book's title. First, *Territories of Difference* brings together two important social movement concepts in the Pacific: territory and difference. They are also persistent theoretical concerns in a number of fields. Any territory is a territory of difference in that it entails unique place making and region making, ecologically, culturally, and socially. In cases where different ontologies are involved, the theoretical and political treatment of difference becomes even more important. Second, I have decided to use "life" instead of "nature" or the "environment" in the subtitle since it seems to me that what is involved in situations such as that of the Pacific is the understanding and defense of life itself, in all of its complex manifestations. At issue for movements such as PCN is not just the environment, but the fact of being different and, in the last instance, life itself; hence, territories of difference are also *territorios de vida*, or territories of life, in the activists' conceptualization. Finally, the Spanish *redes*, more than the English term *networks* commonly used to translate it, conveys more powerfully the idea that life and movements are ineluctably produced in and through relations in a dynamic fashion ("assemblages" would be a better translation). Images of *redes* circulated widely in the southern Pacific in the 1990s; represented graphically as drawings of a variety of traditional fishing nets, lacking strict pattern regularity, shaped by use and user, and always being repaired, *redes* referred to a host of entities, including among others social movement organizations, local radio networks, women's associations, and action plans.[19]

1 place

Esta tierra es nuestra
Aquí hemos sido árboles y pájaros
Hemos aprendido
El ritmo de las olas
Para convertirnos en hijos del agua . . .
Esta tierra es nuestra
Como la felicidad
Que hemos inventado

Esta es nuestra tierra
La hemos fundado con dolor y sangre
Es lecho de nuestros sueños libres
Cuna de nuestros anhelos
Y tumba de nuestros viejos
Aquí el agua tiene sabor a nosotros

—From the collection *Esta Tierra es Nuestra*, by the Tumaco artist and popular communications activist Jaime Rivas, produced by Fundación Habla/Scribe, Cali, 1993.

[This land is ours
Here we've been trees and birds
And learned the rhythm of the waves
And become children of the water . . .
This land is ours
As is the happiness
We've invented

This land is ours
We founded it with pain and blood
It is the bed of our free dreams
The cradle of our desires
And the tomb of our elders
The water here tastes like us]

—Translated by John C. Chasteen

Introduction: The Pacific as Place, "Then and Now"

It seems long ago that Sofonías Yacup mentioned in the preface, the liberal politician from Guapi, one of the main towns in the southern Pacific, described the Pacific as a "lethargic and recondite littoral, an absent place entrapped in its own isolation," abandoned by the national government to its own destiny, and in dire need of redemption and progress (1934). Being a nationalist, Yacup had in mind a style of development grounded in local and national conditions. Like most treatises of the period, *Litoral Recóndito* contained a disjointed mixture of scientific observations, ideological injunctions, incipient use of statistics, defense of the *ideal latino* with the concomitant critique of North American materialism, and a catalogue of unclassifiable prescriptions, all of which could be said to constitute, in today's language, a call for an alternative modernity. If one had visited the Pacific in the 1960s or 1970s, one might say that little had changed since Yacup's passionate words of the 1930s; less recondite perhaps, the littoral was still seen by most as lethargic and cursed by its own history, and the era of development had yet to arrive. This situation changed drastically in the 1980s. As a well-known anthropologist eloquently put it,

> New times are announced for the lands of the Pacific corridor. From the tempest are born the new ideologues who, like demiurges, no longer summon the idiom of geographical determinism as the obstacle to the region's development. Again the *Mar del Sur* [the Pacific Ocean] awakens the amazement of the colonizer; this time, however, it is not Balboa who with his feverish dreams of golden particles contemplates the region in awe from the height of the Darién mountains. Now, it is the [scientific] gaze that has quantified the landscape, inventoried the forests, classified the species, measured the depth of the bays and which, in doing so, has seemingly lifted the veil that in an illusory manner portrayed this territory as a world populated by jungles, insalubrious places, rivers where the heaviness of heat made thinking impossible, and where only indians and blacks could dwell in their primitive spatiality. (William Villa, cited in Vargas 1993: 293)

In short, if for most of its history the Pacific was imagined as a faraway place doomed to backwardness by its very natural conditions, a place where only the extraction of resources by outsiders was practicable, the situation was to change dramatically in the 1980s and 1990s. What did this sudden change represent in the making of this region? For truly the new scale of concern and action did not erase overnight the natural and social dynamics of yesteryear. The question at stake is the steady incorporation of a given region into modernity, the nation, and the globe. The main processes of incorporation have changed, with development and technoscience occupying a prominent role today. So have the scale and intensity of the transformation. How did the unprecedented conditions of the early 1990s result in the ensemble of processes and practices that transformed "the Pacific"?

This chapter is concerned with what could be termed the making of a socionatural world. By this I mean an understanding of the complexity of relations between the biophysical and human domains (physicochemical, organic, and cultural, broadly speaking) that account for particular configurations of nature and culture, society and nature, landscape and place, as lived-in and deeply historical entities. Anthropology, geography, and ecology have been the disciplines most concerned with this question. In the 1950s and 1960s, cultural ecology approaches saw this complexity in terms of adaptation between humans and the environment. This view was criticized for its functionalism and because it saw the environment as an inert background to which organisms and humans adapt. It gave way to a dialectical view of the relation between organism and the environment, according to which each shapes and produces the other through continuous interactions. In the 1970s, the dialectical relation between humans and the environment was further complicated by setting it in the context of the political and economic forces in which they are immersed. Adaptation became a more encompassing concept entailing biological, social, and political processes, all of them mediated by culture (e.g., Whitten 1986 [1974] for the case of the Colombian Pacific). In the 1980s and 1990s, a new biocultural synthesis spearheaded by biological anthropologists provided an elegant theoretical framework for this view of adaptation, opening the door for infusing the study of nature/society relations with poststructuralist concerns with knowledge, power, gender, and identity (Goodman and Leatherman, eds. 1998; Hvalkof and Escobar 1998; Rocheleau, Thomas-Slater and Wangari, eds. 1996). In some cases, as in the Pacific, the notion of adaptive strategies was used in the 1990s to signal this complexity. In this synthesis, culture and nature are treated as fully historical and constructed, and socionatural worlds become the result of human action even if conditioned by particular environments.

This historicized view of the relation between nature and culture constitutes a further critique of modernity's view of nature as an inert background for the unfolding of the human saga. Worldwide, societies have been ceaselessly constructing bridges between nature and culture (Latour 1993). As we shall see, what types of bridges are built, and how, makes all the difference (see chapter 3). Modern capitalist societies link nature and culture in ways that contrast sharply with how black and indigenous communities do it. But this is getting ahead of the story. In this chapter, I am interested in building a view of the region called "Chocó biogeográfico" by biologists and planners and "Pacífico geográfico" or "region-territory" by social movement activists as constructed through geobiological, human, and technoscientific processes operating at many levels, from the microbiological to the geological and from the local to the transnational. Above all, I am interested in elaborating a view of the region as a place. Why place? Because place continues to be a crucial dimension of the making not only of local and regional worlds, but also of hegemonies and resistance to them. The tendency today is to state that globalization has rendered place irrelevant, meaningless, or at least secondary in the constitution of places and regions. But is this so?

If anything has characterized social science debates since 1990 it is the concern with globalization. These debates have been characterized by a pervasive asymmetry by which the global is equated with space, capital, and the capacity to transform while the local is associated with place, labor, tradition, and hence with what will inevitably give way to more powerful forces (e.g., Dirlik 2001; Escobar 2001; Harcourt and Escobar, eds. 2005). This marginalization of place has had profound consequences for our understanding of culture, nature, and economy, all of which are now seen as determined almost exclusively by global forces. It is time to reverse this asymmetry by focusing anew on the continued vitality of place in the creation of culture, nature, and economy. Place continues to be important in the lives of most people, if by place we mean the engagement with and experience of a particular location with some measure of groundedness (however unstable), boundaries (however permeable), and connections to everyday life, even if its identity is constructed and never fixed. There is an implacement that counts for more than we might want to acknowledge, which suggests the need for "getting back into place," to use the philosopher Edward Casey's expression (1993). This seems to be, indeed, an increasingly felt need of those working at the intersection of environment, culture, and development. Scholars in this field are often confronted not only with social movements that retain strong attachments to place, but also with the realization that any alternative course of action must take into account place-based (although not place-bound) models of culture, identity, nature, and economy.[1]

I have found it useful to think about the production of a region such as the Pacific in terms of six distinct, although interrelated, historical processes. This chapter will be mostly devoted to the first two processes. Subsequent chapters will focus on the remaining four.

1. Historical processes of geological and biological formation. Geologists and paleoscientists discuss the region's specificity, particularly its high levels of endemism and biological diversity, in terms of geological and evolutionary processes. Following theories of complexity, this geological and biological history can be seen partly in terms of self-organization of nonorganic and organic forms of life. Only the most basic elements of this history will be presented in this chapter.
2. Historical processes constituted by the daily practices of the local black, indigenous, and mestizo groups. This is the domain of history and anthropology. Through their daily practices of being, knowing, and doing, local groups have actively constructed, though in the midst of other forces, their socionatural worlds for several centuries. This chapter will highlight the contribution of the black river communities to this process.
3. Historical processes of capital accumulation, from the local to the global. Capital is doubtless one of the most powerful forces constructing most rain forest regions of the planet. Nevertheless, the construction of the Pacific cannot be explained solely in terms of capital. Indeed, it could be posited that forms of noncapitalism exist and are actually being created today out of the dynamics of place-based cultural and ecological practices, even if this occurs in the engagement with capital, modernity, and the state (see chapter 2).
4. Historical processes of incorporation of the region into the state, particularly through development and conservation representations and strategies. In the early 1980s, the Colombian Pacific was represented for the first time by state discourses as an entity susceptible to development. In the 1990s, this representation took the form of an ambitious sustainable development strategy, still under implementation today. Capital and development constitute a two-pronged strategy for the reterritorialization of the Pacific as a modern space of thought and action (see chapters 3, 4).
5. The cultural-political practices of social movements. After 1990, black and indigenous movements became important players in the representation and construction of the Pacific as region-territory. These movements set into motion a cultural politics which operated chiefly through the ethnicization of identity in close connection with ecological and alternative development concerns. By positing the notion of the *Pacífico geográfico* as region-territory of ethnic groups, the social movements of black and indigenous communities made visible the cultural, ecological, and economic place-making strategies of the communities (see chapter 5 and this chapter; chapter 6 for supra-place strategies).

6. The discourses and practices of technoscience, particularly in the areas of biodiversity conservation and sustainability. Since the early 1990s, biodiversity has become a powerful discourse in environmental and international development circles; it originated a network of sites that embraces significant domains of cultural and ecological action. As a network, biodiversity exemplifies the role of technoscience in the making of socionatural worlds. This network is confronted by networks of heterogeneous actors that include ecosystems, social movements, and NGOs; all of these networks became an important element in the struggle over the Colombian Pacific as territory (see chapters 3, 6).

In very schematic fashion, these processes can be further divided into two overall strategies, strategies that are not bounded and discrete but overlapping and coproduced:

1. Strategies of localization by capital, the state, and technoscience. Capital, state, and technoscience engage in a politics of scale that attempts to shift the production of locality in their favor. Nevertheless, to the extent that these strategies are not place-based (even if locally articulated), they inevitably induce a delocalizing effect with respect to places. This effect is in keeping with the overall thrust of modernity of sundering place from space (Giddens 1990) and deterritorializing social and ecological life (Virilio 1999).
2. Subaltern strategies of localization by communities and social movements. These are of two kinds: place-based strategies that rely on the attachment to territory and culture; and network strategies that enable social movements to enact a politics of scale from below. In the Pacific, this entails engagement by local movements with biodiversity networks, on the one hand, and with other place-based actors and struggles, on the other. In this way, social movements develop a political practice that can be described as place-based yet transnationalized (Harcourt and Escobar, eds. 2005; Escobar 1999b, 2001). There are localization strategies by other groups that do not fit easily into these two categories, such as those of armed groups and drug cartels. I will end the chapter with a brief discussion of the devastating effects of strategies of displacement and deterritorialization by armed actors.

The first part of the chapter provides an overview of the physical and economic geography of the Colombian Pacific. This is continued in part II with a broad discussion of the history of the region's settlement and change, particularly by black groups. In these two parts, I take the existing geographical, historical, and anthropological literatures as the material to be analyzed. Part III starts the discussion of the strategies of social movements for the defense of the Pacific as place. This initial discussion, to be continued throughout the book, introduces an important conceptual innovation produced by the social movement of black communities in the second half of the 1990s, namely, a political ecology framework articulated around the notion of the Pacific as region-territory of ethnic groups. Part IV, finally, discusses the current situation, particularly the impact of forced displacement on the strategies of place. The conclusion restates the notion of the politics of place and points toward chapters on capital, nature, development, and identity that will examine these dimensions of place in greater depth. As an exercise in political ecology, the aim of this chapter will be, as Dianne Rocheleau well put it, "to incorporate multiple past and present stories of places and peoples before attempting to 'solve' their 'problems'" (1995a: 1047). This goes against much ahistorical development and conservation intervention today and speaks of the value of place-based political ecologies.

I ◈ Notes on the Geological and Biological History of the *Pacífico Biogeográfico*

The *Pacífico biogeográfico* provokes images of excess: "The most spectacular rainforest of the world" (Palacios Santamaría 1993: 363); "one of the places on the planet where water is most abundant" (Lobo-Guerrero 1993: 122); "the wettest section of the New World" (West 1957: 22) with a mangrove forest that is "one of the most luxuriant of the world" (West 1957: 62), and with "the most beautiful coconut trees the Expert had ever seen, including those seen in Asia and Africa" (Ferrand 1959: 14); "perhaps one of the most complex ecosystems in the world" (Whitten 1986 [1974]: 25). It is repeatedly said that the Colombian Pacific is one of the richest regions of the world in terms of biological diversity, some places having "the highest degree of endemism in South America and probably in the entire world" (Gentry 1993: 201). In a more poetic vein, the Pacific is likened to a "cosmic singularity" where "the laws of nature all seem to be altered and where the natural and the supernatural haunt fishermen, miners, gold artisans and even the professionals born to this jungle place. Rivers of gold, earthquakes, tsunamis, and interminable rain all serve as the context for human passions" (Ramírez 1991: 15). Purely scientific description would not do, even if most often this is what we get. As Taussig says, with places like the Colombian Pacific in mind, "I am perplexed as to the absence of heat in movies set in these torrid zones. And not only in the frivolous medium of film, but in our ever so serious anthropological monographs as well as in the stories we tell. Nature has its 'special effects', as do movies, but some, it appears, are more equal than others." And he continues: "The humidity and heat bend your will to live," but with the important caution of being "mindful of the fact that African slavery in the new world was in part based on a notion that those slaves took the heat a whole lot better than Europeans" (2000: 1, 2).

The reference to physical elements calls for a phenomenological description of place in terms of "special effects" and the experiences that are central to regions such as this one, including heat, water, rain, and of course the forest. It could be a phenomenology of substances and daydreams, à la Bachelard: gold, trees, earth, visions, and all kinds of animals, plants, and beings ranging from those of the sea and the rivers to those of the forests and the infra- and supra-worlds. Taussig is right in pointing at the paucity of such descriptions; this study is not going to fill the gap. Local poets and *cantadoras* (women singers) of the Pacific are more likely guides to a phenomenology of the place, and some experts in the oral tradition have tried their hand at it with a good measure of success (see especially Friedemann and Vanín 1991; Pedrosa and Vanín 1994). For now, I am more interested in restaging, as it were, the various realist descriptions of the Pacific given by natural scientists, anthropologists, and geographers so as to convey a notion of the place. What I hope will emerge at the end of the chapter is a view of what scientists call the Pacífico biogeográfico as a region constituted by historical processes that entail geological, biological, political, and sociocultural dimensions.

Complexity theorists argue that all the structures that surround us—from mountains and ecosystems to social institutions—are the product of historical processes. History is a property not only of human and biological processes (evolution), but also of physics and chemistry; according to some biologists, language and meaning are a property of all living beings (Markos 2002; Goodwin 2007). In the last instance, history is a feature of the intensity of matter and energy as they become actualized, producing the particular forms we see in the world. Some argue that the same basic mechanisms are involved in the historicity of geological, biological, and social structures, even if they operate in different forms and take different shape in the three domains. Often these structures show emergent properties that are the result of unpredictable interaction between parts. Intensifications of matter and energy foster nonlinear dynamics that result in the generation of novel structures and processes, different kinds of accumulation of materials, shaped and hardened by history. Two of the main structures that result from nonlinear dynamics are self-organizing meshworks of diverse elements and hierarchies of uniform elements. Meshworks and hierarchies (discussed in more detail in chapter 6) coexist and intermingle and might give rise to one another. Rain forest ecosystems and small-town markets that arose spontaneously in many parts of the world up to the eighteenth century are examples of meshworks; bureaucracies and organizational structures with conscious goals and overt control mechanisms exemplify hierarchies. While hierarchies result in the formation of strata (e.g., sedimentary rocks, species, social classes), meshworks produce self-consistent aggregates out of the articulation of heterogeneous elements in terms of functional complementarities (e.g., granite rocks, ecosystems, certain networks of social movements). This new materialist philosophy promises to give us a view of the commonalities among the nonorganic, organic, and social worlds; its aim is to explain the construction of a historical universe in which self-organizing processes, showing emergent properties, play an important role (e.g., de Landa 1997, 2002; see also Prigonine and Stengers 1984; Kaufman 1995; Solé and Goodwin 2000; and Camazine et al. 2001 for accounts of complexity and self-organization in science, particularly biology).

This way of thinking about nonorganic, organic, and social life is useful for analyzing the production of place in its various dimensions. I begin by considering at some length the description of the long geological history of the Pacific. At the level of discourse, it is not difficult to notice the assumption of life in the metaphors employed: [2]

> Much before the Chocó geographic area existed, South America was populated by plants and animals originated in the southern supercontinent of Gondwana, made up of Africa, Australia, India, and South America. After the fragmentation of Gondwana due to the movement of tectonic plates, South America separated to become an island-continent. This separation started from the south, between Africa and South America, continuing northward and giving place to the southern Atlantic ocean. At the beginning of the Cenozoic, 65 million years ago, South America had completely separated from Africa and slowly drifted apart towards the northeast. This isolation lasted for over 50 million years, during which time the fauna changed through evolution according to the climatic and ecological changes that were taking place. . . . Finally, the South American and Pacific plates collided provoking seismic and volcanic activity that caused the rise of the Andean cordillera along the Westernmost border. In a similar manner, the collision between the North American and Pacific plates resulted in a series of orogenetic processes [the lifting of mountains] in the southern and eastern parts of North America, creating a series of islands

between the Central American nucleus of Nicaragua-Honduras and the northwestern tip of South America, what today is known as Colombia. The same tectonic movements that caused the rising of the Andes also caused geological foldings in Western Chocó. In the Eocene, beginning 55 million years ago, before the joining of the two American continents, there occurred an orogenetic process to the West of the Andes. Rocks from this period (and from the Miocene) have been found all along the coastal cordillera, from Panama to Cabo Corrientes, including in the Gorgona Island and the northwest of Ecuador [see map of the Pacific at the beginning of the book]. In this way there appeared a mountain chain from the sea bottom west of South America or at least a series of islands bordering the depression known as the geosynclinal Bolívar. This depression was at that time one of the marine connections between the Atlantic and the Pacific, keeping apart the landmass of the two continents which were on a collision course. (Alberico 1993: 236)

The closing down of the gap between the two continents is relatively recent in geological terms, dating from the early Pliocene (3.7 to 3.1 million years ago), and is considered a landmark event in terms of the biotic connection between north and south. The partial emergence of the Panama Isthmus as a series of islands had started by the middle Miocene, about 12 million years ago. The importance of these changes in biological terms, as paleoscientists explain, was enormous and accounts for the peculiar biological makeup of the Pacífico biogeográfico. Among the relevant facts are the following: the terrestrial exchange of faunas between 9.3 and 8 million years ago, increasing rapidly after the sealing of the isthmus in the Pliocene; the growing diversification of northern genera and species in the new environments (e.g., frogs, toads, serpents, deer); conversely, the colonization of Central American tropical forests by species from the south; the establishment of barriers that affected the distribution of certain species (e.g., rodents of the *Orthogeomys* genus and of certain freshwater crabs of the *Potamocarcinus* genus). These changes were, in turn, affected by transformations caused by glaciations occurring during the past 500,000 years which resulted in the joining of many islands with the continent given the drastic fall in the sea level (e.g., Gorgona island, off the coast of present-day Guapi in the southern Pacific, which separated again from the continent about 10,000 years ago). The alternation of glacial and interglacial periods fostered fragmentation and reunification of the neotropical forests, which in part is believed to have acted as a "motor of speciation," thus explaining their high degree of diversity. Another factor contributing to diversity was the emergence of the Andes, which conditioned the isolation of the Pacific lowlands from the Amazon lowlands, thus stopping the interchange of species between the two regions. The result of isolation and further speciation was the high degree of endemism that biologists mention when discussing the Pacific (Bernal and Galeano 1993; Gentry 1993; Duque-Caro 1993; Alberico 1993).

The geology of the region has been poorly studied. Whatever is known comes from research done by government agencies and multinational mining corporations interested in the region's rich mineral deposits, including oil. The geologists' map of the Pacífico biogeográfico resembles a well-drawn quilt in which bright colors represent relatively well-differentiated bands—geomorphological regions (e.g., alluvial sediments, basalts, volcanic rocks, auriferous areas)—according to geophysical properties, the type of mineral deposit, and the geological time of their formation. Equally imposing is the stratigraphic representation (stratigraphic columns) that shows successive and changing bands of mineral materials, from the Cenozoic to the Pleistocene, arranged according to families (e.g., the Uva and Naya-Naipí families, south of Buenaventura) and formations (e.g., the Darién or Dagua formations), all of which are characterized by particular mineral composition and strata (see, e.g., Galvis and Mojica 1993). Of great importance in this representation is the discussion of the historical processes that resulted in such geomorphologies; these range from tectonic movements and magmatic and telluric activity to geological evolution caused by the action of the ocean, rivers, and climate. These forces yielded the structures and strata which, through self-organizing processes, came to constitute the matrix for the development of biological and social life. The physical landscape, including the layout of the land and the various types of landscape units that provide the context for biological and human action, is of central importance.

West's (1957) study of the lowland areas of the Pacific is still taken as a landmark study, although much has been done over the past two decades to qualify and go beyond this pioneering work. West distinguished three landscape types: the lowlands of recent alluvium, formed by marine sedimentation along the Bolívar geosyncline; these are followed eastward by hill lands of dissected tertiary rock; and finally the mountain areas of Mesozoic origin (see also Pedrosa 1996a; Martínez 1993). The Atrato river basin (two hundred miles long) constitutes the best example of alluvial morphology in the lowlands. Equally important are the deltaic areas along the littoral from Cabo Corrientes to Ecuador, which form the characteristic landforms of this landscape, including natural levees, swamps, and shallow lakes, or *ciénagas*. This alluvial strip is five to fifteen miles wide, although it widens to thirty-five miles in the Patía area north of Tumaco. In the southern part of the region, besides large rivers like the Patía and Mira, there are many smaller rivers, ranging from twenty-five to fifty miles long and often no more than ten miles apart, flowing from the Andean cordillera toward the ocean. Natural levees and freshwater back swamps give way to mangroves, which become predominant in the vicinity of the sea. The alluvial plain is bordered by a band of hills, which, in the southern Pacific, extends two hundred miles from Buenaventura to Ecuador. This hilly area constitutes a belt twenty to forty miles wide between the relatively narrow alluvium band and the Western cordillera of the Andes. Close to the rivers, the landscape is characterized by terraces which are usually the site of habitation and cultivation. As one moves away from the rivers, the terrain slopes into jungle-covered hills one hundred to two hundred feet in elevation. Valleys between the hills may harbor swamps with various kinds of palm trees. The slopes of the Andean Western cordillera are also covered by dense forest; as in the hills, an intricate system of shallow roots prevents major erosion. More disaggregated analyses identify a larger number of formations, from the ocean, beaches, and mangroves to various types of neotropical forests, including lowland, sub-Andean, and Andean forests, all of which present markedly diverse characteristics (e.g., Pinto-Escobar 1993).

West introduces a distinction which is useful in visualizing the layout of the landscape, as far of the littoral itself is concerned: "four geographic belts arranged in sequence from the sea inland":

> (1) a belt of shoal water and mud flats immediately off the coast; (2) a series of discontinuous sand beaches, interrupted by tidal inlets, estuaries and wide mud flats; (3) a zone of mangrove forest, usually one-half to three miles wide; (4) a belt of fresh-water tidal swamp, lying immediately back of the brackish-water mangroves. Inland from the tidal swamps on slightly higher ground lies the equatorial rain forest that covers the greater part of the Pacific lowlands. Each of the three latter littoral zones—the discontinuous beach, the mangrove, and the fresh-water swamp—is distinguished by a given arrangement of certain types of landforms and vegetation associations. Each also presents peculiar problems in terms of human travel and subsistence. (West 1957: 53; see also Pedrosa 1996a: 36)

On the northern coast the littoral is made up of cliffs, while south of Cabo Corrientes it is often characterized by sandy beaches interspersed with estuaries, small rivers, tidal ponds, and often flanked by coconut trees. The beaches are molded by the tides and ocean currents; while in many places this has meant an advance of the sea inland at the expense of the beach and the forest, in others the opposite process has taken place. It seems to be true that many beaches have experienced significant retrogression, a fact that the old people lament. This retrogression has been due to the slow action of waves and currents, but also to catastrophic events such as tsunamis (West 1957: 57; Pedrosa 1996a: 35). The great tsunami of January of 1906, its epicenter in northwest Ecuador, radically transformed the beaches, villages, and mangrove areas for a stretch of sixty miles around the Tumaco area. It is said that in one of the neighboring islands one of the waves reached a height of seventy meters.

The mangrove forest is itself seen as an adaptive response, particularly the legendary system of aerial roots that characterize it (Cantera and Contreras 1993; Rangel and Lowy 1993; Prahl, Cantera, and Contreras 1990). This adaptation to high levels of salinity and unstable grounds entails the exchange of gases in anaerobic substrates and a particular type of seed reproduction. The most common mangrove species in the southern Pacific are the so-called red mangrove (*Rhizoporaceae* family) and the black mangrove (*Avicenniaceae*). This amphibian ecosystem occupies large extensions in this part of the Pacific. Some of the trees can reach heights of over one hundred feet, with diameters greater than three feet. At low tide, the root system can be ten to fifteen feet high, forming "an almost impenetrable maze of vegetation" (West 1957: 63). This structure explains why activities in the mangrove are largely regulated by the tides. Moreover, the mangrove swamps create a system of meandering channels known as esteros that establish a network essential to transportation between villages, rivers, and beaches. Besides its value in regulating a number of important functions in the maintenance of aquatic life (see chapter 2), the mangroves afford rich resources to the local inhabitants, from wood for house construction and tannin extraction to a wide variety of fish and shellfish (local crabs, oysters, clams, and many species of fish). Well known are the women who brave the harsh conditions, their legs submerged in mud and their bodies under attack by swarms of mosquitoes they try to keep at bay by burning cow dung, to collect shellfish and mollusks from the swamp.

In what Whitten came to describe as "the wet littoral" (from the Esmeraldas river in Ecuador to the San Juán river in Colombia, roughly what I call the southern Pacific), "the mangrove swamp seems to crystallize the buzz of life, conditioned by the sea and backed by jungle resources" (1986 [1974]: 34). But this is not all, since "the mangrove swamp itself provides a major contact point for a good many international transactions, ranging from lumber booms to possible international political intrigue [and including tannin extraction from the tree bark, long in desuetude]. *Such transactions are a crucial part of the socio-economic environment of Afro-Hispanic culture, which is caught up in, though not altogether dependent upon, international economy*" (34). Today, such ties entail the destruction of the mangroves for the establishment of large-scale shrimp cultivation for export (see chapter 2).

Moving inland from the mangrove is a narrow belt of freshwater swamp, often formed by the overflow of rivers under the action of the tide. An important tree predominates in this belt, the *nato* (*Mora megistosperma*). Two other species of great economic importance to local people and to timber extractors are the *sajo* (*Campnosperma panamensis*) and the *cuángare* (*Virola*, spp.). All of these are large trees. The freshwater swamp is also home to the famous *naidí* palm tree (*Euterpe oleracea*), the tips of which are harvested, canned, and shipped abroad under the product name "heart of palms." This beautiful palm tree, which adorns the *vegas* (riparian areas) of the rivers with its tall, slender figure, has been decimated in many parts of the littoral. Palm trees are very important all over the Pacific; biologists proudly mention the thirty endemic species that are found nowhere else in the world (Bernal and Galeano 1993: 222). Very well known, besides the *naidí*, is the *chontaduro* palm tree, which gives a date fruit of the same name sold in many cities of the interior by black women who set up their stalls on busy street corners, most conspicuously in the city of Cali. Palm trees provide many products to the locals, from dates and oil-rich foods to material for roof thatching.

The rivers and the forest are the most immediately striking elements of the Pacífico biogeográfico. To the north, the large rivers Atrato and San Juán, along with the Cordillera Occidental and another important mountain chain, the Serranía del Baudó, constitute the most important elements of the topography and ecology of the region. To the south, a large number of rivers (the Mira, Mataje, and Patía being the most important) flow westward from the Andes to the ocean. Five rivers flow into the Tumaco ensenada, creating a proliferation of estuaries that are utilized by locals for transportation and fishing, always attentive to the rhythm of the tides; a similar river system makes up the rural areas of the port of Buenaventura. All along the rivers, levees or vegas of various widths provide the space for human settlements and the growing of such crops as maize, coconut, cocoa, and plantains. Traveling up and down the rivers of the Tumaco ensenada, for instance, one sees at every so many bends of the river a cluster of a few houses facing the river, which on occasion give way to a larger cluster or town. As we shall see, anthropologists discuss this pattern of settlements in terms of social and ecological articulations among various parts of the river, the forest, and the various worlds (see chapter 3).

The rain forests occupy the largest part of the region (77 percent, or about eight million hectares [a hectare is 2.47 acres], over 50 percent of which, owing to the low intensity of traditional production systems, still has continuous cover with low levels of intervention; see, e.g., Sánchez and Roldán, 2001). As in other tropical humid forests, the Pacific is characterized by several layers, including tall trees sixty to one hundred feet high that might form a canopy, a medium stratum of shorter tress and palms (twenty to thirty feet high), and a ground cover of ferns, vines, tree seedlings, and other vegetation. Yet there are structural and physiognomic factors that differentiate the Pacific rain forest from other similar forests, for example, a high density of small and medium-size trees, which in part explains the unusually high levels of plant diversity. The elevated rain level is another factor contributing to this diversity. In fact, the Pacífico biogeográfico "claims the world record for plant diversity" (Gentry 1993: 208). Despite the inadequate taxonomic knowledge (much regretted by botanists), this claim seems undisputed. The predominant families (*Leguminosae, Rubiaceae, Palmae, Annonaceae, Melastomataceae, Sapotaceae*, and *Guttiferae*) are similar to those of other neotropical forests, but here too there are significant specificities, such as the preponderance of *Guttiferae* and *Melastomataceae* species. To add to this list of unique features and "world records" recorded with pride by biologists—often for strategic conservation purposes—a number of families present the largest leaves known in the world in their respective categories; the largest of these have been described as "tissue masses 1 meter long and 50 cm wide, probably the largest individual leaves of any dicotyledon in the world" (Gentry 1993: 218). For the conservation biologists, as we shall see in subsequent chapters, the consequences could not be more clear. One of the best-regarded biologists to work in the Colombian Pacific wrote,

> In this context, the Chocó region, where even the low areas show marked patterns of cloudy forests, is in all likelihood one of the most interesting

spots on earth from the perspective of the evolutive "hot spots." It is thus that the Chocó gets to be situated at the center of the growing concern with the preservation of the world's biodiversity. Moreover, to the extent that biodiversity might be directly translated into an economic patrimony, a significant part of the global capital of biodiversity could reside in the endemic plants of the Chocó forests. (Gentry 1993: 218)[3]

Many constructivists may find the above geological and biological description beside the point. It is humans, to be sure, who make the distinctions I have been tracing (mountains, swamps, forests) and who endow them with meaning. Indeed, it is part of the making of place that humans make such distinctions. Yet this constructivist perspective of place and region misses some important points. First, the landscapes these distinctions demarcate are by no means irrelevant to the kinds of lives people and other beings make in them (lives are not the same in a desert as in a rain forest, to use two extreme examples). Second, and conversely, the external world is certainly not irrelevant to the kinds of distinctions humans make. Local people might—and indeed do in some instances—make partially different distinctions from those of the scientists. Ethnobiologists have discussed at length the relation between "folk taxonomies" and scientific ones, some arguing for a single "map of nature" that human groups label differently, others favoring instead a radical incommensurability among worlds (Berlin 1992). Third, different places have different things to offer humans to work with and live in, and this has everything to do with how humans construct places. Places are thus coproductions between people and the environment. Sea, forests, mountains certainly constitute marks in the territory, even if for the constructivists they might not constitute " 'natural' markers." As Ingold (1993) so aptly put it, boundaries may be drawn in the landscape only in relation to the activities (the "taskscapes") of the people (or animals) for whom it is recognized or experienced as such. This, however, begs a question of importance to ecologists: Why are certain features of the landscape more readily identified as boundaries than others? And in what ways? The "tasks" that forms of matter and energy besides the human perform by and on themselves are crucial.

Also of importance to ecologists is the effect of certain boundaries on the kinds of biological lives that emerge in divergent landscapes. Biologists have no doubt that the Andean mountain range, for instance, acted historically as a natural barrier between the Amazon lowlands and the Pacific. The entire exercise of the ecological zoning of the Pacific for conservation purposes in the 1990s was based on the premise of the existence of such barriers and natural landscape units (*unidades de paisaje*) (e.g., Ministerio del Medio Ambiente/IGAC 2000). True, any boundary both separates and connects. This is even truer of the Andes, which connects and creates ecological interdependencies between the Pacific lowlands and inter-Andean valleys such as the Cauca Valley, for instance, in such a way that they could be seen as two parts of the same larger ecoregion. But the differences are also real. In the last instance, what is at stake here is the difference between a system and its environment, a theme that has been persistently addressed by systems theory since its inception in the 1950s (see, e.g., Churchman 1971). The most recent solutions to the system/environment puzzle emphasize both the self-referential, self-organizing, or autopoietic character of the system and the system's structural relation or coupling with its environment (e.g., Jantsch 1980; Maturana and Varela 1980, 1987; see chapter 6). In the Pacific, the recursiveness of biophysical and cultural processes has effected and tended to maintain certain basic relations and boundaries, whereas the more contemporary social dynamics tend to shift them, thus making more difficult, if not impossible, the region's functioning as an autopoietic whole.

In the Colombian Pacific, to sum up, an entire world opens up between the tall mountain range to the east and the littoral to the west. Like many other socionatural worlds, it is profoundly unique. To reiterate, its boundaries are constructed and are in fact being transformed by whites and mestizos who are moving in large numbers to the Pacific, with very deleterious effects. In recent decades, this region has been named in various ways by different actors. This means neither that the rain forest is an irrelevant or inert background for the meaning-making activities of humans, nor that it can be understood independently of the complex history so sketchily outlined above. The most relevant question in terms of my purpose of examining the Pacific as place continues to be that of what happens when humans arrive on the scene. Historians and anthropologists have described this process.[4]

II ◈ Settlement, Habitats, and Peoples of the Colombian Pacific

As things Pacific became an object of discussion, the settlement of the lowlands since the Spanish conquest became an important focus for historians and other scholars, particularly the spatial and demographic expansion of the black population, its relation to economic changes introduced by white people, its effects on preconquest indigenous peoples, and its relation to and impact on the coastal, mangrove, river, and forest ecosystems. Historians like Germán Colmenares (1979) have contributed to the history of the region with studies of the interaction between slave-owning haciendas in the neighboring inter-Andean regions and the structure of gold mining in the Pacific. In recent years, a handful of historians have started the laborious historical investigation of the region, representing the processes by which it arrived at its present form. Under the encompassing title "Settlement, Habitats, and Peoples of the Pacific," the French-born and longtime Colombian resident Jacques Aprile-Gniset (1993) advanced an interesting hypothesis about the Afro-Colombian expansion in the Pacific that culminated in the late nineteenth century and can still be seen at play in some ways today. Like Aprile-Gniset, the Cali-based historian Mario Diego Romero (1995) has written an early and crucial chapter of this history under the suggestive title "Settlement and Society in the Colombian Pacific, XVI–XVIII Centuries," namely, the advance of the gold-mining frontier based on a particular organization of the enslaved population. Romero (1997) continued this task by researching the ethnohistory of a single river basin, the Naya river, south of Buenaventura, from the perspective of the territorial appropriation by black communities that became so debated in the context of the territorial and cultural rights law of 1993, Ley 70. Studies in the 1980s and 1990s gave an overall idea of the slow construction of the Pacific as a region by focusing on specific localities and on aspects of the underlying logics shaping the region. What follows is a synthetic account of the emerging views based on a variety of sources.[5]

Aprile-Gniset distinguishes two cycles in the settlement of the Colombian Pacific. The first was that carried out by aboriginal groups. Despite some evidence of early habitation of coastal areas in the northern Gulf of Urabá as far back as eight thousand years ago, little is known about the intervening period until the early pre-Hispanic period (about 1000 B.C.); even after this period the archaeology of the Pacific is scant at best (see Stemper and Salgado 1995 for a review). According to some, the region was integrated to other parts of the Americas many centuries before the arrival of the Spaniards through its many important river deltas, such as the Baudó, San Juan, and the Patía, that conquistadors like Pizarro discovered in their voyages. To the south, the Tumaco cultures of La Tola were important two thousand years ago (A.D. 500 to 500 B.C.). Based on the abundant ceramic record they left, some researchers hypothesized that the vitality of this culture was due to contact and migration with other parts of the Americas and possibly East Asia; its disappearance is thought to have been caused by natural disasters, perhaps tsunamis and earthquakes that pushed the survivors to migrate out of the area (Errázuriz 1980; Pedrosa 1996b). But the settlement of the Pacific by native groups was substantial. For at least a millennium, indigenous communities occupied the territory with particular forms of appropriation of the environment—flora, fauna, rivers, sea, and the use of gold and platinum—practicing until recent decades a type of subsistence livelihood appropriate to the humid forest, without the domestication of animals or significant sedentarization (Pedrosa 1996b; Ulloa, Rubio, and Campos 1996). Today, indigenous groups make up about 5 percent of the population, with demographically, culturally, and politically important groups such as the Embera and Wounaan in the northern Chocó province. [6]

Aprile-Gniset (1993) labels the second cycle Afro-American. It articulated with the Spanish conquest and expansion, gained momentum in the seventeenth century, became strengthened in the eighteenth century, and reached its maximum territorial and demographic expression after the end of the nineteenth century. This cycle had in turn several phases: a partial expansion, restricted to the accessible mining enclaves; this phase was advanced through military conquest and the subjugation of indigenous inhabitants, often met with fierce resistance; as another historian put it, the Indians' "hostility, high death rate, inability to produce quantities of bullion, and conflicting Spanish ideologies concerning their treatment and religious conversion soon necessitated the use of blacks to fill the labor shortage" (Sharp 1970: 24). This phase gave origin to a purely extractive economy with rudimentary mining techniques based on enslaved labor (stream placering or panning and other types of gold extraction, such as pit placering or horizontal drafts in interfluvial areas or high terraces). Romero examined this process in detail for the southern Pacific. In his view, the fundamental dynamic was characterized by a simultaneous process of adaptation and resistance to enslavement, articulated around the mining groups, or *cuadrillas*, created by the slave owners as production units. The cuadrillas developed cultural and social forms of organization that resulted in domestic relations in which women provided internal group cohesion while men took on the relations with the white society. Over time, the cuadrillas gained mobility and expanded their field of relations, thus creating conditions in which they could obtain their freedom. In the interstices of the slave-based mines there appeared settlements of free blacks devoted to traditional mining based on kinship relations; these groups used their own resources as well as resources borrowed or learned from indigenous groups for their cultural and ecological adaptation to the environment under conditions of greater autonomy.

This dynamic was at the basis of the occupation of territories linked by the black communities (Romero 1995), coinciding with Aprile-Gniset's second phase. This phase is characterized as

> extensive and peaceful, fostered initially by maroons and subsequently by manumitted blacks; it is an agrarian phase and of independent mining that starts in the late colonial period, prospers after manumission [1852], reaches its full territorial development in the early years of the twentieth century, and continues to exhibit some of its dynamics even today. To sum up, if the first phase is characterized by an external colonialist project, the second takes up the form of an endogenous agrarian colonization. Where the military enterprise of the Spanish soldier failed there succeeded, three centuries later, the work of the peaceful black colonist cultivating plantain, rice, manioc, maize, and coconut palm trees . . . an intense circulation resulted in a prodigious territorial expansion through the maximal dispersal of the population and the establishment of numerous agrarian habitats. (Aprile-Gniset 1993: 12, 13)

This process was punctuated by a host of factors in myriad ways. To mention a few of these: the precarious establishment of the mining enclaves (*reales de minas*) did not take place until the first half of the seventeenth century, when indigenous groups were subdued and the first enclaves appeared in rivers like the Telembí, Timbiquí, Guapi, Iscuandé, and Maguí. These mining centers maintained a significant relation with cities like Cali, Popayán, Pasto, and even Guayaquil and Lima for the supply of slaves and miscellaneous provisions. There was a circulation of knowledge among blacks and indigenous groups around mining and agricultural techniques; the system of "slash and mulch" (as opposed to the more common slash-and-burn agriculture), as West (1957) has called it, was an indigenous practice that blacks adopted to their own ends. Interethnic marriages, especially between black men and indigenous women, were common in certain areas, and there were occasional alliances against slave owners and invaders. Indeed, indigenous groups, organized in the well-known *encomiendas*, were often forced to supply food for the cuadrillas. Slaves obtained their freedom in a variety of ways, including self-manumission through the purchasing of their freedom with the proceeds from independent mining they conducted on Sundays and festivities. *Cimarronismo* (maroonism) was also important as a source of freedom, leading to the consolidation of free settlements known as *palenques*, where processes of cultural, demographic, and military resistance and reconstitution took place. In the late colonial period and well before manumission, there thus appeared a significant number of free blacks (*libres*) who, without breaking completely with the Spanish mining enclaves, started the exploration and colonization of other areas of the Pacific. While articulated with the enclave system, the settlements of libres did create conflicts with the slave-based system.

Race was a central organizing factor in the process of settlement and the region's economy. For some researchers, nature, landscape, economy, and race constituted an integrated system in the assembling of the Pacific. The view of the region as a sort of pantry or cornucopia of riches to be extracted was inextricably linked to the harnessing of black labor (from colonial slave mining to today's African palm cultivation), not infrequently through representations of race that depicted blacks in natural terms. Rural areas and the peoples who lived in them were seen as lazy, savage, and primitive, and only blacks were seen as able to withstand the humid jungle. Under these conditions, blackness functioned largely as a cultural category, although phenotypic characteristics were also important (Leal 2003). From colonial times, the Pacific has been seen as the black region par excellence in Colombia; however, until recently the study of race and racism in the Pacific was largely restricted to white–black relations in towns and to arguments about the poverty of the region as a whole. This made sense up to a point: as we shall see (chapter 5), the category of "black" was not predominant until recently in black riverine settlements, and in the 1990s most scholars and movement activists discussed the Pacific in terms of cultural difference and rights more than of race and racism, even if these were often in the background of the discussion. The exception has been black urban movements, for which race and racism were much more salient concerns.

This perspective has changed over the past few years, particularly through research in ethnohistory and historical anthropology and geography, a great deal of it by Afro-Colombian scholars. This research looks at ethnicity, region, and nation as interwoven processes involving the Pacific, the larger region (the *suroccidente colombiano*, or Colombian southwest), and the nation. These newer works highlight the relations of power between the Andean centers and their "white" inhabitants and lowlands such as the Pacific. What this historical research shows is the production of what the Cartagena historian Alfonso Múnera (2005) has named "hierarchical geographies of race." Although with variations, these geographies were clearly established by scientists, politicians, writers, and authorities in colonial times and adopted without significant alterations by republican leaders after independence. In these racialized geographies, writers established a direct connection between climate and territory and racial groups: the Andean plateaus were inhabited by the civilized and good white people of European origin; the humid, tropical lowlands by blacks and indigenous groups incapable of reason and progress. The most progressive views after independence envisioned the mixing of races but endorsing the ineluctable inferiority of blacks and Indians when constrained to their territories. These recent historical studies emphasize the relations of dominance and subalternity mediated by ecological and ethnic processes, besides the more commonly studied economic considerations.[7] They also make clear the spatial-cultural bases of coloniality anchored in notions of territory, nature, and race.

The Cali architect and urbanist Gilma Mosquera has developed a useful account of the progressive transformation of settlements along the rivers of the Pacific since the nineteenth century. Mosquera's model starts with an empty space that is progressively settled, resulting in distinct spatial formations. Until the beginning of the twentieth century, river settlements were characterized by dispersed family units and groups of families brought together by labor or kinship relations. Small multifamily nuclei started to arise following the decision by a group of dispersed families in a river segment to "construct a village" (*hacer pueblo*). These neighborhoods typically started with three to ten units that were production and living units and evolved into nuclei of eleven to thirty homes that included isolated and joined units. With the advent of newcomers, this protovillage became a village, usually a one-street town with a tendency to the gradual separation of production site and living quarters (thirty to two hundred units). Subsequent phases turned the village into a town proper, with a more marked separation between living and production sectors, some commerce and tertiary activities, and greater social differentiation. If in the first stages one or two family trunks predominated, in the towns there was greater differentiation. However, each habitat type reflects a tight link between kinship relations, production relations, the organization of space, and cultural practices; the family continued to be the organizing axis of residential and production space. After the 1920s, this system of relations "entered into contradiction with the dominant national ideology, slowing down the definitive insertion of the local communities in the capitalist regime of production, from which they have only benefited in a marginal way" (Mosquera 1999: 55).

The pressures for dispersed settlements to nuclearize contributed to the breakdown of the family-based traditional production and authority systems. These pressures ranged from natural forces to in- and out-migration and production opportunities (e.g., timber, mining, agriculture, tourism in the coastal towns). By the 1940s, demographic expansion in much of the Pacific required the adaptation of new land for food production, including maize, cassava, plantain, rice, sugarcane; it also entailed taking advantage of external demand for products such as rubber and vegetable ivory. This growing integration of village, river, and region has been at play in the rivers surrounding Tumaco ever since. Many villages were established at the end of the nineteenth century by a couple arriving from other parts of Nariño or Ecuador in search of land or fleeing political violence; slowly, this founding couple brought a number of relatives and friends to start a village. Through kinship strategies encompassing the village, the river, and neighboring rivers these villages diversified according to social, production, and spatial strategies. For some researchers, this pattern was geared toward multiplying access to precarious resources and territories. Since the 1950s, with the increase in the pace of modernization, the final narrowing down of the territorial frontier, and closer links to the town of Tumaco, this goal became less important as migration and labor opportunities in the city increased. Migration to Tumaco often resulted in a network of circulation of people, products, and information between the city and the river, and it continues to be so today (even if significantly disrupted by paramilitary control in some areas). Kinship relations and nonconsanguineal residential alliances are more geared to ensuring a symbolic-territorial affiliation to the river community than access to material resources per se. As affiliation by river residence becomes less practicable, it is the combination of types of social relations that confer dynamism to particular places and or territorial identities.[8] In general, however, it is in these towns, most of which have no more than two or three thousand inhabitants even today, where Euro-Andean values, the state, the market economy, and armed actors are making inroads with full force (Villa 1996).

Socioterritorial affiliation is accomplished in multiple ways in contemporary contexts. What should be highlighted is, first, the continued importance assigned to place-related identities (see chapter 6); and, second, that while other regional and national markers of identity are used, until recently the river affiliation predominated locally. "People living on a given river consider themselves as a single community," wrote West (1957: 88) in the 1950s. The sense of belonging engendered by river affiliation has weakened, yet in many places, the river continues to be a vital space of social interactions through which people actualize opportunities afforded by either the local or the larger regional or transnational contexts. To be sure, in recent decades the pace of environmental degradation, the transformation of production relations, and the presence of urban imaginaries have taken a toll on river-based cultural and ecological practices. Yet there continues to be an "aquatic space" (Oslender 1999, for the Guapi river) that bespeaks the continued importance of aquatic environments for many local groups. This space includes coasts and beaches, the rich, flooded *guandal* forests, and the networks of rivers and estuaries largely regulated by the tides for up to twenty kilometers upstream; it also involves the construction of houses on piles along the river margins, the heat and humidity, and the many visions that inhabit the aquatic spaces and the forest.

After 1993, "the river" became once again an important cultural and political referent. Along some of the rivers of the southern Pacific activists of the social movement of black communities proposed that collective territories be conceived of on the basis of the dynamics of rivers, particularly the river basin since the river is "the space where one lives and acts" (Cortés 1999: 135). This model was not considered practicable around the larger rivers, such as the Patía and San Juán; moreover, some activists considered that a strict river focus could be too provincial and limiting for regional identity and organization. Nevertheless, the discourse about the river as a central cultural-spatial matrix of territory and identity is significant in itself. Needless to say, the settlement patterns have changed. West's map of the Rosario river (1952) shows a "traditional" pattern similar to Mosquera's (West 1957: 115). By 1993, when I visited this river, the pattern of dispersed settlement was still dominant, although it was undergoing significant changes, particularly because of the expansion of a nearby African palm plantation.

These introductory remarks on settlement, territory, and identity should give an idea of the most salient elements in the making of place.[9] Don Porfirio Angulo (don Po), a local community leader in his sixties, summarized his own recollection and experience of the settlement process he witnessed over many decades in the Caunapí river, and it is fitting to end this section with his words:

> The first settlers of the Caunapí were grandchildren of slaves in Barbacoas. They arrived in the early years of the century fleeing the civil war. One of the first to arrive was Pedro Damián Castillo Cortés; he discovered the Caunapí when he was hunting, and he found out that it was good for hunting and also for agriculture, particularly plantain, maize and rice. These promising lands attracted others, who came to build their huts and try out the territory [*hacían rancho y ensayaban el territorio*]. Bernardo Cortés Castillo founded La Espriella, named after the engineer that drew the railway lines. My grandfather, Luis Angulo, was the fourth founder. There were no roads, and everything was done through the river. The first external factor was the railway, which produced significant destruction of the forest. That was in 1926. Soon after there was an increase in the exploitation of rubber and *tagua* [vegetable ivory], followed by the cultivation of cacao. But local people did not care about money. They exchanged these products for kerosene, salt, and clothes. People only cleared up what they were going to plant, the rest they kept as jungle, even if in those days we did not know about "biodiversity." Work was collective [*mingas* and labor groups] and family-based. I built my house in that way, without a single nail. In the 1950s, Colonel Rojas Pinilla declared entire areas of the Mira and Caunapí as lands for colonization. From the Mataje to the Patía, this was still open forest used for hunting and agriculture, with the railway in between. But this changed in the 1950s with the arrival of the first Japanese to the Mira river to plant bananas,[10] and even more when the Spaniards arrived with their cattle. One single man, Valentín García, cleared 5,000 hectares of forest between 1958 and 1962. That was the time also when the railway was abandoned to construct a road, to the detriment of the people. With the road came the present avalanche of people from elsewhere to grab land, until the arrival of the *palmiculturas*. Right now there is growing consciousness that the forest is disappearing and we are going to be finished [*se nos está yendo el bosque y nos estamos acabando*]. As you can see, the history of the Caunipí is beautiful. There were great *marimberos* [marimba players]. Not any more.[11]

Don Po's narrative, influenced by the discourses of the 1990s and the mobilizations for collective land titling in which he was participating at the time of the interview, gives an idea of the process of settlement discussed above. According to Villa (1996), most regions of the Pacific had an open frontier until about 1950. Villa believes that the model of territorial appropriation remained viable until this time despite the commodity cycles of boom and bust. After the 1950s, with increasing colonization, technological changes in mining, fishing, and timber extraction (for example, the use of high-power pumps for mining, gasoline-powered saws for timber, and dynamite and large nylon nets, or *trasmayos*, for fishing), and great land concessions to foreign companies, the frontier finally closed down. With the arrival of the big oil palm companies in the 1980s, large-scale technology, development, and market forces, more than ecological dynamics or local cultural practices, began to shape the territorial transformation of the entire Tumaco region. But this is getting ahead of the story.

III ◈ Place Making and Localization Strategies in the 1990s: The Pacific as Region-Territory of Ethnic Groups

> *Tierra puede tener cualquiera, pero no territorio.*
> [Anybody can have land, but territory is another matter.]
> —don Porfirio Angulo, Tumaco, August 1998

Don Po's succinct assertion reveals the significant conceptual, cultural, and political changes that swept over the Pacific throughout the 1990s. From the 1950s on, peasant struggles all over Latin America were carried out in the name of land. *La tierra pa'l que la trabaja* (land to the tiller) was the slogan that fueled peasant movements, land takeovers, and agrarian reforms. In Colombia, the National Peasant Users' Association (ANUC) led the struggle with varying degrees of success (Zamocs 1986). Although ANUC had its heyday in the 1960s and 1970s, it was still of some importance in the 1980s in regions such as rural Tumaco, even if the imaginary of "land" and "peasant" was to prove patently insufficient to respond to the changed institutional, economic, and political context. By the time Ley 70 was sanctioned, the categories of land and peasant to refer to black people were on their way out, to be replaced by "black community" and "territory." I deal in this section with the emergence of "territory" in the Pacific in the late 1980s and 1990s, leaving the account of what Restrepo (2002) has aptly called "the ethnicization of black identity"—enabled by the concept of "black community" enshrined in Ley 70—for a subsequent chapter. Here I will examine the concepts of territory and region-territory produced by black and indigenous organizations over the 1990s as subaltern strategies of localization.

The concept of territory has a short but complicated history in the Pacific, and it has been influenced by international factors I will not discuss here.[12] It has its roots in a series of factors that converged first in the northern Chocó province in the 1980s, including the following: the crisis of the traditional settlement model after 1950; the declaration in 1959 by government law of the Pacific lowlands as *baldíos* (public empty lands), which enabled a faster pace of extraction of resources by people external to the region; and the increasing presence, since the 1960s and 1970s, of missionaries, state agencies, and development projects, such as the Dutch-funded DIAR project for peasant development. Its most immediate source, however, was the development in the 1980s of a strategy of defense by the large black peasant communities of the middle Atrato river against the predatory activities of timber companies.

Starting in 1985, these communities, with the help of Catholic groups inspired by liberation theology, began to delimit their territory and develop strategies for natural resource use. This strategy inaugurated a new form of property "whereby a conception of a collective territory without borders or limits gave way to the territory as delimited in a map and which allowed a census of its owners" (Villa 1998: 439). There thus appeared an increasing number of local organizations, grouped under the name Asociación Campesina Integral del Atrato (ACIA, Integral Peasant Association of the Atrato River), intended for a measure of self-governance. Other similar organizations followed by the end of the 1980s and beginning of the 1990s all over the Pacific, including other river organizations (e.g., in the San Juan and Baudó rivers, with the organizations Acadesán and Acaba, respectively); another important factor was the consolidation of indigenous resguardos (official territories) following the emergence of the strong indigenous organization OREWA (Organización Regional Embera-Wounaan), grouping the Embera and Wounaan peoples of the Chocó. These resguardos were demarcated without the participation of the black communities, even in those places where they had practiced traditional forms of occupation, so that often black peasants were relegated to the status of landless *colonos* (colonists). It was in this context that ACIA began to articulate a concept of territory for the black communities, with a focus on ethnicity and cultural rights. By 1987, ACIA had reached an incredible goal in its negotiation with the government: the demarcation of six hundred thousand hectares which, although still not legally titled to the organization, were put under a special management plan based on the cultural practices of the communities. For Villa, this agreement "was the beginning of a new territorial order in the Pacific. It demonstrated the efficacy of a political discourse that articulated black cultural experience to the specific pattern of appropriation of the territory" (1998: 441). But not until ten years later (1997) did ACIA, by then representing about 120 organizations, receive legal title to almost eight hundred thousand hectares under Ley 70. By this time, however, many of its beneficiaries had left the area, fleeing the violent actions of paramilitary, army, and guerrilla groups that became so prevalent in the middle Atrato starting in the mid-1990s (Wouters 2001; Agudelo 2000).

The ACIA process fostered the pioneering development of a series of concepts and technologies that had illustrious careers throughout the 1990s. The idea that "the territory" was fundamental to the physical and cultural survival of the communities, and the argument that these communities have unique ways, rooted in culture, of using the diverse spaces constituted by forest, river, mangrove, hills, and ocean were the two most important conceptual innovations. The experiences of the middle Atrato were important in other respects, such as the development of the technology of territorial and cultural workshops (*talleres*) that were to become ubiquitous throughout the rivers and towns of the Pacific with Transitory Article 55 (AT-55, issued in 1991 as part of the constitutional reform process; it would give way to Ley 70 in 1993), often with the support of the progressive Catholic Church (the Pastoral Afrocolombiana). Villa describes the reverberation of activity up and down the rivers of the Pacific that took place around the AT-55 and then Ley 70 in the early to mid-1990s:

> In the rivers, the people now have other reasons to get together in order to perform their chants, dances and playful activities besides the rituals of the Saints or the dead. Now the *decimero* [local poet] arrives to the meeting to recollect the birth of the river organization, or to evoke the trip taken by some in the community to Bogotá in order to show others how they lived in the Pacific and to teach them about the territory. Chants and dances become an integral part of political meetings, the elders narrate the story of the peopling of the river, the original places of settlement are marked on the maps, somebody teaches the history of slaves and masters, others about foods and fiestas of the past, and yet others tell about the history of encounters between Indians and blacks. Yet the voyages through the river are not only an oral exercise but a real one, such as the trip taken for a number of days by the people of the San Juán in 1992 starting out at the delta in Itsmina. Hundreds of people from the Asociación Campesina del San Juán took to their boats, stopping in each town and enlivening the trip at each stop with music and dances. *The voyage was a geographical recognizance of a territory that they now learned as their own.* Along the parallel rivers

of the Pacific, organizations now come together and the traditional orality of the people becomes uncontainable as they narrate the histories of their rivers. . . . As the ignominious history of extraction of resources is woven into the encounters, those from the north learn that their history is not different from those from the south, and in the penuries of the past and the certainty of a common future a collective identity is thus forged. *It is here that a concept of region becomes manifest and is learned as a lived-in process.* (1998: 444, 445; emphasis added)

What, then, were the particular landmarks in the emergence of the territory as articulatory discourse in the southern Pacific? As I mentioned, a number of social and economic processes under way since the early 1980s helped provide the context for this emergence. Capital stepped up its operations throughout the decade, setting in place a brutal accumulation model that greatly affected territory and place (see chapter 2). State development plans were being implemented for the first time with the entire region in mind (see chapter 4). International organizations were beginning to pay attention to the region's biodiversity. These factors contributed to creating the notion of the Pacific as a region. And then there was the irruption of popular mobilization along the rivers centered at this time on those segments of the river that had a higher degree of family-based production practices. If ACIA had been a salient factor in the emergence of the concept of territory—indeed, ACIA's experience shaped indelibly the AT-55—it was the flurry of organizations and activities around AT-55 that finally brought this concept to life at a regionwide level.

The discovery of the territory linked to AT-55/Ley 70 in the early 1990s thus had two moments. In the first, local organizations developed a series of technologies, such as *monteos*, literally traversing the *monte* or forest/territory with the entire community to recognize places of habitation, cultivation, and hunting and gathering, past and present; the collective drawing of color maps on large sheets of paper, not infrequently done by the young under the instructions of old people; the gathering of oral histories and traditions; and the numerous meetings among organizations to develop a common strategy and political discourse for the implementation of Ley 70 and beyond (see chapter 5). Villa's description captures this moment well. The process was of great political importance since it brought together ethnoterritorial organizations and communities in a discussion about territory, culture, and history. In a second moment, the participation of external NGOs, academics, experts, and international organizations became more prominent. Two methodologies became intertwined: satellite and GIS-based cartography and participatory workshops to elicit local representations of the territory. Informed by a landscape ecology framework, this social cartography often took place in the context of government programs, such as the Ecological Zoning Project carried out by the IGAC (see Ministerio del Medio Ambiente 2000); it involved interdisciplinary teams of natural and social scientists working with local communities in the elaboration of the communities' "mental maps," subsequently contrasting them with the maps afforded by modern cartography. While the technical maps centered on relatively homogeneous production systems and landscape units (in terms of geomorphology and vegetation), the social cartography or participatory mapping with communities tried to get at a whole set of social and ecological aspects, including the *espacios de uso* (use spaces, such as the various types of *monte* or forest, river, mangrove, sea, village, home), production systems, history of settlement, tenancy, local projects, hunting and gathering, visions of the territory, local knowledge of plants and animals, informal borders, commerce, and the like. In the best cases, these exercises incorporated an awareness of the difficulties of comparing these two systems of knowledge.[13]

As a political ecology tool, maps, whether drawn by experts or communities, may reveal or hide important aspects. First, feminist researchers have used community maps of land use and cover in relation to local knowledge to unveil gender relations and make visible women's contributions and assets (Rocheleau 1995b). Who counts, draws, and narrates and how is of decisive importance in these cases; there was some awareness of these issues in the Pacific in the 1990s. Second, maps contribute to the production of territoriality (the *territorios de comunidades negras*) in particular ways; whether this corresponded to *long-standing* territorial logics is a subject of debate. For Oslender, the mapping exercises involved "an imposition of fixed boundaries onto local epistemologies of fluid boundaries and tolerant territorialities, forcing local communities to translate their territorial aspirations onto maps which Western-style institutions will accept as legitimate documentation to accompany their land right claims" (2001: 253; cited in Restrepo 2006: 83). In other words, mapping, no matter how participatory, introduced a new form of representing spatiality and of thinking about the territory. For some activists and intellectuals, there was consistency between the long-standing practices of territoriality and the conception of it in Ley 70; in other words, there was an articulation between the historically constructed territoriality of the communities, the collective titling process, and the expectations of the communities, with the law operating as an instrument for the defense of historical, ethnic, and cultural identities. In this view, juridical rights entail the recognition of ancestral rights and articulate with the responsibilities of the communities to protect the territory for the *renacientes* (present and subsequent generations) (see Cassiani, Achipiz, and Umaña 2002).

The titling process became an important focus of the work of social movement throughout the 1990s, often imbricated with a variety of government programs, such as the Biodiversity Conservation Project (Proyecto Biopacífico, PBP) (see discussion in chapter 4), and the Ministry of the Environment's Natural Resources Management Program, PMNR, funded by the World Bank and implemented by two government agencies—the Agrarian Reform Institute (INCORA) and the Social Solidarity Network (Red de Solidaridad Social); INCORA was directly in charge of coordinating most of the titling process after 1994. The guidelines for this program had been negotiated by government agencies and black and indigenous organizations in 1992 (Sánchez and Roldán 2001; Offen 2003). As far as the social movements are concerned, a landmark meeting took place on June 18–22, 1995, in the facilities of an alternative development NGO in Perico Negro, located in a predominantly black region south of Cali. Convened as a meeting (*encuentro*) of black and indigenous communities and peoples (*pueblos*) of the Pacific and under the rubric "Territory, Ethnicity, Culture, and Research in the Colombian Pacific," this meeting—which I mentioned briefly in the introduction—brought together for the first time the great majority of black and indigenous organizations of the region. Chief among the achievements of this meeting were the establishment of guidelines for negotiations among black and indigenous communities and between these and the state, and a theorization of the concept of territory according to the movement concept of *cosmovisión* (worldview) of the local communities.[14] That the organizations gave primary importance to research and knowledge production as part of their political struggle was reflected in the fact that they devoted significant effort to drafting the statutes of the newly created Instituto de Investigaciones Ambientales del Pacífico (Institute for Pacific Environmental Research). This process revealed the various and at times conflicting interests of the organizations of the region, while making clear the extent to which knowledge production and even research had become a central practice of the social movements themselves.[15]

But the most critical achievement of the Perico Negro *encuentro* was the discussion of the *cosmovisión* and conception of territory of the various movement organizations. Indigenous representatives explained the *cosmovisión* of their respective communities (Embera, Wounaan, Eperara-Siapidara, Awa, Catío, Chamí, Zenú, and Tule peoples). They emphasized the integration of people and nature, traditional management practices, the role of traditional authorities, and the resulting conservation of the environment. Their declaration of principles focused on the defense of resources and traditional knowledge through strategies articulated around the principles of unity, territory, cultural identity, autonomy, and self-sufficiency. The presentations by black organizations similarly focused on "the manifold logics of appropriation of the territory" by the communities. Among the elements emphasized were the river-based models of appropriation with their longitudinal and transversal dynamics (see chapter 3), the organizational process for the defense of the territory, and the local knowledge, patterns of mobility, kinship, and gender relations. Their declaration of principles highlighted their right to a territorial strategy that built on traditional appropriation models in order to resist the onslaught by capital and the dominant culture. This strategy would start by recognizing that the Pacific is a territory of ethnic groups, thus requiring a special policy regime:

> In the same way that it is impossible to separate the territory from culture and ethnicity, thus the knowledge we have about biodiversity is a cultural knowledge that requires particular treatment, including appropriate property regimes. . . . It seems to us that in the definition of the relation between the region-territory and the rest of the country it is necessary to be clear about the relation between the organizational processes and the institutional agents [state and NGOs]. . . . *The identity that needs to be constructed today at the heart of the black communities is one based not on race but on the defense of the territory*; it underlies our conceptualization of life and the world and a set of cultural aspirations . . . that is, the development of our own cultural dynamic from the perspective of a logic of resistance. (Fundación Habla/Scribe 1995: 52, 53, 54; emphasis added)

While these debates surely reflected the salience of anthropological and conservation discourses (including the issue of intellectual property rights linked to traditional knowledge that was by then becoming prominent in Colombia and around the world), activists nevertheless gave their imprint to these concepts and pushed them for the first time into an articulated, elegant conceptualization that was to be of great political expediency in the context of the organizations' encounter with the state and technoscience. Above all, at this meeting, it is possible to find, for the first time, discussion of the concept of the Pacific as region-territory of ethnic groups. This concept was to emerge in full flower over the next two to three years and to become an essential point of reference for rethinking development, sustainability, and conservation—certainly throughout the 1990s and in a general way up to the present. As we shall see in greater detail in chapter 3, the territory came to be defined as the space of effective appropriation of ecosystems by a given community, while the notion of the Pacific as region-territory of ethnic groups was seen as a political construction for the defense of territories. If the territory embodies the life project of the community, the region-territory articulates the life project of the community with the political project of the social movement.

The region-territory also emerged as a category of interethnic relations pointing toward the construction of alternative models of life and society. In other words, the region-territory was both a conceptual innovation and a political project, what could be called a subaltern strategy of localization. It located biological diversity within an endogenous perspective of the ecocultural logic of the Pacific. Even if produced by social movements, the concept of territory did not fully emerge out of the longstanding practices of the communities, where rights to land have been customarily allocated on a different basis (according to kin, tradition of occupation, etc.). In practice, the hope was that, once collective tenure was secured, particular rights to land would be negotiated internally by communities. Some observers saw the emphasis on collective territories as a mistake of the movement based on a misperception of their strength at the time. Be that as it may, from an ecological perspective the concepts of territory and region-territory can plausibly be seen as a representation of the collective ecocultural practices.

In the PCN conceptualization of the 1990s, the territory was considered a challenge to developing local economies and forms of governability capable of supporting its effective defense. The strengthening and transformation of traditional production systems and local economies, the need to press on with the collective titling process, and working toward organizational strengthening and territorial governability were all important components of this region-centered strategy. Besides being influenced by larger debates and interactions with academics, other activists, state experts, and NGOs, PCN's conceptual framework was based on its own analysis of the situation along many of the rivers and on intensive interaction with communities and their leaders. Workshops conducted during the 1990s with river community leaders invariably suggested a number

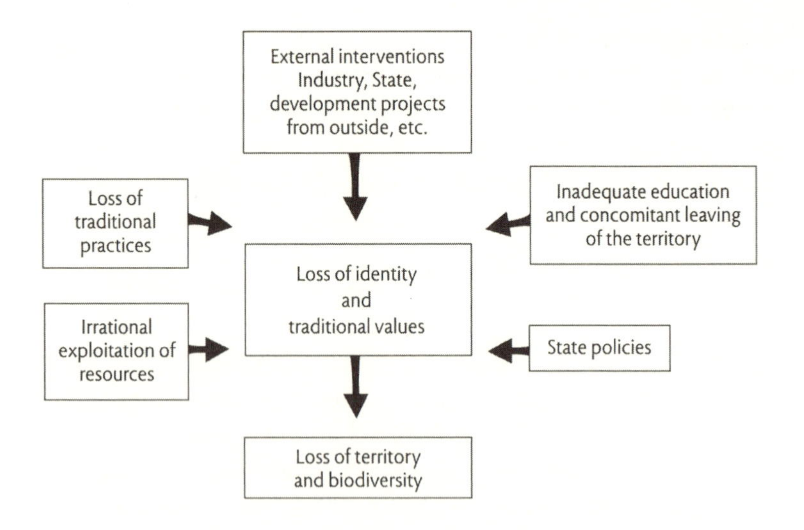

Figure 1. System that generates loss of territory, according to river community leaders.
Source: Workshop on Ecological River Basin Design, Buenaventura, August 1998

of factors linked to the "loss of territory," including "loss of traditional values and identity," loss of traditional production practices, excessive exploitation of resources, state development policies, increased pace of industrial extraction, and alienating educational models for the young people (see figure 1). In more substantial discussions with community leaders and social movement activists, a series of other factors linked to the loss of territory emerged, including the spread of plantations and specialization of productive activities; changes in production systems; internal conflicts in the communities; the cultural impact of national media; out-migration and the arrival of people foreign to the region espousing the ethics of capitalism and extraction; and inadequate development policies, the neoliberal opening to world markets, and the demands of the global economy.[16]

Less developed at this point was the gender dimension of the territory and region-territory. The main attempt by women activists and some anthropologists was to use the language of gender complementarity of tasks (vernacular gender) to explain the spatiality of territorial appropriation. This notion suggests a gender division of labor in terms of differential utilization of spaces, tasks, and forms of knowledge about health and the environment. In this conception, since they tend to remain in the river settlements, whereas men move more broadly in spatial terms, women are seen as a stronger referent of belonging to place, and this belonging is strengthened through a woman's descendants and her kinship group, often involving a network in several rivers. Women and men thus effect different but equally important appropriations of use spaces. If men demarcate the territory in their mobilization for production, women are seen as consolidating it through socialization processes and the construction of identities through a panoply of practices of food, healing, and production. When these practices are disrupted (e.g., by the introduction of market foods), the links to territory start to weaken. Hence the crucial gender dimension to effective territorial appropriation, which has become broadly recognized by black community activists. This recognition is accompanied by an awareness of inequality in many spheres, particularly the domestic and the political. For many black women's organizations at present, while the struggle for ethnocultural and territorial rights should be the leading goal, this needs to be advanced simultaneously with mobilization for the needs and interests of black women as members of the black community (Grueso and Arroyo 2005; see chapter 5 for further discussion).[17]

The accomplishments of the collective titling process have been very significant, among the most accomplished in Latin America in terms of ethnic territories. An evaluation of the titling program conducted in 2001 (Sánchez and Roldán 2001) showed that indigenous territories were augmented by 324,288 hectares, corresponding to forty-four new *resguardos* and the expansion of twelve more, for a total of 1,994,599 hectares. The results for black communities were most impressive: 2,359,204 million hectares titled to over twenty-two thousand families organized into fifty-eight community councils; the collective territories for the black communities had been increased to about 4.8 million hectares by 2003. The area thus titled represents 52 percent of the total region. Throughout the 1990s, besides the territorial titling achievements, the new legal instruments had a number of beneficial aspects; despite difficulties and tensions, they fostered interethnic collaboration as well as collaboration between ethnic groups and the state; enabled local organization development; were essential to placing the question of black and indigenous identities under discussion; and enabled for a time the construction of an alternative imaginary of territoriality and development. The long-term results in terms of biodiversity conservation and local autonomies would have been noticeable if the process had not been disrupted drastically after 2000, as we shall see shortly.[18]

To sum up, the concept of territory produced by a number of movements in the 1990s articulated a place-based framework linking history, culture, environment, and social life; it evinced the development of a spatial consciousness among movement activists and, to some extent, within the riverine communities themselves. This conception resonates with academic frameworks in which nature and culture are seen as interconnected in overlapping webs of humans and other beings, and communities are seen as multiply located—they are simultaneously place-based and networked across places. What takes place is an encounter between self-organizing ecosystems and people from below, on the one hand, ("rooted networks," in Rocheleau's counterintuitive description, 2000: 3), and hierarchical organizations of various sorts (e.g., capital and the state), on the other. "The Pacific" may thus be seen as a complex matrix of processes in which people, territory, species, and so forth are held in relation within a social context; social movements attempt to steer this dynamic away from the social disruption and ecological degradation at local and larger scales, by building on the self-organizing tendencies and relations of solidarity that exist or that might be created or re-created. In their view, for this strategy to have a chance, it must rely on the social and ecological relations that have existed in place, even when confronted with processes of spatial restructuring. As complexity theory suggests, "structures" do not necessarily predetermine the kinds of connections people make. New possibilities and categories may arise and lead to emergent worlds. One may conclude, with Rocheleau in her place analysis of Kathama in the famous Machakos district of Kenya, by suggesting that we "take away from these stories the lesson that the community organizations and membership were their own sources of surprise, rather than simply mechanisms to weather or respond to external surprise. In view of this we might think of communities as crucial and foundational categories for entities that are themselves complex and contingent" (2000: 20).

This is to say that without the region-territory, communities would be even more vulnerable to being reconstructed by capital and the state. Despite its support for the titling process, the state has its own project of region construction, in which the Pacific would be but a link in a strategy of Andean-style development for the entire *occidente colombiano* (Western Colombia). This state project has become increasingly inseparable from the use of force in a complicated dynamics with other armed actors. This process, which has gained ascendancy in the southern Pacific since the late 1990s, has radically transformed the politics and production of place.

IV ◈ Displacement and the Territorial and Cultural Imperatives of Development and Modernity

Oh, my God, what scary things / are they saying from afar
where they fell, burn, and kill / without a question even asked. . . .
Oh, my God, what scary things / has told me Don Severino
that the land's no longer useful / for either coconut or farm
Oh, my God, what will become / of the life of our children
if the land which is their future / is being turned into a whirlwind?[19]

Sebastião Salgado's photographic exhibit *Exodos* (2000), an impressive and deeply human six-year-long project of portraying displaced people worldwide, leaves one with the impression of both the ineluctability of displacement on a large scale and the crisis dimensions it has taken on in recent times.[20] Few countries validate Salgado's vision as clearly as Colombia, with the Pacific as one of the main theaters of displacement. The figures are "a scary thing," to be sure: close to three million internally displaced people by 2002, tens of thousands of them from the Pacific and a disproportionate percentage from indigenous and black groups. The "peace refuge" that was the Pacific until the mid-1990s became just another battlefield in a country at war (Wouters 2001; Agudelo 2000). The targets have been the territorial, social, and cultural integrity of the black and indigenous groups; the organizing processes, including systematic threats against activists; and the control of natural resources (e.g., timber, gold, African palm) with total disregard for environmental regulation and the rights of their inhabitants. The goal of this "cruel little war" (Joxe 2002) and "geography of terror" (Oslender 2004), as activists see it, is the disappearance of the ethnic groups of the Pacific as distinct cultures.[21]

Take the case of the Pacífico nariñense.[22] Despite intense capitalist activity since the early 1980s (see chapter 2), black organizing around the culture-territory equation had been strong since the early 1990s. Everything seemed to suggest that ethnic organizations were advancing a new model of territorial and political ordering. With the arrival of right-wing paramilitaries en masse after 2000, however, this project changed dramatically. Through a combination of terrorizing strategies that included threats against and assassinations of community leaders and activists, the explicit prohibition of the community councils formed around the PMRN and Ley 70, selective displacement of communities and leaders from particularly coveted lands (particularly for African oil palm but increasingly for coca cultivation), "emplacement" of people in other cases (restricting people's mobility in some places), and capturing of state funds through Plan Colombia through phony NGOs, this particular armed actor was able to demobilize the movements, weaken the institutional frameworks established over the previous eight years, install a criminal economy based on coca production, and expand noticeably the oil palm frontier with the government's blessing. Additional consequences have been the exodus of activists, the emergence of a cadre of young black leaders more willing to work within the capitalist model being implemented by force, the disarticulation of the territorial practices of communities, and a necessary transformation of the organizing strategy from an emphasis on the territorial-cultural project to the defense of human rights and strategies for dealing with displacement.[23] What appeared in Tumaco has been characterized as a "landscape of fear" (Oslender 2004; Cortés Severino 2007) and as a new economic model that brings back some of the savage extractive practices of the past, combining them with illicit, criminal, and parastatal practices that articulate well the demands of transnational capital (Almario 2004, Restrepo 2005). This is an important aspect of imperial globality, and will be discussed further in the book's conclusion.

To put it bluntly, the Pacific is being subjected to the territorial and cultural imperatives of imperial globality; this project must be seen as a *simultaneously economic, ecological, and cultural reconversion*, a reconfiguration of the biophysical and cultural landscapes of the Pacific. Not infrequently, one finds in some areas a river controlled by guerrillas next to another controlled by paramilitaries, both pushing people to plant coca, while the army keeps watch a few kilometers downstream.[24] From the perspective of the local social movements, all of the external actors—including guerrillas, paramilitaries, capitalists, and the state—share the same project, namely, the control of peoples, territories, and resources; activists are very clear that *this project does not coincide with the interests and reality of the black and indigenous communities*; this project is a planned process related to the historical experience of racism and coloniality. In cases such as the Pacific, activists emphasize strengthening people's capacity to withstand the traumas of capitalist modernity (from poverty to war) *in place*, building on people's struggles for the defense of place and culture, and fostering people's autonomy over their territories. Food security and cultural and territorial rights are central to this goal.[25]

Consistent with this view of the problem, the Primer Encuentro Nacional de Afrocolombianos Desplazados (First National Convention of Displaced Afro-Colombians) held in Bogotá in October 2000 identified a *principle of return* as a general policy for all ethnic groups of the Pacific, given their particular culture and relation to the territory. To the extent possible, resettlement should be seen as the exception, not the rule, and as a temporary measure. The strategy of the social movements is the effective declaration of the Pacific as a *territorio de vida, alegría, y libertad* (a territory of peace, happiness, and freedom), free of all forms of armed violence. Also considered were an effective system of early warning and prevention of displacement (many displacement events have been announced with ample anticipation, with no preventive action taken by the state; on the contrary, there has been a proven correlation between guerrilla presence followed by army presence and followed finally by paramilitaries, who effect the previously announced displacement, not infrequently accompanied by killings); and "socioeconomic stabilization" (a term coined by the Social Solidarity Network)—that is, fulfillment of the state's duty to guarantee the full exercise of social, cultural, and economic rights to all communities.

The seeming *ensañamiento* (obsessive targeting) of the Pacific and its peoples cannot be explained without reference to the racism enthroned by Euro-Andean modernity, on the one hand, and to the obsession with development, on the other. As I have explained elsewhere (2003a; see also chapter 4), displacement is an integral element of Eurocentric modernity and development. Modernity and development are spatial-cultural projects that require the continuous conquest of territories and peoples and their ecological and cultural transformation along the lines of a logocentric order. The massive displacement observed today worldwide, whether relatively voluntary or forced, is the end result of processes that started at the dawn of modernity and crystallized at the end of the eighteenth century, when capitalist modernity became consolidated as a cultural-economic project. Today, given the intensification of neoliberal globalization, it is possible to assert that *the gap between modernity's displacement-producing tendencies and displacement-preventing (or correcting) mechanisms is widening and becoming untenable*. The crisis could assume such magnitude that the resettlement schemes and refugee camps of today might come to be seen as pilot projects for the future. Displacement can also be seen as the most visible manifestation of the regime of social fascism on the rise (Santos 2002).

As the case of the Pacific suggests, strategies for preventing displacement and enabling return should take as a point of departure an understanding of resisting, returning, and re-placing that is contextual with respect to local practices, building upon movements for identity, territory, and autonomy wherever they exist. These strategies should foster alternative development for livelihood and food security as a required minimum (more on this in my discussion of projects in the Yurumanguí river in the book's conclusion). As a Nasa indigenous leader put it, the minimum condition for resisting in place and defending culture is to "deglobalize the belly." Modern institutions (state, UN system, humanitarian aid) have an important role to play in this regard, but they need to be approached from the vantage point of these criteria.

Conclusion: Globality, Coloniality, and the Politics of Place

I started with a view of the Pacific as place, that is, in terms of historically constituted individual entities and processes that, while operating at different levels, interacted in complex ways to form a whole (an assemblage, as I will call these entities in the chapter on networks), the Pacífico biogeográfico. I highlighted the production of a political ecology framework by social movement activists centered on the territory-culture equation. This framework, as Rocheleau (1995a: 1039) would put it, helped us "to understand the broader political outlines that bind local ecologies, economies, communities, and cultures into regional and global systems." I suggested that, while many of the long-standing biophysical and cultural dynamics that have produced the region are still in place, many of them have also been deeply affected, if not outright destroyed. Under these conditions, what happens to the defense of place? I leave this question pending until the concluding chapter.

Things do not look good for the Pacific. The massive displacement witnessed there makes newly visible the coloniality of power specific to the current phase of imperial globality. If until 2000 the global design implanted in the southern Pacific could to some extent be counterbalanced by the place-based actions of ecosystems, movements, and communities, the regime of terror that came about in the late 1990s imposes clear limits on such a strategy. As we saw briefly, the social movements are maintaining minimum strategies of resistance. These strategies are still enabled by the construction of identities and territories of the past decade, although it is clear that such practices are no longer sufficient and some perhaps not even appropriate to the moment. A lot will depend on what happens in Colombia and internationally, as I will discuss in the conclusion. For now, it is sufficient to highlight the tremendous deterritorialization effected by armed actors, although with an underlying basis in national and transnational processes.

Why does a politics of place continue to be important? First, the struggle of the social movements of black communities of the Colombian Pacific illustrates, at least until the late 1990s, the viability of a politics of place in the context of imperial globality. In its encounter with state agents, experts, NGOs, international biodiversity networks, and others, the movement developed a unique cultural-political and ecological approach that articulates the life project of river communities with the political vision of the social movement. To this extent, the movement can be interpreted in terms of the defense of practices of cultural, economic, and ecological difference. More generally, the goal of many of today's struggles is the defense of place-based conceptions of the world and practices of world making—more precisely, a defense of particular constructions of place, including the reorganizations of place that might be deemed necessary according to the power struggles within place. These struggles are place based, yet transnationalized (Harcourt and Escobar, eds. 2005; Escobar 2001). Although the current situation in the Pacific makes a defense of place in the style of the 1990s largely impossible, the struggle over place has not become meaningless and has not ended.

The politics of place can be seen as an emergent form of politics, a novel political imaginary in that it asserts a logic of difference and possibility that builds on the multiplicity of actions at the level of everyday life. Places are the site of dynamic cultures, economies, and environments rather than just nodes in a global capitalist system. For Gibson-Graham, this politics of place—often favored by women, environmentalists, and those struggling for alternative forms of livelihood—is a lucid response to the type of "politics of empire" that is common on the Left and that requires that empire be confronted at the same level of totality, thus devaluing localized actions by reducing them to accommodation or reformism. "Places always fail to be fully capitalist [and modern, in the Eurocentric sense], and herein lie[s] their potential to become something other" (Gibson-Graham 2003: 15; see also 2006). The politics of place should also be an important ingredient of imaging after the third world and its developmentalisms (fears of "localisms" notwithstanding, but of course taking all the risks into account). Politics of place is a discourse of desire and possibility that builds on subaltern practices of difference for the construction of alternative socionatural worlds; it is an apt imaginary for thinking about the "problem-space" defined by imperial globality and global coloniality (Scott 1999).

If territory is to be thought of as "the ensemble of projects and representations where a whole series of behaviors and investments can pragmatically emerge, in time and in social, cultural, aesthetic and cognitive space," that is, as an existential space of self-reference where "dissident subjectivities" can emerge (Guattari 1995: 23, 24), it is clear that this project has been under development by the social movements of the Pacific. The defense of the territory entails the defense of an intricate pattern of place-based social relations and cultural constructions; it also implies the creation of a novel sense of belonging linked to the political construction of a collective life project. At stake with Ley 70, as we saw, is not just land but the concept of territoriality itself as a central element in the political construction of place on the basis of black and indigenous cultural experiences. The struggle for territory is thus a cultural struggle for autonomy and self-determination. This explains why for many black people of the Pacific the loss of territory would amount to a return to slavery or to becoming "common citizens."

I have introduced here the rudiments of a framework for understanding the production of the Pacific as place. This production takes place through a series of historical processes—biological, cultural, social, economic, political, technoscientific. To the strategies of production of locality by capital—themselves delocalizing in relation to place-based dynamical processes, from those of ecosystems to culture—social movements oppose strategies of localization which focus on the defense of territory and culture. The idioms of biodiversity, sustainability, traditional production systems, cultural rights, and ethnic identities are all interwoven by movement activists into a discourse for the defense of place. Each chapter that follows will develop further a given aspect of the dominant and subaltern strategies for the production of place in the Pacific. One of the most readily visible and salient strategies is that of capital (see chapter 2). Not until the last chapter, on networks, however, will we have a full view of the manifold processes that account for that unique place that is variously called the *Pacífico biogeográfico*, *Chocó biogeografico*, or, simply, *la costa Pacífica*.

2 capital

We are the descendents of the slave trade. Our papers say: "Afro-descendents: descendents of the Africans brought to America with the transatlantic slave trade." What do I personally think? If the slave trade is at the basis of capital accumulation, then inequality and racism are also at the basis of the same process. I can make headway on the problem of territory, or ethnoeducation, up to a certain point, but if I do not solve the fundamental problem I do not solve anything. Often times what makes one feel pleased are those fights we haven't lost; the situation is so complex, and we fight in the midst of such adversity and against so many powerful enemies—economically, materially, politically—that even when we play them to a tie we feel satisfied [since] most of the times they win [smiling], but we have tied in some instances. These give us the courage to go on.

—Carlos Rosero, PCN activist, quoted in *Landscapes of Terror: Between Hope and Memory*, a documentary by Catalina Cortés Severino, 2007

Introduction: The Arrival of *Palma Africana* in the Pacific

From July 1, 1958, to June 30, 1959, Maurice Ferrand, a French expert from the UN Food and Agricultural Organization (FAO), visited Colombia with the intention of advising the government on a program for "the promotion and development of the production of African oil palm (*Elaeis guineensis*)." The program was to include "the identification of favorable regions for the establishment of new plantations, the adoption of modern techniques, and the initiation of a research program to insure the success of the new varieties" (Ferrand 1959: 1). Ferrand spent a good part of his time "in the backward region of Tumaco," where he found "soils that are the richest and most fertile one can find in the tropics" (14). Such soils were found particularly in the area between the Rosario and

Mira rivers. Following the railway that connected Tumaco with the small town of Candelillas to the east, he noticed "a band of mangroves 15–20 km wide, useless for cultivation," followed by a band with a low water table where he found "the most beautiful coconut trees the Expert had ever seen, including those seen in Asia and Africa. After passing the km. 30–35 of the railway the land elevation changes and the water table goes down to 2–4 meters. There one finds large areas ideal for the cultivation of the oil palm, and an equally perfect climate with an annual rainfall of 3 meters well distributed throughout the year." He concluded, "The entire region is of first order quality for oil palm, especially where the water table does not go higher than 2 m. from the ground" (14).

Couched in terms of the promotion of a "colonial product" in the Tumaco region, the history of this product, despite the fact that its introduction in Colombia did not take place until the end of the era of the colonial empires, bears many of the marks that have been associated with the biological trade characteristic of the colonial enterprise. For while its introduction to a number of other regions in the American continent (for example, in Bahia in Brazil, the Antilles, the Guyanas, and Haiti, where it became a subsistence crop) seems to have occurred in parallel with the introduction of slaves, it was not until the 1920s that the product arrived on Colombian soil. But even this late chapter in the history of the plant was infused with colonial overtones. After a couple of failed attempts by the French, a Belgian botanist and the director of the Brussels Museum and Botanical Garden, Florentino Claes, successfully brought into the country the seeds that were to find their way to Tumaco and other areas. Claes came to the country with the intention of doing research on tropical plants and, having observed the ideal conditions present in many regions, contacted the missionaries in charge of running what at that time were still considered *territorios de misiones* (Catholic mission territories), such as the Pacific, with a proposal for bringing oil palm seeds from the Congo. The *Elaeis guineensis* thus made it to Colombia from the Eala Botanical Garden in the Belgian Congo (Patiño 1948).

By the early 1930s, the seeds had given rise to elegant plants in many disparate places, from the towns of the Putumayo, such as Puerto Asís, whose central plaza was adorned with twenty-seven palm trees planted by Capuchin missionaries in 1933, to some of the main urban avenues, such as the Avenida Roosevelt in the city of Cali, which was experiencing a period of growth after the opening of the Panama Canal. More important, the seeds made their way to the Agricultural Experiment Station in Palmira, just north of Cali. In the 1930s and 1940s, other oil palm seeds were introduced to other regions: for instance, by the United Fruit Company into its banana-growing areas in the Magdalena region; twenty-one thousand seeds from the Institut de Recherches pour les Huiles et Oleagineaux in Paris; seeds from Nigeria and the Ivory Coast, purchased by the Colombian Institute for the Promotion of Cotton; and four hundred seeds from Sumatra and the Dutch Antilles received from Honduras by the biologist and researcher Victor Manuel Patiño "by courtesy of Dr. V. C. Dunlap, of the Tola Railroad Co, thanks to the good offices of Dr. Wilson Popenoe, director of the Pan American Agricultural School" (Patiño 1948: 13). By 1945, Patiño had established a small plantation in the Agroforestry Experiment Station in Calima, between Cali and Buenaventura. By the time of Ferrand's arrival, the selection of mother trees for the production of good quality seeds had already started, a step that the expert considered indispensable to the development of the sector. Thus concluded successfully another chapter in the circulation of seeds through diverse colonial and neocolonial regions. From Africa to America, in this case, the biological material was transported according to a process that involved colonial authorities, metropolitan and local researchers and scientific institutions, botanical gardens, and international organizations like FAO.[1]

In subsequent decades Ferrand's mission assumed tremendous importance in radically transforming the natural and social landscapes of the Tumaco region. Together with industrial shrimp farming, which started in earnest in the second half of the 1980s, African palm was to become the arrowhead of capital and state action in the development of the region and its much-touted insertion into the Pacific Basin economies. What were the landmarks of this process that has resulted in the appropriation of great extensions of forest previously devoted for the most part to traditional production systems, in the dislocation of thousands of black men and women, and in the sizeable accumulation of capital by investors from Cali, Medellín, Bogotá, and abroad? Along with the apparatus of development (see chapter 4), capital has been the most important operator both of modernity and of the spatial and sociocultural reconstitution of the region. State, capital, and development stratify and hierarchize and transform the nature of labor, the function of land, and the role of money (e.g., Polanyi 1957). State, capital, and development transform gender and ethnic identities, not infrequently generating unprecedented forms of violence. They are "apparatuses of capture" (Deleuze and Guattari 1987) that effect a significant reconversion of local territories, economies, and cultures in the image of Euro-Andean modernity, even if, crucially, this transformation is far from complete and does not operate in a dualistic fashion (tradition versus modernity; capitalist versus subsistence practices), as these introductory remarks might suggest. Today, the spread of the oil palm is effected by force through violence and displacement; indeed, along with coca cultivation, the oil palm has become the primary motivation for displacement in many parts of the Pacific.

This chapter deals with the actions and effects of capital in the region of Tumaco over the past two to three decades. Part I focuses on African palm and industrial shrimp production. At first sight, these processes of capital accumulation seem to be susceptible to being described through classical political economy, that is, as the implantation of a capitalist mode of production with the subsequent proletarianization of large segments of the population and the crude exploitation of natural resources. This kind of analysis indeed describes to a great extent what has been happening in the region. The fate of much of the Colombian Pacific in recent decades can be seen as the steady incorporation of the region into the world capitalist system, a process that started with the organization of gold mining by the Spaniards in the first half of the seventeenth century (Colmenares 1973; Romero 1995) and is described by political economists in terms of boom and bust cycles (of gold, platinum, precious woods, *tagua*, or vegetable ivory, rubber, timber, and now even genes). This analysis is necessary but no longer sufficient, and in the first part of the chapter I give a broader view of this process. Part II shifts to a theoretical register to discuss current conceptualizations of the relations between capital, nature, and the economy in order to make visible other dynamics at play in the action of capital over natural and human life. Part III presents local attempts at alternative capitalist development in the Tumaco region, particularly community shrimp farming; by introducing a framework for looking at "diverse economies"—following the work of J. K. Gibson-Graham (1996, 2006)—this section asks whether these communal forms could be considered noncapitalist forms of economy. Finally, the last section raises the question of whether it is possible to rethink production from an ecological perspective, theoretically and based on the lessons of the southern Pacific.

In recent years, researchers have interpreted the history of the Pacific in terms of a long-standing extractive economic model inaugurated by the expansion of the gold mining frontier since the seventeenth century. In this view, the region's social and natural landscapes have been primarily shaped by systems of production that depend on the exploitation of natural resources and the exportation of the surplus, so that the benefits of the economic activity do not stay in region. The nonextractive local systems developed by the black and indigenous communities do not participate significantly in the generation of surplus, even if they are crucial to the food security of the local inhabitants. This fact has had indelible consequences for all aspects of Pacific development, from the landscape to local settlement patterns, and from its articulation to larger economies to the local modes of appropriation of territory and resources (Leal 2004; Sánchez and Leal 1995). This proposal is important in understanding the history of territory and place in the Pacific even if, as we will see, it leaves important aspects out of the picture.

The underlying issue raised in this chapter is that of the character of capital and the economy in the age of imperial globality and global coloniality. Needless to say, I do not expect to give a comprehensive answer to this question here but point at some trends that enable one to go beyond established political economy conceptions that both take "the economy" for granted and perform an analysis of capital largely in terms of structures and devoid of culture. The analysis of capital has been amplified in recent decades; the most significant transformations in the framework of historical materialism have come from the engagement of Marxism with culture, feminism, ecology, and poststructuralism.[2] Two of the most useful trends for my purposes are the cultural analysis of the economy in economic anthropology, on the one hand, and the critique of the "capitalocentrism" of political economy by poststructuralist Marxist feminists, on the other. Both trends can be traced back to the germinal work of Karl Polanyi (1957; Polanyi, Arensberg and Pearson, eds. 1944), and are best represented, respectively, by the works of Stephen Gudeman (1986; 2001; Gudeman and Rivera 1990) and Julie Graham and Katherine Gibson (Gibson-Graham 1996, 2006). Polanyi argued that the economy is an "instituted" process and that the market is only one way of organizing the economy, even if it became predominant over the past two centuries. Others have described at length the long process through which the economy became a distinct category in the Western experience, separated from morality, politics, society, and culture and thus able to provide the basis for a positive science (Dumont 1977; Foucault 1973). With these notions in mind, Gudeman argues that what is known as the science of economics is just one historical discourse constructing a particular view of the economy, even if it has come to be considered a universal model. Those who use Western economic frameworks to investigate the economic practices of peasant communities in Latin America, for instance, fall into the trap of utilizing a reconstructive methodology "by which observed economic practices and beliefs are first restated in the formal language [of economics] and then deduced or assessed with respect to core criteria [of the same universal model] such as utility, labor, or exploitation" (Gudeman 1986: 28). They thus discover in the empirical data their own assumptions. Gudeman sees this epistemological trap at work in all allegedly universal models of the economy, whether neoclassical or Marxist.

That capitalism naturalizes its market model of the economy is understandable; the idea of a world organized in terms of individuals, self-regulating markets, and commodities is necessary to the functioning of the system. Less understandable is a certain naturalization of capital as *the* central organizing principle of social life in the critical discourses of political economy. Yet this is precisely what Gibson-Graham argue has taken place. In their view, most variants of this discourse have endowed capitalism with such dominance and hegemony that it has become impossible to conceptualize social reality differently. All other realities (for example, subsistence economies, cooperatives and other local initiatives, barter and solidarity economies, even feudal or neofeudal economies) are thus seen as opposite, subordinate, or complementary to capitalism, never as economic practices in their own right or as sources of economic difference. By highlighting how in globalization descriptions "all forms of non-capitalism become damaged, violated, fallen, subordinated to capitalism" (1996: 125), they refuse to accept the death of noncapitalist forms of economy.[3]

The project that emerges from these critiques is that of liberating the economic imaginary from its sole reliance on the languages of capital, individuals, markets, and the like. What needs to be done is to make visible what Gibson-Graham call "discourses of economic difference," that is, other ways of thinking about and organizing material and social life. For Gudeman, this entails the investigation of local models of the economy; these are deeply informed by local culture, even though they exist not in isolation but in their engagement with dominant models. As the local models of nature to be discussed in the following chapter, these models are an important dimension of place making. For Gibson-Graham, what is at stake is the possibility of finding noncapitalist forms of economy at play in concrete situations, such as the so-called traditional productions systems of the Pacific. Noncapitalism is defined not in terms of the control over the means of production but in terms of the circulation and appropriation of surplus labor and surplus value. We will see the relevance of this shift in emphasis from control over the means of production to control over the circulation of the surplus for some local economic efforts in the Pacific. Suffice it to say for now that with their poststructuralist notion of "economic difference" and the possibility of alternative capitalist and noncapitalist economies Gibson-Graham are not only creating fissures in the well-sutured frameworks of political economy but also inviting their readers on a journey of revisioning the material and cultural reality of most people on the planet.

I ◈ The Conventional But Not So Conventional Form of Capital: *Palmicultoras* and *Camaroneras* in Tumaco

The Arrival and Consolidation of African Oil Palm ◈ The history of capital in the Pacific does not start with the arrival of the *palma africana*. While I will not deal here with other important domains of capital in the Pacific—such as timber, commercial fishing, gold mining, canning of hearts of palm, tourism, and, lately, coca—some of these have been of paramount importance in introducing a capitalist regime of culture and economy. Since the 1940s, timber has been particularly important in the Pacífico nariñense in this regard. After a cycle of export of *tagua* (1885–1935), a timber industry proper emerged in the 1940s. This first wave was initiated by North American, German, and Spanish capital, and its purpose was the export of timber for construction. Some of the processing was done locally, thus incorporating a portion of the population into the activity. If there ever was a moment in the history of Tumaco that can be described in conventional terms of capital and the formation of a proletarian labor force, it is this period; however, the labor force involved only the small number of workers employed by the factories, since most of the people involved in this activity in the river areas had no workers' contract or standard work conditions. This phase ended abruptly in the 1970s owing to a combination of factors, including the decimation of trees, government policies that prohibited the export of unprocessed timber, and the unionization of workers. National capital took over after the exit of foreign capital, and the extraction strategy changed dramatically. Over the past few decades, extraction and processing have been articulated around a network of small sawmills on a number of rivers that ensure the supply of timber through mechanisms of indebtedness. Production shifted to being entirely oriented toward the national market, bypassing Tumaco to the benefit of Buenaventura to the north. From there, the timber is sent all over the country, first, for construction, and, second, for the paper industry and other uses (Leal and Restrepo 2003). A wave of tannin extraction from the bark of mangrove trees in the 1950s and 1960s was also important in creating conditions for the development of the timber and oil palm industries later on.[4]

The arrival of the *palma africana* was a slow process. Initially in charge of implementing Ferrand's proposal was the Cotton Promotion Institute (Instituto de Fomento Algodonero, IFA), which was replaced in 1969 by the more modern Colombian Institute for Agricultural Research (Instituto Colombiano Agropecuario, ICA). Ferrand had identified the area of Tangarial, some twenty-seven kilometers from Tumaco, as the site for an agricultural experiment station (*granja experimental*) for the promotion of palma africana and other crops, like coconut. According to the official history, the Tangarial granja was established through the donation of two thousand hectares by the government. In fact, it was the result of a combination of measures in which don Primitivo Caicedo, a local *agricultor* (farmer), played a somewhat significant role. Don Primitivo was an enterprising man who had once met General Gustavo Rojas Pinilla, the only military dictator to rule Colombia in the twentieth century, on the general's visit to Tumaco in the mid-1950s. On that occasion, don Primitivo and some of his friends, including don Segundo, Pedro Angulo, and doña Bertilda Preciado, followed the general from Tumaco to Candelillas, then an important stopping point on the railway line, with a petition in *papel sellado* (an official paper with government seal). In the petition, the agricultores asked the general to create a branch of the Agrarian Bank (Caja Agraria) in Tumaco. Feasting on roasted pig with Rojas Pinilla in Candelillas, they extracted a promise from the general to this effect, one which he ultimately fulfilled. It was this Caja Agraria that was set on fire by disenchanted and disgruntled agriculturalists many decades later, sparking the famous Tumacazo of 1979, by all accounts the most crucial popular uprisings in the Pacific in modern times.[5]

Don Primitivo had worked on the railway until it shut down before the end of the 1960s for alleged high costs of operation. After a year of unemployment which he spent on his farm, don Primitivo was hired by IFA to help in the demarcation of the granja experimental at Tangarial. As he recounts, "Given that most of these lands were occupied in small parcels, it was impossible for anybody else to get in; I did the counting of the parcels, and ICA went on to buy them and took whatever was still under open public lands [*baldíos*]." Two years later, feeling proud of his role in helping the granja take root, he expressed a wish to go back to his small farm, and he was rewarded with enough seeds to plant ten hectares in palm. He cleared fourteen of the forty hectares of forest (*puro monte*) he owned, which also contained plantain and cocoa trees, in order to plant the palm. Not many local farmers followed his example. Of the few who did, including, according to don Primitivo, Emiliano Caicedo, Domitilo Sevillano, and Arquimedes Valencia, it was said they were sowing weeds and they would end up like beggars since they would not have even the plantain to eat. To this, don Primitivo answered, "If I lose everything here, I will move back to Chilbío, since that's my place, or I'll go to Dos Quebradas, since my wife is from there, or to the Coast, where I have my father's inheritance."

Don Primitivo was a highly dignified man who had read the book of Charles the Great, believed in an ethics of work, and had established several rural schools during his lifetime. Like most of the region's people of his generation, don Primitivo was a trusting man who believed in the goodness of people and was an advocate of learning and of knowing the law; he used to read the legal codes in the local library and at the mayor's office because "in Colombia there is no law for dispossession, but he who doesn't know lets himself be dispossessed"; however, he learned to be suspicious of those who have knowledge since "those who study become deceitful and just want to live at the expense of others." He also articulated very well the sentiment of many of his peers about people's growing dependence on money. "Before," he said, "there was more and better *comida* [food]. Somebody would kill a rabbit, another a deer, and it was possible to eat; but today if you do not have money you don't eat." His judgment of money was drastic: "Today, even to be a beggar you have to have money" (Como decían Emeterio y Felipe, ahora hasta para pedir limosna hay que tener plata). In commenting on how in the years of the rapid spread of palm cultivation people would so easily sell their plots at the sight of a bundle of money, he decried the fact that "people let themselves be fooled by money; those who have not known money are capable of killing or going mad for it, and that's why we lost." There is more at stake here than a nostalgic "before" and "after" lamentation. As Taussig (1980) put it in the case of the black people of the Cauca River Valley, at issue is the wholesale attack by the world of the commodity on a life-world and worldview in which the commodity has played a relatively restricted role. This is an important aspect of the transformation of the Tumaco region for a good number of decades now, and it has become accentuated with the massive displacement that has accompanied the arrival of the African palm and shrimp industries (and, of course, coca).

It is in terms of displacement and the transformation of landscapes, labor, and production practices that one can analyze, in the first instance, the development of commercial agriculture in the area. Although don Primitivo and a handful of other small agricultores did relatively well with their small parcels of palma (they made enough for the farmer to "buy his shirt and food for the family, so they can eat, not as rich people but at least eat," as don Primitivo put it), it is clear that, from the outset of Ferrand's conception, the industry was meant for large capital investment. Ferrand recommended a social distribution for the development of the crop, later adopted by ICA: "Farmers with large financial means should be directed to the best soils and terrains, and each of them should exploit a minimum of 1,000 hectares," he wrote in his report. "This minimum, which can easily be surpassed, should enable the installation of an oil processing plant under optimal economic conditions" (1959: 38). But he added a complementary measure: "Next to the industrial plantations, it is of great social and economic interest for Colombia to establish, under well thought out conditions, colonization schemes based on the cultivation of oil palm" (39). He envisaged a dual agricultural system of large-scale holdings next to small-scale farming that has come to characterize Colombian agriculture for decades (Fajardo, ed. 1991). As we have seen, ICA's early attempt to lure local black agriculturalists into oil palm production did not meet with much success. Besides economies of scale, small-scale production faced limitations in terms of the commercialization of the crop, which small agriculturalists have to sell for arbitrarily low prices to the few enterprises that have processing plants. A cooperative of small producers spearheaded by don Primitivo and his friends and headed in the 1980s and early 1990s by don Primitivo's son, Rubén, a college graduate with expertise in the management of cooperatives, did make some inroads into the market, but they were no competition for the mighty enterprises from the country's interior, which had vast amounts of capital at their disposal. In fact, in an episode that demonstrated the utter imbalance of power, in the late 1980s the cooperative lost most of its land (over one thousand hectares) to the Varela group from Cali, who appropriated it after a trial by paying the cooperative the ridiculously low compensation of five million pesos, with which the small farmers were able to buy fewer that thirty hectares shortly thereafter.[6]

The concentration of land caused by this situation was staggering. According to official statistics, fourteen thousand hectares were planted in African palm in 1989. By 1996, the area controlled by agribusiness enterprises devoted to palm (locally known as *palmicultoras*) had increased to thirty thousand hectares, although not all of it was under cultivation; most companies had plans for further expansion. According to some estimates, 60 percent of this land was developed at the expense of primary forest. Land acquisition has relied on a combination of measures, ranging from the buying of land at prices that would be considered high by local farmers to less agreeable means, including outright coercion and threats. Over time, many black agriculturalists learned not to sell their land so readily, realizing that the money they obtained from its sale would not last long in the precarious Tumaco neighborhood, a city which saw its urban population almost double in the 1990s as a result of the massive migration caused by the rapid pace of land concentration. Resistance to selling was visible in the 1990s along the Rosario river, for instance. Since one of the main palma-growing areas is located between the Caunapí river, the right margin of the Rosario river, and the Tumaco–Pasto highway, this resistance is significant. In the mid-1990s, El Sande, a community of about fifty people located between the Mira river and the Ecuadorean border, refused to sell its land to one of the largest palmicultoras, Palmas de Tumaco, which almost surrounded it. The members of the community refused to sell in the belief that their land is their only means of subsistence and survival, even if that means that all of their commerce had to be done with towns in Ecuador under disadvantageous conditions. As don Primitivo remarked in 1993, "We are small agricultores; foreigners are grabbing most of the land."[7]

By the late 1990s, there were about two dozen large palmiculturas, five of them over 1,000 hectares in size.[8] All of them were controlled by large capitalist groups from the interior, particularly Cali and Bogotá. This crop has increased in other parts of Colombia as well, and as a result the country became the fifth largest world producer in the 1990s, after Malaysia, Indonesia, Nigeria, and the Ivory Coast. There were about 120,000 hectares under cultivation in 1992 in the country as a whole, which yielded about 285,000 tons of oil. African palm represented close to 7 percent of permanent crops in the country and 3 percent of the agricultural gross domestic product. By 1999, there were about 150,000 hectares in cultivation, yielding about 440,000 tons of oil, of which 94,000 were exported, chiefly to the European Union (CEGA 1999). Forty per cent of all exported palm oil came from the Tumaco area, even though only about 13 percent of the total land under cultivation was located there. Although in the early part of the first decade of the twenty-first century there was a program to increase production by small farmers by 4,000 hectares, credit and technical assistance has been directed almost exclusively to the large growers (see Castaño 1996 for details). With government support, large growers have been able to bring to the country the latest agronomic, genetic, and technological developments from the world leaders in Southeast Asia, especially Malaysia; as a result, in the late 1990s Colombia claimed the second highest yield of tons of crude oil per hectare (3.67). Moreover, FEDEPALMA (the Palm Growers' Association) advocated for and secured a "new pact" with the state which would make of the palm oil industry an important "engine of integral development" for the country as a whole (FEDEPALMA 1993; Mesa Dishington 1993; CEGA 1999). Job creation is touted as an important reason for the expansion of the industry, although it is by no means as substantial as its proponents make it out to be (see below). Palmicultoras increased the pace of land appropriation after 1999, with the help of paramilitaries and the resources of the largely U.S.-funded Plan Colombia. The sector was again heralded as a development pole by the second administration of President Álvaro Uribe (2006–10). Today, the industry is experiencing a boost because of the discourse of biofuels; in this growing world market, often couched as a sustainable alternative to fossil fuels, African palm appears as a main contender. Environmentalists and indigenous and black activists are rightly warning about this development, since again it spells out an even faster destruction of the forest and displacement of local groups.[9] But the strategy of expanding the cultivation of oil palm, in many cases by forcing people out of their territories, has continued until today, often times with government blessing.

Needless to say, the negative social impact the plantation system has had on local communities and ecosystems does not enter into FEDEPALMA's or the government's facile calculations. Traversing on foot one of the largest plantations in 1993—Astorga, which extends between the Caunapí and Rosario rivers and the Pasto–Tumaco highway—I witnessed a situation barely conceivable fifteen years earlier. Hundreds of Afro-Colombian workers could be seen busy at work along interminable rows of palm trees, planted regularly alongside a two-hundred-kilometer grid of side and main roads built to transport the product out for processing. On the plantation, women are in charge of keeping the area under each tree free of weeds, while men apply fertilizer and collect and transport the fruit to the extraction plant located at the center of the plantation. A vast reticular system of roads ensures speedy transport of the fruits to the processing plants. Most workers are hired by intermediaries and are paid the legal minimum wage with no benefits or security of any kind.

The discipline of the landscape—row after row of green trees, literally as far as the eye can see—corresponds to that of the workers. The end of the working day at five o'clock is preceded by the erratic noise of the trucks that transport the workers either to the riverbank, where they will row down to their homes in their canoes, or to the road that connects the plantation with Tumaco and smaller towns nearby; some of the smaller towns have seen their population grow significantly in an effort by the new labor force to keep their expenses low. At dawn the next day the same traffic occurs, only in the opposite direction.

The local landscape and pace of life before the plantations' arrival was very different. Those who still have farms along the rivers constitute a mirror in which the plantation workers can see their recent past. The new disciplined and simplified landscape is an effect of the apparatuses of capture of capital and the state as they exert their action on territory, humans, and money. The results are land (private property); disciplined labor (for the extraction of surplus value), although this cannot be defined in terms of straight proletarianization since labor arrangements are of many kinds, including subcontracting, piecework, indebtedness, kinship networks; and a sense of money as a generalized means of exchange for commodities. The steady march of the plantation has taken place at the expense of the local forests and agricultural plots and of the rivers, now seriously contaminated with the effluents from the plantation processing plants, which include pesticide and herbicide residues. Local production systems and, hence, subsistence have also been significantly transformed, as have black communities and cultures, many of them have now become sources of cheap labor for the plantations in a land that used to be theirs. To the polysemy of places and cultures, the plantation opposes the monocultures of mind and space of modern agriculture and social relations (Shiva 1993). The networked, partly self-organizing world of forest, rivers, farms, and communities enters the process of simplification and hierarchical structuring, by sheer force if need be, as in recent years.

This story, however, does not necessarily entail an ineluctable line resulting in an unqualified capitalist triumph. In the late 1990s, many black workers maintained the hope of returning to small farming, either temporarily or permanently, in order to spend the bulk of their life on the rivers. This suggests that black consciousness is far from being completely transformed even if, as the old people are wont to lament, their offspring are no longer interested in traditions and much less in rural life. There is a space between the colonization of consciousness by capital and modernity and the consciousness of that colonization where the process is actively contested and significantly reshaped by those undergoing it (Comaroff and Comaroff 1991). Even the palm cooperative of small producers represented for a time a warning against the common idea that capitalism can only beget more capitalism. While the hold of the capitalist commodity sector seems to be increasing dramatically in this area of the Pacific, it would be necessary to investigate what is happening in the noncommodity sector (independent, communal, and collective territories) and in household strategies in both rural and urban areas in order to ascertain the full character of economic regimes in this complex landscape of forms and practices. As I will discuss below, certain collective forms of the economy that emerged in the mid-1990s cannot be reduced to strict capitalist terms.

The production of palm by small farmers is by no means negligible. In 1999, there were 1,427 farms with fewer than 5 hectares planted in palm, and 350 with between 5 and 20, amounting to about 3,500 hectares, representing 18 percent of the total area under cultivation and between 10 and 15 percent of the total production (FEDEPALMA 1999). But cultivation practices there are entirely different. Rather than planted in regular rows, the palm trees are dispersed throughout the farm, interspersed with other food crops, such as plantain and cocoa. This does not mean that the palm trees are planted randomly because agricultores have an accurate mental map of the entire farm as well as an understanding of the associations among the various trees and crops. Farmers take into account not only the type of seeds and soils, but also the regimes of shade and temperature around each plant, including weeds (e.g., vines). Black farmers thus tend to reproduce in their plots the heterogeneity and complexity found in primary or secondary forests. For government technocrats, such complexity appears irrational, disordered, and ignorant (e.g., CEGA 1999: 35–37). For the farmers, it is a way of ensuring food supply and the reproduction of the domestic group, while generating some surplus for the market. At issue here are two different models of land and nature (see chapter 3). For now, suffice it to say that while the plantations follow the demands of capital, local agriculturalists have been able to configure a rhythm of production which does not conform entirely to the rationality of capital; they retain a certain level of autonomy regarding the disciplining of labor, landscape, and time that is an imperative of capitalist production. Since the late 1990s, however, capitalists, local politicians, and development agencies have pushed to transform small farm practices to resemble those of the large plantations. They have done so chiefly through a strategy spearheaded by a regional mixed corporation, Cordeagropaz, set up with the collaboration of the large growers.[10] If successful, this program will result in a significant rupture in customary cultivation practices.

How is one to gauge the completeness or incompleteness of capitalist development, avoiding the extreme of seeing capital everywhere without falling into the trap of seeing resistance everywhere? There probably are as many answers to this question as there are Marxist, post-Marxist, and poststructuralist frameworks. It should be clear by now that the emphasis in this section is neither on a detailed economics of the operations nor on deciding the character of the industry once and for all. Above all, I want to bracket the assumption that these enterprises can be described in purely capitalist terms, but to settle this matter would require more detailed study. Suffice it to say that even the largest palmicultoras hire a relatively small number of employees on a year-round basis—about two hundred for the larger ones, that is, those with two thousand to four thousand hectares under cultivation. The rest of the workers are hired only when cyclical activities (e.g., weed control and fertilizer application) are required. By the mid-1990s, it was estimated that the palm-growing industry was responsible for about ten thousand direct jobs for the close to twenty thousand hectares under cultivation (Angulo 1996). In terms of labor, there are two main types of situations. First, most of the administrative and technical personnel have a permanent contract; most of them are also from outside the region and are not black. Second, there is an amalgam of labor relations for the labor-intensive operations, particularly the harvesting and transporting of the fruit. These usually operate on the basis of the constitution of a work group by a contractor; the contractor in turn uses various forms, including indebtedness (*endeude*), daily wage (*jornal*), piece rate (*a destajo*), or a given agreement to share costs and benefits (*sociedad*). Kinship, symbolic, and friendship relations are often involved in these arrangements. The process is also gendered: of the three operations involved in harvesting—cutting of the bunches of fruit off the palm tree with a long tool, done by a *tallero*; collecting the fallen fruit, or *pepas*, from the ground, the work of the *peperas*; and transporting the fruit, usually by mule, to the nearest road, done by the *muleros*—the first is done by adult males, the second by women, frequently aided by their young children, and the third by younger men. Peperas are paid half as much as the talleros and one-third less than the muleros. Women are paid a destajo for this work, a common practice throughout the world. In other words, the organization of labor is a gendered, racialized order (this aspect of the coloniality of power is discussed in chapter 4).

The interpretation of this constellation of labor arrangements is not easy. One explanation sees at play a double strategy—processes of proletarianization and of peasantization—articulated to the vagaries of world markets (Whitten 1986 [1974]). Others see the Tumaco economy as a polyphony of pragmatic activities determined by the need to adapt to an uncertain environment (Arocha 1991). From a Marxist perspective, the articulation by capital of various forms of labor, land tenure, and production practices—and between capitalist and noncapitalist forms—is geared toward the expanded reproduction of capital. In other words, while not strictly capitalist, black labor is subsumed to and articulated with capital. More recent approaches in this tradition see black labor and farms as articulated with capital but nevertheless retaining some margin of autonomy. Capital is seen as locally inscribed according to the specific social and cultural conditions, and this inscription increasingly transforms local rationalities toward the market and commodity forms (Restrepo 2004). This view sees capital as a heterogeneous set of practices, even if it is still articulating many forms of labor and production (see also Quijano 2000). At the end of the chapter, I will introduce a more diverse panorama of economic forms that enable a less capitalocentric reading of palm and shrimp production.

Finally, one must keep in mind that oil palm (and shrimp aquaculture: see below) are global operations. Palm oil is the major cooking oil in many Asian countries, such as China and India, the second most common edible oil in Europe, and now a main crop in the global biofuel economy. The crop has expanded hugely over the past fifty years, experiencing a 32 percent increase in global consumption in the second half of the 1990s (Wakker 1998, cited in Gibson 2000). Despite the increasing criticism of its rapid expansion in tropical rain forests, where it destroys indigenous livelihoods and biodiversity, the prospects for the growth of the industry seem bright again, even more so in light of biofuels. Thus in the mid- to late 1990s the Malaysian government drew up plans for the development of one and a half million hectares for oil palm in Sarawak, most of them located in traditional lands of the indigenous Dayak. Following in these footsteps, the then-president of Colombia, Andrés Pastrana, announced in March 2001 his intention to devote three million hectares to large-scale plantations, in units of twenty thousand hectares or more. He presented this project before the Asian Strategy Leadership Institute conference in

Malaysia, luring Malaysian industrialists in particular to invest in Colombia so that "progress and social development can reach large areas of Colombia that are ready to join in the cultivation and processing of this primary commodity."[11] The continued expansion of the palma Africana has become almost an obsession of the second Uribe administration, which seems heedless of its dramatic human and ecological costs.

Local Mangroves vs. Global Shrimp ◈ Mangroves are a vital part of the social, cultural, and ecological life of the Pacific. Over half of the Pacific mangrove forests are located in Nariño, nearing 150,000 hectares. Biologists say 70 percent of the mangroves of the Pacífico nariñense has experienced significant intervention, particularly after the wave of industrial tannin extraction from the bark, which finally eased in the early 1970s. Nevertheless, they continue to be a vital source of sustenance for local people. Mangroves play a crucial role in the maintenance and reproduction of river and ocean life. Biologists emphasize the fact that mangroves are one of the most productive ecosystems on earth, recycling large amounts of nutrients and maintaining water quality, a fact that some mangrove-based black communities in nearby Ecuador have used as the basis for creative statistical calculations to counter the facile claims of large-scale shrimp operations. When the total productivity of the mangrove is considered (in terms of embodied energy, or "emergy," for instance), shrimp production appears as an extremely poor alternative (Martínez Alier 2002). Mangrove forests house a large number of species of flora and fauna, including fish, mollusks, and crustaceans; they are the site of complex interactions among aquatic species, insects, and birds and thus are crucial to the local biodiversity. In addition, mangroves are essential to the early reproductive stages of many species. Shrimp eggs, for instance, are laid in the open seas near coastal areas, but the larvae slowly move to the mangroves, where they achieve a good part of their growth before returning to the sea. Mangroves protect coastal areas from erosion and encroachment by the sea, and they provide a livelihood to many people in the surrounding areas. Among the resources that locals obtain from mangroves are timber for house construction, aquatic species from fish to mollusks, wood for charcoal production, tannin for dying handicrafts, and some fruits and medicinal plants. A good deal of movement and transportation also take place along the estuaries that exist in the mangrove forests. So, while profoundly localized, mangroves are not just local in that, ecologically speaking, they fulfill many functions of importance beyond locality.[12]

Why these rich and by most accounts beautiful socionatural worlds could be under rapid destruction is not easy to understand, but this is precisely what has been happening since the mid-1980s. By then, industrial fleet fishing of ocean shrimp had come under attack by environmentalists because of its destructive impact on seabed habitats and high level of by-catch of many other species, including turtles. Some trawler fleets dump up to fifteen tons of fish for every ton of shrimp, as reported by Shiva (2000: 40). Fishing communities throughout the world had been calling for a ban on mechanized trawlers since the 1970s. With a degraded marine ecology came a significant decrease in shrimp and prawn catch. Coupled with a seemingly inexhaustible appetite for shrimp in the United States, Japan, and Europe (which in 1993 consumed two billion pounds; Filose 1995), this situation led to the dream of large-scale aquaculture as a solution to both the ecological crisis caused by overfishing and the problem of supply. Always eager to help, the international development establishment provided the missing link in terms of ample financing and technical assistance for industrial aquaculture. Between 1988 and 1995, the World Bank and the regional development banks provided 69 percent of external financing to 40 percent of the projects worldwide. Other bilateral or multilateral lending institutions funded a significant percentage of the remaining projects. Many of these projects had technical assistance and some funding from the Food and Agricultural Organization (FAO) and the United Nations Development Program (UNDP) (Tirado 1998). Thus started one of the most important, and questionable, chapters of the so-called blue revolution (Shiva 2000). In many regions this development is taking place at the expense of mangrove forests; in order to cut costs and have a quick high return on investment the industry has relied on uprooting the mangroves to make room for the construction of vast ponds. By 1990, over 26 percent of world shrimp production came from culture systems, compared with barely 2 percent ten years earlier. Growing consumption and scientific and technological advances (such as advances in seed production and genetics, particularly hatcheries for the production of seed stock to replace the unreliable collection of natural post-larvae; pond management; and nutrition; see, for instance, Fast and Lester 1992) suggest that the future growth of the industry seems guaranteed—if the ecological and social problems around it can be controlled.

In Latin America, Ecuador took the lead in the development of shrimp farming and is in fact the second largest world producer of shrimp in captivity. The industry has been severely affected by several factors in recent times, from viral diseases that kill the shrimp in its early stages and water pollution to a government ban on new pond construction because of the alarming level of mangrove destruction and increasing community opposition. In Colombia, the industry took off in the mid-1980s thanks to generous government subsidies and support, including paid tours for interested industrialists to the main producing countries in Asia. Colombia is still a relatively small producer (10,000 metric tons/year produced in 1997 on about 3,000 hectares) compared to Ecuador (130,000 tons/year in 180,000 hectares) or Thailand (150,000 tons in 70,000 hectares). However, its contribution to capital accumulation in the region is not negligible. In 1994 shrimp aquaculture represented a value of fifty-five million dollars in exports, with production in the Tumaco region representing over 30 percent of the total. Among the factors cited as favorable to the development of the sector in this region are the excellent quality of the Tumaco Bay waters, which enabled a high level of productivity (although contamination is growing every year with the effluents of both palm and shrimp industries); government subsidies; credit programs totaling fifty-two million dollars for the period 1989–95 for the aquaculture sector; growing global demand; the possibility of using mangroves to lower construction costs; and a steady program of government-supported research and technical assistance. This final factor became crucial after 1992, when the industry faced its first crisis as a result of viral contamination of the shrimp, a decrease in global demand, and a revaluation of the national currency. Several industries were forced to end operations during the 1990s.[13]

As in the case of the palmicultoras, shrimp farms were established largely by capitalists from Cali, Bogotá, and Medellín. The largest farm, Maragrícola, had 265 hectares under shrimp cultivation and 45 in tilapia in the mid-1990s. It was set up by a well-known capitalist group from Cali (Mayaguez), whose main line of investment is in sugarcane.[14] The shrimp industry employs very few workers directly (Maragrícola had only 220 workers in the mid-1990s, and all of the qualified employees were from the interior). A larger number of jobs were created indirectly in the fish- and shrimp-packaging plants in the city of Tumaco. It is mostly women who are hired for these jobs; they start lining up at 5:00 a.m. outside several fish- and shrimp-processing plants hoping to get a job for the day. In most of these plants working conditions are poor. The workers stand on a wet floor for the entire day, are paid *a destajo*, that is, according to the weight of the shrimp they are able to process (clean, behead, peel, and devein), and often end up with severe back pain and bloody hands. The largest processing plant, established by several of the *camaroneras* (shrimp farms), has better working conditions, providing the workers with rubber boots and aprons.[15] As in other parts of the world where sweatshop conditions prevail, the sexual division of labor here, while providing women with a meager source of income, not only relies on sexist ideologies about women's manual dexterity and obedience, but also represents little in terms of a betterment of living and social conditions vis-á-vis men and society at large (see, for instance, Ong 1987; Mies 1986). Work for men in the camaroneras is more sporadic. Intermediaries hire groups of men to perform certain activities, such as the harvesting and maintenance of the ponds every four months. The harvest is conducted at night, and the men have to be on the water for most of the evening. The work lasts for a week at most, after which the men either become unemployed or return to their customary fishing and agricultural activities, which a significant number of them are still able to perform.

Besides the destruction of many hectares of mangrove forest (2,600 hectares in Nariño by the mid-1990s; Castaño 1996: 34), with the concomitant impact on people's livelihoods, the camaroneras contribute to the environmental degradation of the area. These problems originate first in the fact that the industry uses a variety of inputs intensively—from a fish-meal-based concentrate as feed to fertilizers, antibiotics, and chemicals for regulating the pH of the water—and, second, in the acids and sulfates produced by the operation, given the inadequacy of the soils for this type of aquaculture (Tirado 1998; Shiva 2000). Large amounts of organic residues, decomposed by microorganisms, contribute to the production of phytoplankton, which demands large quantities of oxygen upon decomposition. The effluents of the farms, with their high concentration of organic and inorganic residues, cause further problems of hypernutrification and eutriphication of the receiving waters, thus altering the equilibrium of the mangrove and estuarine ecosystems. This effect is made worse by the practice of partial recycling of pond waters to relieve stress on the shrimp and to reduce their susceptibility to diseases. There are many other ecological effects, from salinization of the water table and mortality of biota in the estuaries to the introduction of viruses and non-autochthonous species. Some alternative methods have been suggested to lessen some of these problems, the main one being the construction of farms inland in zones of higher elevation, thus sparing the mangroves. This is doubly important given the difficulty and estimated high costs of restoring the mangrove forest; these have been calculated as ranging from $5,250 to $125,000 per hectare, depending on the situation, with an average of $50,000 per hectare (Tirado 1998: 26).[16]

Not surprisingly, the destruction of mangroves for shrimp farming has met determined opposition from local communities and local and international environmental groups worldwide. Local movements linked up in 1997 to create a network, the Industrial Shrimp Action Network (ISA Net, www.shrimpaction.org) to campaign against industrial shrimp farming and assist local organizations and communities in dealing with its impact. Fourteen countries participated in the founding, including Honduras, another large producer, Ecuador, and Colombia, the latter represented by PCN. In October 1996, delegates from twenty-one nongovernmental, community, and social movement organizations met in Choluteca, Honduras, to discuss the worldwide spread of shrimp farming. In what became a well-known communication called the Choluteca Declaration (see Acción Ecológica et al. 1997), this group deplored the destruction and contamination of mangrove forests, estuaries, and swamps; the displacement and marginalization of local communities by shrimp farming; and a host of unsustainable practices, such as the introduction of nonnative species and the capture of shrimp larvae in ways that adversely affect the survival of other species. Among other things, it proposed a global moratorium on the expansion of shrimp farming; sustainable practices compatible with the structure and functioning of natural ecosystems and the cultural practices and socioeconomic interests of coastal communities; and the active involvement and participation of local communities. Among the NGOs and social movement organizations present were Acción Ecológica and Fundación Ecológica de Muisne (Fundecol) from Ecuador, the PCN from Colombia, and Greenpeace.

Struggles of this sort often find eloquent expression in the narratives of local people, as in the case of a black woman reported by a local grassroots group in Ecuador, Fundecol, in its call for a mobilization against an illegal camaronera in Muisne in 1999 (cited in Martinez Alier 2002: 83):

> We have always been able to cope with everything, and now more than ever, but they want to humiliate us because we are black, because we are poor, but one does not choose the race into which one is born, nor does one choose not to have anything to eat, nor to be ill. But I am proud of my race and of being a *conchera* [shell fish collector] because it is my race which gives me strength to struggle in defense of what my parents were, and my children will inherit; proud of being a *conchera* because I have never stolen anything from anyone, I have never taken anybody's bread from his mouth to fill mine,

because I have never crawled on my knees asking anybody for money, and I have always lived standing up. Now we are struggling for something which is ours, our ecosystem, but not because we are professional ecologists but because we must remain alive, because if the mangroves disappear, a whole people disappears, we all disappear, we shall no longer be part of the history of Muisne, we shall exist no longer, and this we do not wish to accept. . . . I do not know what will happen to us if the mangroves disappear, we shall eat garbage in the outskirts of the city of Esmeraldas or in Guayaquil, we shall become prostitutes, I do not know what will happen to us if the mangroves disappear. . . . What I know is that I shall die for my mangroves, even if everything falls down my mangroves will remain, and my children will also remain with me, and I shall fight to give them a better life than I have had. . . . We think, if the *camaroneros* who are not the rightful owners, nevertheless now prevent us and the *carboneros* [charcoal makers] from getting through the lands they have taken, not allowing us to get across the *esteros* [estuaries], shouting and shooting at us, what will happen next, when the government gives them the land, will they put big "Private Property" signs, will they even kill us with the blessing of the President?

This was exactly what happened in the Tumaco area in the early 1990s when a large camaronera enclosed seven hundred hectares of mangrove forest and put up "Private Property" signs to prevent local *concheras*, *carboneros*, and *leñeteros* (charcoal makers and firewood collectors) from using the mangroves and estuaries in their customary way. In the declaration by the black woman of Muisne can be seen the same expression of *dignidad* (dignity) voiced by don Primitivo Caicedo and doña Ester Caicedo, one of the leaders in the communal shrimp-farming projects in Tumaco (see below). Further, a number of features in her narrative deserve to be highlighted from the perspective of a place-based political ecology: the distrust of the money-based world, which suggests a semiotic resistance to the reconversion of the environment into an economic resource; the defense of culture and place in the conviction that losing them amounts to blacks' disappearance as a people; and a notion of power and of what lies beyond place (e.g., life in Esmeraldas and Guayaquil) that, even if not articulated as such, suggests a critical posture toward the state and the larger modern world. It is precisely these aspects that are at the basis of much of the environmentalism of the poor worldwide. Far from relying on a static notion of culture, the narrative reflects a profound engagement with a world that, though frequently menacing, encourages local people to develop transformative economic and ecological projects.[17]

Shrimp farming has also contributed to the transformation of the landscape. The "mirror of water"—as the multihectare pond surface is known in the trade—became a new vista for the local inhabitants, accustomed as they were to the fractal dimensional landscape of estuaries and mangroves, with their unique aerial roots, or to the ever-fluctuating mirror of the sea. The undulating, smooth space of the sea, which the fishermen (*pescadores artesanales*, or traditional fishermen in the language of development) learn to navigate under the guidance of the winds and the moon, and the fractal space of the mangrove forest are now accompanied by the striated and regularized space of modernity requiring a host of technosigns: the points at which the fish meal concentrate is to be poured into the ponds or the pH measured; the tower for watchmen and the regular formation of light poles casting their shadowy light from dusk to dawn to prevent theft; the fences and signs of private property, odious to the local inhabitants; the scheduled times of seeding and of harvesting of the product. Even the shrimp are processed and classified according to an internationally established standard of twelve sizes, which determines not only the price of the product but the satisfaction of the northern consumer. Very different is the knowledge that regulates the practice of concheras, leñeteros, and pescadores artesanales in their customary environments; for some, this knowledge is closer to art, pragmatic improvisation, or performance than to a scientific reading of signs afforded by instrumental rationality or the reliance on a body of context-free knowledge. The drastic transformation of the landscape effected by capital in this way also effects a transformation in the interrelated modes of knowing, being, and doing that often remains inadequately explained.[18]

II ◈ The Contradictions Between Capital, Nature, and the Economy

Certain trends in the social theory of capital and nature help one understand the transformation being brought about by capital in the Tumaco region, beyond the most obvious effects on the landscape, social relations, and subjectivities. The result of the engagement of liberal theories of the economy with the environment has yielded the field of environmental economics. The main idea behind this field is the internalization of the hidden environmental costs of capitalist production (the so-called externalities) into economic calculations; according to this framework, economists would need to take into account the environmental costs of oil palm and shrimp production in the Pacific (contamination of rivers, destruction of mangroves, opportunity costs of the ecosystem) to arrive at more real costs of production. This approach, although a step forward in some ways, is limited because it leaves unquestioned the capitalization of nature and labor, upholds the plantations as a practice, and leaves untouched the market framework. An altogether different approach, ecological economics, constitutes a more farsighted proposal. This approach seeks to re-embed the economy in social structures (conflict and power), the ecosystem (biophysical processes and laws), and social perception of the environment (culture; see the table in the introduction). A major claim of ecological economics is that socioenvironmental processes cannot be reduced to market values. On the one hand, there is no single value standard to which all values can be reduced (e.g., money), as in conventional and environmental economics; on the other hand, it is often the case that in environmental struggles there is incommensurability of values (e.g., between how corporations and communities value mangroves) so that it is impossible to find a unified value standard.

For these reasons the problem of valuation that is behind environmental conflicts is central to ecological economics. Approaches such as multicriteria evaluation provides a means for bringing contrasting value standards into dialogue and possible resolutions that are not based on a common language (e.g., market) or value (money; see Martínez Alier 2002; Munda 1995). The case of mangroves is a clear example of an ecological distribution conflict in its various dimensions: it entails a defense of the mangroves as a source of livelihood; it pits poor, place-based communities against capitalist enterprises with ties to world markets; it brings to the fore conflicts over the valuation of resources, with a clash between simple monetary valuation on the part of shrimp-farming advocates and a multicriteria valuation on the part of ecologists, communities, and NGOs that includes a variety of values (including livelihood and subsistence, biomass production, food security, cultural values, coastal defense, landscape value, aesthetic values, biodiversity); and it highlights the cultural and place-based nature of the environment, given that the ecosystem is seen in starkly different terms by the communities that live in it.[19]

From a Marxist perspective, palmiculturas and camaroneras are putting in place a very peculiar set of productive forces and production relations, to the benefit of some and the detriment of others, including the natural environment. That this is well understood by many local black agriculturalists is attested by their repeated protests against the fact that the *gente de afuera* (people from outside the region) are grabbing and monopolizing the land, by their continued conviction that it is better to produce use values and items for market exchange in one's own land than to work for the *empresas* (enterprises), and by their incontrovertible judgment that all those who leave their lands end up badly in Tumaco. Or one could invoke Marx in an attempt to explain why a Barbie doll found in a poor household in the Rosario river hangs on the wall of the main room, its crystal paper wrap intact, along with photographs of dead relatives, stamps of the Virgin Mary, or a magazine cutout portraying an electric blender, in a small hamlet with no electricity. In all likelihood, this Barbie was brought here by a relative from Cali in one of his or her not-too-infrequent return trips to the river. That the Barbie does not circulate but is rather made to stand for a still somewhat remote modernity is another reminder of the vagaries of the commodity form in a place that is not yet fully committed to it and where, as Taussig (1980) would put it, existence has not yet been fully disenchanted by the commodity.

There are other questions to be explored from an environmental perspective. As is well known, there was no explicit ecological treatment in Marx, and the cultural analysis was undeveloped. The challenge for ecological Marxism has been to extend historical materialism into physical nature (how natural history influences human history and vice versa). Since the 1980s this extension has led to an active debate about what is called the second contradiction of capitalism. Briefly, this concept builds on the notion that capitalism, given its unplanned character, is a crisis-ridden and crisis-dependent system which results in intraclass and interclass conflict and hence in repeated crises. The starting point of Marxist crisis theory is the well-known contradiction between productive forces and production relations or between the production and realization of value and surplus value, leading to realization crises. Capitalism continuously restructures itself through these crises, in ways that are themselves contradictory. But there is a second contradiction of capitalism that has become pressing with the aggravation of ecological crises. This contradiction brings to the fore the conditions of production, insufficiently theorized by Marx but placed at the center of the inquiry by Polanyi (1957). A condition of production is defined as those factors that are not produced as commodities, that is, according to the law of value, even if they are treated as such; this includes those aspects that Polanyi called "fictitious commodities," such as land (nature), labor (human life), space, and many general and communal conditions of production. Modern capitalism has brought about the progressive capitalization of production conditions. Palm trees produced on plantations, privatized land and water rights, industrially produced shrimp, and genetically engineered seeds are all examples of the capitalization of nature and human life as production conditions.[20]

This process is mediated by the state, which regulates the production of production conditions and makes them available to capital in the form, for example, of subsidies, technical assistance, and control of labor on behalf of the capitalist sector in Tumaco. This regulation is often highly bureaucratized and increasingly governmentalized (e.g., Brosius 1999; Escobar 1999a; Luke 1999). The state legitimizes its control over production conditions in terms of the general interest, including, for example, progress and development and economic growth. But this only pushes capital and the state toward more social, hence more visible, forms of control over production conditions via more flexible planning and planned flexibility, including, as we have seen, the increasing use of violence. The World Trade Organization and the struggles to combat it are a case in point. The key issue here is that capitalism subverts itself when it shifts to more social and ideological forms. There is a contradiction between productive forces and relations, on one side, and production conditions, on the other. In other words, capital tends to create its own barrier by destroying production conditions. This becomes the basis for the action of social movements, such as ethnic, women's, and environmental movements and new coalitions between labor and other groups. Their struggle has two sides: the defense of production conditions (e.g., human life, the body, nature, space, and place) against capital's recklessness and excesses; and control over the crisis-induced restructuring of production conditions (e.g., the drive to expand the oil palm frontier through a combination of force, rights violations, and the support of the government and Plan Colombia). This is another dimension of what Guha and Martínez Alier (1997) describe as the environmentalism of the poor.

In the Tumaco region, the capitalization of nature in the shrimp and oil palm sectors creates a contradiction between these two types of capital (the growing contamination of the rivers by the palmicultoras influences negatively the shrimp sector, for instance). There are thus contradictions between individual capitals and capital as a whole. Over the past two decades, the state has favored the palmicultoras over the shrimp industry, given the larger political clout of the plantation growers and the more manifest ecological devastation caused by shrimp farming. The inauguration in late 1993 of a state-of-the-art fishing port facility for large-scale processing of shrimp and ocean fish gave some impetus to the shrimp sector. Attempts by this sector to control its own production conditions (e.g., through the local production of larvae to avoid the problems that affected the industry in Ecuador) did not prevent a crisis in the sector again by the late 1990s, when several farms were forced to close down their operations. Alliances between the oil palm sector, local black elites, and local and regional state institutions seem to have gained ascendancy over the reproduction of production conditions in the present decade, including labor. These are opposed by social movements and, at times, by local elite movements for regional, autocentered capitalist development.[21]

In conclusion, the outcome of struggles within and between capital, the state, and place-based and social movement organizations affects the production and regulation of production conditions in complicated ways. Whether state policies pertaining to the conditions of supply and regulation of environment, labor power, infrastructure, and space will enable the reproduction of capital as a whole has yet to be proven; since 2000 the instrument of choice has been force and new legal apparatuses (see the conclusion). The use of force might mean that capital will continue to destroy its own conditions rapidly and that neither state nor capital will be able to rationally reconstruct those conditions on a long-term basis. This brings to the fore a series of questions, from the eventual (and ongoing) reconstruction of production conditions by capital, the state, or social movements (or a combination of these actors) to a possible redefinition of production, which I will address in the conclusion of the chapter. Before I do so, however, one final piece of the economic puzzle needs to be discusseed: the possibility of gleaning noncapitalist practices in some community projects in the Tumaco region.

III ◈ Beyond Capital: Collective Shrimp Farming as a Noncapitalist Practice

Doña Ester Caicedo was one of the most important pillars of the *camaronera comunitaria* (community shrimp farm) of the barrio Unión Victoria in Tumaco from its inception in the early 1990s. Many years ago, while working as a maid for a famous television personality in Bogotá, she dreamed of being a television star, and she almost succeeded. But she fell in love with a man from Tumaco and, upon becoming pregnant, returned to her native region. Back in the littoral, she found that her only alternative was selling fish on the highway near Tangareal. After her lover left her, she worked for a time as the manager of a small local hotel that belonged to a friend. With her meager savings, she acquired some land and, going to the mangrove to cut the main poles, she built a house for herself and her young daughter with her own hands. With the house finished, she became a good catch in the eyes of the local men. Although she longed for a companion and more children, she thought, why bring a man into the house she had built with her own *fuerza* (strength) only to find that sooner or later he would feel as if he could boss her around? She wasn't one to put up with that anymore. Besides, as she says, she was born for a different fate: she was destined to be a leader, to be as free as the wind, and to devote her life to more important things, including her children.

Doña Ester was born on the Mira river and as a child lived in many places throughout the Pacífico nariñense. As a young girl she sold plantains. Her aunt took her to Tumaco for some schooling, and later she traveled to the big cities of the interior to work as a maid. By the late 1980s, now in her late thirties, she still felt that she wanted to transform her life and re-create herself, so she found a new trade as a charcoal maker; it was a hard job done mostly by men, but the income was good. Soon, however, the announcement of the arrival of electricity to the area threatened her new occupation. The news caused her to lose sleep, but she only worked harder. In the midst of these changes she met José Joaquín ("don Chepe") Castro, a veteran carbonero and natural-born leader. In the late 1980s, as the first wave of camaroneras began operations, doña Ester and don José Joaquín, with the help of Luciano Castillo, who several years earlier had led a successful land takeover in another part of the city and had learned community organizing, founded the Asociación de Carboneros y Leñeteros (Charcoal Makers and Firewood Collectors' Association, ASOCARLET). Since don José Joaquín had worked with INPA (Instituto Nacional de Pesca y Acuacultura) in the past and had some experience in traditional fish farming, it did not take long for the new partners to start thinking about opening their own camaronera. The idea became an obsession, a powerful dream that, against all odds and the best judgment of most, they decided to make real.

The mangrove forests customarily used by locals had always been considered *tierras baldías*, or empty public lands. Even the mangroves within Tumaco's urban perimeter, such as those in the barrios Unión Victoria and Porvenir, where the camaroneras comunitarias were established in the early to mid-1990s, were regarded as such. As the adjacent mangrove areas began to be claimed by the large companies, they were enclosed and made off-limits even to the carboneros and leñeteros who had lived off them all their lives. This was the situation when, in the fall of 1993, the twenty-five ASOCARLET members who originally had decided to embark upon the building of a shrimp farm were about to finish constructing their first one-hectare *estanque* (pond); they constructed the estanque entirely by hand, a titanic endeavor at which the entire membership worked three days a week for a whole year, devoting the rest of their time to income-generating activities in order to survive. No sooner were they declaring their enterprise a success, however, than they were told that the land on which they had so lovingly and laboriously carried out their task already had an owner. The owner in question was a large camaronera called Maja de Colombia, of unknown capital origin, that was ready to expel them from the land. This camaronera had claimed seven hundred hectares of mangroves and estuaries. One day a fateful Private Property sign was placed at the very entrance to the community pond by Maja's arrogant paid guards. After much negotiation and intermediation by the regional development corporation, Corponariño, Maja agreed to cede fifty hectares to the community, on which they continued to expand their enterprise. By 1998, ASOCARLET's camaronera had five estanques in operation. Throughout the construction process, they had some support from the city government, mostly to pay for food while the construction lasted, and received technical assistance from a sympathetic INPA biologist and from Plan Padrinos (then called Foster Parents International). Notwithstanding the importance of this help and the fact that it resulted indirectly from the politicking of local political leaders, the camaronera was decidedly the result of a tenacious community effort that lasted a number of years.

To be sure, no project in a context like that of Tumaco in the 1990s—saturated as it was with institutions, discourses, and organizations of the state, the private sector, and civil and political society concerning development, the environment, black culture, participation, and so forth, as we will see in detail in chapter 5—could be said to be simply the result of community initiative. For local black politicians and state bureaucrats, the experience of ASOCARLET was made possible by the early support of institutions like Plan Padrinos and INPA, which were trying to come up with strategies that went beyond the use of black people as mere cheap manual labor for development projects directed by outsiders.[22] Yet by all accounts the local leadership was definitive. Over the years the group developed an increasingly sophisticated approach to their activities—indeed, a political ecology approach. Among the features emphasized in this approach were the strengthening of black culture, the defense of the territory, the sustainable use of the mangrove, and the protection of biodiversity. Again, these features were thoroughly permeated by expert discourses, including those of the black movement. Next to them were discourses of other provenance, such as that of *convivencia solidaria* (living in solidarity), now extended to the natural environment, and *economía solidaria*, which derived from the cooperative movement, with some roots in the Pacific since the 1950s. Also present were long-standing conceptions of forest, mangroves, and nature in general.[23]

There are many vantage points from which to assess this experience. To start with, there is the perspective of an Afro-Colombian intellectual and activist from Cali, Harold Moreno, who, attracted by the debate about the black communities then raging throughout the region in the wake of AT-55, arrived in the region in the early 1990s. Moreno had obtained his bachelor of arts degree in the well-known communications program (*Comunicación Social*) at the Universidad del Valle in Cali in 1986 and was eager to employ his knowledge in the service of the growing popular communications movement that swept over large parts of the Pacific and resulted in a vibrant movement for popular radio stations and literacy programs (see chapter 4). Although not an activist in the movement, he was part of what Sonia E. Alvarez (1998) has called the "social movement web," that is, the larger set of actors and organizations that constitute a politico-discursive field in which people concerned with black cultural politics act and circulate, thus contributing, often in contradictory ways, to the meaning of black cultural politics. Upon arriving in Tumaco, Moreno participated actively in workshops on AT-55 with river communities. He read anthropological texts on black culture avidly, worked with the progressive local priests engaged in the same process, learned much from what the old people had to say in the workshops, and came to believe that all work in the Pacific had to rely on solidarity, friendship, and reciprocity, although without minimizing the importance of expert knowledge.

Moreno joined the ASOCARLET project as an advisor and teacher of sorts. He helped the group with *capacitación comunitaria* (community training and animation), the overall planning process, the preparation of documents, and research in the case against Maja. Working on oral histories with ASOCARLET members, he found that the community's active presence in the area went back at least sixty years. Moreno was very much aware of the fact that local people were being rapidly marginalized by *paisas*, the whites from the Antioquia region, so he became committed to showing that the black people of the region had entrepreneurial capacity. And this community certainly had it. By themselves they had formulated schemes for branching out into other activities, from the production of certain endemic flower species to tannin, including other cooperative businesses, from hairdressing to carpentry. And these people had become ecologically minded; they not only planned on reforesting part of the mangrove, but also became interested, as the process moved along, in employing alternative shrimp-feeding sources, minimizing the use of chemicals, and using naturally existing larvae instead of laboratory-raised specimens. They were also mindful of social issues. They discussed and revised the project's bylaws collectively and debated the best ways to share power among the different groups so that power did not lie solely with the group devoted to shrimp farming; as the volume of operations increased, they also wondered if women would be rewarded at the same level as men. Above all, ASOCARLET people cared about their children's future: they wanted them to inherit a workable camaronera and a healthy mangrove, which they thus should actively conserve. This suggested they were aware that certain development models could drastically change their culture, and that awareness explains why in the last instance they wished money could be used "in the same way that they use toothpaste,"[24] that is, as a simple means of exchange.

Moreno believed in the synergistic value of modern and ecological technologies, modern as well as traditional economic practices. His view of an alternative modernity was based primarily on a belief in the ability of local people to craft their own projects. In this he was of the same mind as people like doña Ester and don Chepe Castro. They are not alone in this conviction, of course. What some philosophers (Spinosa, Flores, and Dreyfus 1997) have recently called "the retrieval of history making skills" could be said to be central to the practice that groups like ASOCARLET have been trying to develop. Through their phenomenological argumentation, these philosophers seek to bring back a contextualized and situated notion of human practice that contrasts with the detached view of people and things fostered by Cartesianism and hierarchical models. People live at their best, they maintain, when engaged in acts of history making, meaning the ability to engage in the ontological act of disclosing new ways of being and transforming the ways in which they deal with themselves and the world. This, in turn, requires rootedness in particular places (immersion in the cultural background practices of a collectivity with the risks entailed by speaking out on the basis of this background). In this context, place-based activists and common people ("reticent citizens," 94–115) do not act as detached contributors to public debate but are able to articulate the concerns of their community in such a way that the relevant background practices are changed. This life of *skillful disclosing*—which can be said to characterize doña Ester and don José Joaquín as well as the PCN activists I discuss in chapter 5—is possible only through intense engagement with a place. These philosophers find this practice at play in three domains above all: entrepreneurship, as the skill of cultural innovation; democracy, as the skill of interpretive speaking; and solidarity, as the skill for creating grounds for meaningful community. While these authors believe that the skillful way of being human is being eroded, they also believe it is still sufficiently significant for the recuperation of place-based skills and practices. ASOCARLET is one of those cases in which such retrieval is actively being pursued.[25]

The place-based practice of skillful disclosing is important as a first corrective to the understandable tendency to interpret experiences such as the camaroneras comunitarias as no more than by-products of capitalism and state action. Skillful disclosing can be linked to other ways of practicing the economy. Could ASOCARLET be seen as engaged, in their activity of disclosure, in creating noncapitalist or alternative capitalist forms of the economy? "If they can do it, why can't we?" asked doña Ester, referring to the capitalist camaroneras. As I mentioned above, some black agriculturalists around the Tumaco–Pasto highway and in the margins of the Rosario river were still resisting plantation labor in the 1990s by spending the bulk of their life as smallholders. While this avenue might no longer be open to most of those already working in the plantation, it shows that capitalism is not necessarily the condition for more capitalism. This runs contrary to common belief. Could the discourse of entrepreneurship ever be a tool against the capitalization of subjectivity? Could people's economic identity ever get to be other than capitalist, as much political economy assumes? It is just this possibility that is adumbrated by Julie Graham and Katherine Gibson (Gibson-Graham 1996, 2006), whom I mentioned in the introduction to the chapter. It is a different subjectivity that utters the "if they can do it, why can't we" without meaning that theirs is going to be a business of the same kind. In fact, the opposite might be true; through their activity, these people become "economy makers" in the sense that "the economy" is something they do, not that is done to them. If one were to look at Gibson-Graham's criteria for noncapitalism, one would find that in the case of the two camaroneras comunitarias in the Tumaco area, the process of appropriation and distribution of surplus labor takes place in a very different way. These cooperative efforts lack capitalist investors or a top-heavy (e.g., expert, patriarchal) authority structure in charge of distributing the surplus without the workers' say; instead there is a process of governance that favors a fair distribution of wealth and surplus. These efforts are informed less by hierarchical principles than by the logic of nonhierarchy and the respect for the self-organization of nature itself. Part of the community's goal is to reproduce the base (that is, the material and cultural stock that makes the enterprise possible; see Gudeman and Rivera 1990) and expand it whenever possible into a set of ecological enterprises.

Gibson-Graham's conceptualization of the diverse economy has other features. First, conventional representations of women in the development literature present them as engaged in market-led capitalist projects, whether already capitalist or driven by a (feminine) desire to become so (Gibson 2002; Escobar 1995: 177–92). Instead of a monolithic conceptualization of the economy such as this, it is possible, if one attempts to make visible the variety of forms of transaction, labor, and organization in operation in many localities, to arrive at a representation of the economy as a diverse landscape of forms and possibilities. The ensuing view of a "diverse economy" (see table 2) would decenter what is usually considered the sole dimension of the economy—market transactions, wage labor, capitalist organization—making visible the vast, complex sea of other economic relations, including alternative capitalist, alternative market, and noncapitalist. Gibson-Graham refers to these dimensions as the community economy. In this economy "we see women and place-based economic activism as having an important impact" (Gibson 2002: 76). What about experiences such as the camaroneras comunitarias of the Pacific? They can be seen in terms of J. K. Gibson-Graham's general principle that "while each can be seen as small and insignificant interventions in the face of capitalist globalization, seen through the lens of a diverse economy these projects are exciting and significant attempts to develop the unique specificity of non-capitalist places" (77). Despite their fragility, these efforts demonstrate that a noncapitalist politics of place focused on livelihood is indeed an existing and, at least in some cases, a viable practice.[26]

The framework of the diverse economy, as much as the example of the camaroneras comunitarias, helps one resist the tendency to see all forms of economic activity as becoming capitalist merely because they interact with capitalism. By focusing on processes of appropriation and

Table 2. The Diverse Economy Framework

Transactions	Labor	Enterprise
Alternative market	*Alternative paid*	*Alternative capitalist*
Sale of public goods	Cooperative	State enterprise
Ethical 'fair-trade' markets	Self-employed	Green capitalist
Local trading systems	Reciprocal labor	Socially responsible firm
Alternative currencies	In kind	Nonprofit
Underground market	Work for welfare	
Co-op exchange		
Barter		
Nonmarket	*Unpaid*	*Noncapitalist*
Household flows	Housework	Communal
Gift giving	Family care	Independent
Indigenous exchange	Neighborhood work	Feudal
State allocation	Volunteer	Slave
Gleaning	Self-provisioning labor	
Hunting, fishing, gathering	Slave labor	
Theft and poaching	Surplus labor	

Source: Gibson-Graham 2005: 138; see also Gibson-Graham 2006: 71ff.
The table is organized as columns and is not intended to be read across the rows. Note, for instance, that noncapitalist enterprises (bottom row) are engaged in market transactions (top row).

distribution of surplus, Gibson-Graham wishes to make visible those economic forms that might have a different relation to the market. Households, camaroneras comunitarias, subsistence economies, and so forth appear as diverse spaces of production and consumption—a diversity of capitalist and noncapitalist activities that cannot be simply interpreted as being subsumed, contained, complementary, or in opposition to capitalism. An antiessentialist view of the economy, in which the economy constitutes a realm of heterogeneity and difference rather than a monolithic embodiment of an abstract capitalist essence, makes visible noncapitalist practices and leads to a rethinking of production from cultural and ecological perspectives.[27]

If the economy can be reimagined as a diverse set of transactions, structures, and practices; if ecological rationality, as ecological economists suggest, points to the need to re-embed the economy in natural and social processes; and if the politics of place demands that one take a cultural-political approach to the economy, is it possible to weave these insights into a new and coherent view of the economy? If so, what implications would such an analysis have in regard to the sustainability of environment, culture, and place? I want to review an attempt made in this direction. The attempt centers on the notion of an "alternative productive rationality" that incorporates ecology, culture, and politics in a new view of production. For the Mexican ecologist Enrique Leff (1993, 1995a, 1998a), the potentials of ecological and cultural diversity and of the politics of difference may constitute the basis for such an alternative rationality. Why is this so?

> [Because] sustainable development finds its roots in the conditions of cultural and ecological diversity. These singular and nonreducible processes depend on the functional structures of ecosystems that sustain the production of biotic resources and environmental services; on the energetic efficiency of technological processes; on the symbolic processes and ideological formations underlying the cultural valorization of natural resources; and on the political processes that determine the appropriation of nature. (Leff 1995b: 61)

In other words, sustainability is inextricably entangled in the construction of alternative production paradigms and political orders; they are elements of the same process, and this process is advanced in great part through the cultural politics of social movements and communities in defense of their cultural models of nature. The project of social movements (as I pointed out in brief in chapter 1), is a concrete expression of the search for alternative production and environmental orders also envisioned by political ecologists. A remarkably similar notion appears in the first document drafted by PCN concerning the principles that should orient any development strategy for the black communities:

> The principle of reaffirmation of being should illuminate the construction of our own perspective of the future, a perspective elaborated from our cultural vision and our social forms of being. According to this principle, development plans should be in accord with our productive rationality, our

> forms of organizing, our ways of seeing and living life. In other words, the plans should be *integrated*, not in terms of sectors, as it is usually thought, but because they integrate productive, social, cultural, and political processes toward the search for well-being and human development. . . . It should be emphasized that the principles of compensation, equity, domain (effective control), self-determination, and affirmation of being are constitutive of the principle of sustainability and will enable us to continue to bet on life, peace, and democracy in our country. (PCN 1994: 3, 5)

Leff is adamant about the fact that in the search for a new production paradigm, culture and the environment must be seen as potentials for sustainable development. Throughout history, many peasant and indigenous communities have instilled cultural and ecological conditions in the social relations and productive forces of their societies. Culture was integrated into the conditions of production to the extent that place-based identities and cultural meanings provided the basis of these societies' social and environmental practices. These societies do not produce only goods and services or use values and exchange values; they produce "use-meanings" through the enactment of the ensemble of "meanings-uses" that characterize their cultural models of nature (Escobar 1999a). The environment thus appears as a productive system based on the conditions of stability and productivity of the ecosystem and the cultural styles of the groups that live in it. These observations can be the basis for an alternative theory of production that entails a triple articulation of ecological, cultural, and technological processes. The construction of this alternative productive and environmental rationality is shaped by diverse cultural values, and it inevitably takes place in a field of social conflict. From the specificity of biophysical processes (laws of thermodynamics, ecological productivity, and the balance between negentropic biomass formation through photosynthesis and entropic processes generated by technological transformation of the ecosystem), cultural signification (ensembles of meanings-uses, or meanings and practices of the natural environment), and technological possibilities (in their energetic efficiency) there arises the possibility of an alternative production paradigm. Social movements strive to achieve this alternative paradigm through complex ecological and cultural-political strategies.

In sum, the struggles in the Pacific may be seen as those over production conditions and their restructuring and over unequal patterns of ecological distribution; but they also have to be seen as struggles for culturally specific alternative paradigms of production, development, and sustainability. This is the main lesson derived from an antiessentialist political ecology and view of the diverse economy. Behind production conditions and ecological distribution are power strategies reflecting deeper conflicts over cultural meanings, production paradigms, and environmental rationalities; this is another way of stating the cultural distribution dimension of the framework outlined in the introduction. The search for alternative production and environmental rationalities entails a politics of difference in terms of cultural models of nature (chapter 3), other visions of development (chapter 4), and emerging ethnic identities (chapter 5). In other words, sustainability needs to be rooted in cultural identities and ecological conditions; this is why the struggle for territory and autonomy discussed in chapter 1 is a statement about the sustainability of place and culture. For a number of groups in the Pacific the simple statements "culture makes nature" (as a pure politics of identity that disregards the environment) and "nature makes culture" (a pure politics of conservation that disregards people) are far from representing what they are, whether for the case of activists striving for territorial demarcation as an element in the defense of production conditions against the capitalist onslaught or the community groups engaged in alternative productive projects such as the camaroneras comunitarias. As for the feasibility of these strategies, that must await the discussion of the dynamic of the social movements and the "meshworks" they are creating for the defense of place and culture (see chapters 5 and 6).

Is capital in and of itself capable of generating a significant transformation of the current production systems in the direction of sustainability? It has been argued that capital is undergoing change and developing a conservationist form, centered on the management of the system of capitalized nature (M. O'Connor 1993). This is true of proposals for sustainable development and biodiversity conservation. In the Colombian Pacific, the modern and postmodern forms of capital—conventional and conservationist—are both schizophrenically at play (Escobar 1996). As the region was being brought into the politics of conventional development, discourses and strategies of biodiversity conservation gained some ascendancy. In this way, both forms of capital, exploitative and conservationist, became important for a time, even if the former was by far the predominant form. Conservationist capital attempts a triple resignification: of nature (as "biodiversity reserve"), of local communities (as "stewards of nature"), and of local knowledge (as knowledge of "saving nature"); such a schema constitutes a novel internalization of production conditions, but it does not take nature and places outside the dominant forms of the economy.

Communities are enticed by biodiversity projects to become "stewards of the social and natural 'capitals' whose sustainable management is, henceforth, both their responsibility and the business of the world economy" (M. O'Connor 1993: 5). Once the semiotic conquest of nature is completed, the sustainable and rational use of the environment becomes an imperative. This analysis, however, excludes from consideration the culturally specific and non-market-driven forms of appropriation of nature that characterize local populations.

These forms of appropriation of nature are taken into account in a perspective that sees the environment "as the articulation of cultural, ecological, technological, and economic processes that come together to generate a complex, balanced, and sustained productive system open to a variety of options and development styles" (Leff 1993: 60). A positive theory of production based on this view of the environment would include cultural resistance to the symbolic reconversion of nature by postmodern capital; social and ecological resistance against capitalist destruction and state-driven restructuring of production conditions; socioeconomic proposals for concrete alternative production strategies and development styles; and political organizing to ensure a minimum of local control over the entire process. In the current landscape of hybrid cultures and regimes for the appropriation of nature in the Pacific, this strategy will surely have to combine modern and nonmodern, capitalist and noncapitalist forms and practices. In the long run, the strategy is more likely to succeed if it builds on the self-organizing character of environment and place-based practices (Escobar 1999a).[28]

Conclusion

The situation in the southern Pacific suggests that while capital has made vast inroads into the ecological, social, and cultural life of the region, it is far from achieving a total reconversion, if this were ever possible. Resistance to capitalist-driven labor regimes; attachment to place-based cultural, economic, and ecological practices; varieties of alternative capitalist and noncapitalist forms, whether in the continued existence of aspects of traditional production systems or in recent projects such as the camaroneras comunitarias; and organized strategies of social movements—these are all elements that need to be taken into account in evaluating the actions of capital in this part of the world. They enable one to adumbrate the idea that in this region there exists a complex dynamic of various types of economies, economies one needs to understand without subjecting them to a single capitalist determination. I would like to make some final observations in this regard.

First, even the form taken by capital's strategies of localization is important to the social and ecological outcome of its development. In Papua New Guinea, the oil palm industry has a strong smallholder component, a full 45 percent of the planted area being in this modality. Incentives to promote small farm production have proven successful, given a state specification that any plantation development has to be matched with smallholder estates. As in the Colombian case, smallholders maintain different production practices, devoting some areas to subsistence gardens and redirecting parts of the income to nonmarket, locally valued activities. Even if these trends do not point yet in any definite direction, the differences with the Colombian and Malaysian model are instructive; they make clear the need for comparative studies of global capital (Gibson 2000, and personal communication). Thus it is misleading to talk about capital or even capitalism in the singular (as I have done for the most part in this chapter) without specifying its conditions. In Papua New Guinea, a certain regulation of capital made possible a degree of economic diversity both within the capitalist and the noncapitalist sectors; this did not happen in Colombia, or at least it occurred to a much lesser extent. Similar considerations might apply to aquaculture (e.g., tendencies toward ecological and artisan-type models in some articulation with place-based groups and the state, and playing off the possible range of relations between capitalist and noncapitalist class processes).

This diversity of capitalist forms means, second, that if it is true that capital and modernity are apparatuses of capture that exert their influence on a vast array of heterogeneous forms, the result is not necessarily homogeneous social formations. This observation has been made already in various ways, to my mind, more limited (e.g., the concept of hybridization; Marxist theories of uneven and combined development, peripheral capitalism, and articulation of modes of production). As a set of transnational networks, capital encompasses heterogeneous social formations, such as those of the black groups of the Pacific; it establishes itself in the middle of these formations, becoming part of their horizon. In those regions where cultural and ecological conditions are vastly different, capital operates through this diversity; it requires a certain "peripheral polymorphy" (Deleuze and Guattari 1987: 436). These forms must decidedly not be seen, as in earlier frameworks, as primitive, backward, or transitional forms or as fully existing within a capitalist order. They reflect the existence of different systems of production/signification or meanings-uses that continue to be the source of viable practices. In sum, while as a worldwide process capital tends to organize a "third world" (437), it does so on the basis of a heterogeneity of social formations—which include noncapitalist modes—which, at any given point, might give rise to countertendencies and new forms of economic difference.[29]

Third, in formations such as the Pacific, capital depends on an ongoing activation of representational and, not infrequently, physical violence; this is a requirement of every mechanism of capture as it links locally diverse human and biophysical ecologies to supraplace dynamics.[30] Regimes of representation that exclude and subordinate cultural difference are regimes of violence; this suppression is a foundational element in the political economy of nation building in countries like Colombia (Rojas 2002) and a crucial vector of global coloniality even today. Modern ruling regimes have customarily legitimized themselves on the basis of their own interventions: the more they intervene—in the name of civilization, progress, freedom, democracy, or whatever other goal they deem worthy—the more legitimate they allegedly appear. This feature is even more present in today's imperial globality, albeit perhaps with more fissures. In places like the Pacific, social subjection is accomplished not just through ideology or allegedly legitimate private appropriation of public resources; it requires the open use of force and terror. Forced deterritorialization and reterritorialization are the strategies of choice for two main reasons: on the one hand, the assertive persistence of different regimes of production/signification (black and indigenous cultural practices); and on the other hand, the tremendous rapacity of the ruling groups in social formations such as those in Colombia and extractive economies such as those found in the Pacific.

The case of the Curvaradó and Jiguamiandó rivers in the Chocó province has become well known in this respect. Along these rivers, through a combination of means which includes illegal purchase of land, implacement, displacement, and intimidation, a number of private companies enjoying state subsidies seized over five thousand hectares of land between 2001 and 2004, land belonging to collective territories titled to these rivers' community councils. During the titling process, the communities documented permanence in the region since at least the mid-nineteenth century. Now most of their small towns lie destroyed or deserted, their inhabitants displaced, part of their forests razed to give way to palm and the reticule of private roads that made local movement impossible. The plan drawn by these companies is to extend their holdings to twenty-six thousand hectares in palm and cattle ranching. In some cases, towns have been repopulated by people from outside the region, largely whites from Antioquia, in an example of what some activists call ethnic substitution (displace blacks, bring in whites). This case of imperial globality and global coloniality has been well documented by the Ethnic Affairs Group of the state agency in charge of the titling process, the Instituto Colombiano de Desarrollo Rural, or INCODER (Colombian Institute of Rural Development; see INCODER 2005). Today, through the discourses of development, biofuels, and the free trade agreement—and with some support from the U.S. Agency for International Development—capitalists and the government are promoting African palm in many parts of the Pacific. The expansion of the palm often includes brutal forms of force and cynical measures to entice locals into going along with the schemes. Since 2005, PCN, other black groups, and solidarity organizations in the United States and Europe have repeatedly denounced the palm projects, citing their implied displacement, human rights violations, appropriation of land by private companies, and weakening of the environmental and territorial rights of the communities.[31]

Despite such situations, groups of small agriculturalists, communities engaged in cooperative and semi-intensive aquaculture, and the collective territories can all be seen as effecting unpredictable respatializations oriented at least as much by local operations as by blueprints, planning, or the use of force; in the best of cases, they may enact a model of becoming and multiplicity that contrasts with that of the rational order of modernity.[32] This is to say that some black and indigenous groups of the Pacific might not (or at least not yet) have the same use for the work and accumulation factors, the order factor, and the individualizing factor as dominant modernity does. As in the case of the Zapatista women's insistence on "the right to rest" (Belausteguigoitia 1998), one might speak of the right to no accumulation and to freer labor. These demands might not suffice to reconstitute the local and regional socionatural worlds or to lessen the increasingly strong desire for the commodity among the young, but they might reconfigure the stakes and keep viable other ways of being in place and being in networks, including those created by capital.

The rain forest itself, after all, as a complex meshwork of elements interlocked in their complementarity, also clashes with the domineering linear logic of capitalist modernity. In the long run, the elusive goal of sustainability will depend on the possibility, dim perhaps but not inexistent, of fostering meshworks of activists, local groups, ecosystems, and other actors, such as transnational NGOs and other social movements, that could extract from the dominant logics ever-larger social and ecological spaces. The hope is that they could foster the heterogeneity and diversity that characterize place, nature, and economy. Militating against this project are the most recalcitrant and anachronistic forms of capital, development, and the state. Supporting it are some local groups, practices of difference, the rain forest and mangrove forests themselves, and social movements and their allies.

3 nature

Tierra puede tener cualquiera, pero no territorio.

[Anybody can own land, but territory is something else altogether.]

—Don Porfirio Angulo, "don Po," Tumaco activist and leader, 1998

Introduction: Tales of Nature and the Contest of Epistemologies

Alfredo Vanín, the Timbiquí poet and expert in oral traditions of the Pacific, says that one of the most striking features of the oral literature of the region "is the manner of naming the world, nature, and those beings that accompany women and men in their voyage, be they material beings or spirits of different kinds" (1995: 21). It is said, for instance, that "the world was founded with everything in it: water, air, saints, *visiones y espantos* [visions and scares], trees, serpents, deer and birds. The world exists and functions at many levels, without any entity being more important than any other, save in the kind of power they possess, in which the shaman [*hechicero*] may find nourishment. Men and women grow on the generosity of nature, on its tireless goods and even on those which are unreachable" (21). Visible natural entities like animals, plants, trees, and gold have two contradictory aspects in these narratives: they are both abundant and thus can never be exhausted, but they may also grow distant in space and time and can thus disappear from the reach of people. When pressured too much or put under siege, for example, by too much fishing, hunting, noise, or forest destruction, natural beings go far away or may even transform themselves into *espantos* or *animales de monte* (untamed forest animals). However, they do not die out because in the Pacific, since everything there is perpetually reborn, nothing can go extinct; everything, humans included, is a *renaciente*, the subject of rebirth. Vanín finds it plausible to read environmental significations into some of these stories of abundance and distancing. They are not quite translatable into modern rationality, however, because even though there are narratives of greed that contain a lesson for those prone to fall into cupidity's clutches (especially in relation to acquiring gold) the path to abundance is seen not necessarily in terms of the "wise management of natural beings" but in the holding of a secret that can be learned only if one has been chosen by a saint, a sacred animal, a mermaid, or a spirit of sorts.

This story and others like them that ecological anthropologists and scholars of myth have collected throughout the world should make one aware, in a simple way and even if by contrast, that nature is culturally constructed. The story suggests that there is a deep interconnection between levels or domains of the real—biophysical, human, and supernatural. It also makes clear the imbrication of nature with our knowledge and significations of it. We moderns tend to overlook this fact—nature is one thing, external to us, and our knowledge of it, especially our scientific knowledge, is another thing altogether. This is why one of the most sustained questions moderns have asked is, "What is nature?," and we have done so without much trepidation, expecting to arrive at an unequivocal answer. Indeed, occidental knowledge since the pre-Socratics has persistently attempted to answer this question (Angel 2001). Some current trends in science, including complexity, are claiming that the question What is nature? does not make much sense as it has been historically formulated.

Starting in the 1970s anthropologists have questioned the existence among many non-Western peoples of a category called nature, or at least one that neatly corresponds to our own (e.g., Strathern 1980; Descola and Pálsson, eds. 1996; Restrepo and del Valle, eds. 1996). Black and indigenous groups of the Pacific fall into this group. When activists of the Pacific speak about the *cosmovisión* of the black and indigenous groups or about their ancestral knowledge, they have in mind a way of relating to and signifying the natural world. It is thus important to have an idea about what these local models are, in order to ascertain claims about ecological and cultural difference and about the fact that this difference should inform conservation and development efforts. Part I of this chapter reviews the anthropological literature on the local models of nature of black groups of the Pacific. Part II shifts from ethnographic literature to theory; it discusses schematically the epistemologies of nature in various fields and approaches; in particular, this section reviews the varieties of realism and constructivism that can be gleaned from various debates on the question of nature today. Part III shifts to the ethnographic register again to present the important discussion about the so-called traditional production systems. This discussion was central to the arguments about conservation in the Pacific in the 1990s. These systems may be seen as a particular nature–culture regime or regime for the appropriation of nature that could form the basis of alternative conservation frameworks.

This is followed by an account of the emergence of biodiversity conservation as a global issue. The final section focuses on the discussion of the biodiversity conceptualization developed by activists in their encounter with biologists, planners, developers, and others. It is suggested that this social movement contribution—again, part of the political ecology developed by PCN—constitutes an attempt at gaining some control over the production of nature.

I ◈ The Local Models of Nature of the Black Groups of the Pacific[1]

La ombligada: From the Divine to the Territory ◈ The ritual of *la ombligada* (*ombligo* means "navel") is paradigmatic of the conceptualization of nature of black groups of the Pacific. In the river regions, childbirth takes place in the home, the mother surrounded only by women, including the midwife, who cuts the umbilical cord and receives the placenta. The length to which the cord is cut is decisive for the sexual identity of the child. The midwife buries the placenta and the cord under the house, in between the poles that support it (preferably when the child is a girl) or under a tree by the edge of the forest (for boys). This is seen as being important to the independence of the child later in life. The navel of the newborn is subsequently filled with a natural substance, animal, plant, or mineral, that has been pulverized in such a way as to transmit the substance's properties to the individual. Among commonly used substances are tapir's nails (*Tapirus bairdii*), bones of squirrel or of horned animals, the dry saliva of the eel, a rabbit's foot, the bones of a wild deer, a cock's nail, scorpions or ants, fish bones; the substance may also be gold, a wild plant of ambiguous thermal denomination, or even the sweat of the midwife (Losonczy 1989: 51; 1997: 196–210; Camacho 1998: 38; Velásquez 1957: 245). What is at stake in the ombligada is the metaphorical or metonymic transfer to the body of the child of a property associated with the natural substance. This is done in the belief that the properties are effective in the real world and will shape the personality of the child. The navel thus functions as an interface between the natural and human orders. Hence, by a principle of similitude, it is desired that the child be indomitable like the tapir, fast and fecund like the rabbit, or fertile like certain curative plants; or, metonymically, that the gold in his navel might bring him good luck in mining or that the sweat of the midwife may confer upon him or her the healer's knowledge. Although some substances can be used for either male or female children, there is a marked sexual division in that girls are most often ombligadas with the substance of "hot" or "lukewarm" (*tibias*) curative plants that have been domesticated from semiwild varieties found at the edge of the forest; these plants will transmit their "hot" (*caliente*) character to the girls, essential to their becoming pregnant later in life.

The Belgian anthropologist Ann-Marie Losonczy (1989, 1993, 1997a) has offered a complex explanation of this ritual in terms of various interrelated levels. First, black groups of the Pacific see conception as a divine matter, something coming from *arriba*, above; god mediates between man and woman, and children originate in an extrahuman domain. The actions of the midwife initiate the constitution of the person. The ritual separates the child from the supernatural world from which it comes, joining it to the human world; this is a transference from the mythic place of origin to the humanized world. Second, the ritual grounds the child in the territory of the family and the community and asserts the possession of the territory. Third, there are a number of links and disjunctures that the midwife negotiates: "It is as if for this society being 'a bit of' an animal or plant were constitutive of the actualized human condition, whereas the *disjuncture* between the human being and the supernatural universe from which s/he originates is a necessary condition for her survival. . . . By burying the placenta and the cord, *separated* from the child, the midwife links the child with the communal *territory*" (1989: 53). At the same time, by filling the navel with a substance separated from the natural world, she establishes a link with this world, in what amounts to an inverse and symmetrical operation. In the long run, the aim is to welcome the newborn and humanize him or her, thus attenuating the dangers that the arrival of this ambiguous being entails and the questions it poses about the relations between worlds.

Losonczy's involved explanation leads her to assert that the representational systems of the black groups of the Pacific reveal the existence of "original and coherent cognitive and identitarian strategies" that transform and re-create the imposed sociocultural models, what she calls syncretic "socio-cultural maroonism" (1989: 49). Losonczy's anthropological interpretation is not without precedent in studies from other parts of the world. The study of cultural ways of ordering the world, particularly those aspects related to nature, has come a long way, from studies of ethnoscience and Claude Lévi-Strauss's structuralist explanations of myth and rituals to notions of "cognized models" (Rappaport 1991) and the ethnographic documentation of what today are referred to as "local models of nature" (see Escobar 1999a; Descola and Pálsson, eds. 1996) or ethnoecological frameworks (Toledo 2000a, 2000b). From the perspective of ethnographic constructivism, these models may be seen as culturally specific modes of objectification of the natural world (Descola 1996; Gudeman 2001 for local models of the economy). They constitute what Restrepo (1996a) calls "a grammar of the environment" (*gramática del entorno*) which, according to a certain anthropological tradition, can be studied as a language system. For rural black groups of the Pacific, this grammar "indicates a local model of classification of their surroundings that operates on a nonreflexive basis. . . . as such, it is an anthropological construction, but one that is seen as functioning in the statements, perceptions, and practices of the same people" (Restrepo, personal communication).

The Grammar of the Environment: A Local Model of Nature ◈ Like any human group, the black communities of the Pacific order the real in specific ways, through particular categories, classifications, and relations. This construction is not arrived at in historical isolation. Despite their distinctiveness, the local models of nature of the black groups of the Pacific share important features with those of neighboring groups, such as the Embera (see Ulloa, Rubio, and Campos 1996 for the Embera model; Losonczy 1993 for similarities and differences). Afro-Colombian models are characterized by the centrality of the plant world. Mineral, plant, and animal worlds are opposed in terms of their mobility. Whereas minerals are immobile and animals endowed with mobility, the plant world travels without moving—plants spread horizontally, for instance. Thus plants mediate between what is alive (what moves) and what is not (what does not move). The plant kingdom is associated with the feminine through a complex conceptualization that regulates the demarcation of ethnic territories. Rather than the identification of concrete taxa (as in scientific classifications), the determining classificatory features are thermal. The sun and daylight are the primary sources of heat, while earth, air, and the moon impart cold. Water mixes them both, which is why plants are privileged since they are seen as the fusion and synthesis of the two opposing principles. Organs, cycles, and natural entities can all be classified along an axis that stretches from hot to cold, including the lukewarm and the neutral. The same theory underlies the construction of illness and health, in such a way that whatever is used in healing must balance elements from the various orders. The repertoire of magical and therapeutic plants is vast; most have a wild and a domesticated variety, originating in the space of the forest and in the cultured space of the village, respectively, thus establishing a link between these domains. "Each substance is a reconstruction, a reorganization of order and of the articulation of the world" that takes place "through a creative mixing of different classificatory axes organized around a thermal nucleus" (Losonczy 1993: 43).[2]

This system of classification doubtless creates boundaries between nature and culture, but it does so in a way that differs markedly from modern conceptions. Far from being a rigid binary system, as Losonczy emphasizes, the Afro-Colombian model is characterized by a logic of fragmentation and recombination, "a fluid subtle intellectual strategy" that is based on a principle of mixing that continuously weakens the binary oppositions, since all beings are recognized as having a changing and labile character. The central principle is that of transformation; it pervades nature as much as culture and their interrelations. It is because plants mediate between the natural, the human, and the supernatural—between life and death, masculine and feminine, past and present—that the model enacts a logic of multiplicity and fragmentation.[3] Besides these principles, there are other features that characterize the nature model of the black groups of the Pacific, including further systems of classification and distinctions among various worlds; social divisions, particularly in terms of gender; and a fundamental category already mentioned, that of renacientes. All of the use spaces mentioned above exist in this world (*en este mundo*).[4] This world is in the middle of others that are represented according to an order constructed along the axes above–below and divine–human, each with particular beings and features. There are connections among these worlds, and under certain conditions there can be movement between them. Above this world are several others. Heaven (*cielo*) or the glory (*la gloria*) is the highest world, inhabited by Nuestro Señor Jesucristo (Our Lord Jesus Christ) as well as by saints, angels, virgins, and the *angelitos*, or little angels, children who have died sinless. A significant influence of the Catholic imaginary is at play here, yet Catholic representations are transformed and inscribed in local models of knowledge. Spiritual beings are largely humanized and instrumental. As Quiroga puts it in his study of Afro-Ecuadorians of the Pacific, "Conceptualized within this mythic world, the Saints and Virgins are far from being distant and cold figures. Rather, they are lively, sentient companions of the *pobres* [the poor] and possess many of the same defects and desires" (1994: 62; see also Whitten 1986 [1974]; Urrea and Vanín 1995). Moreover, there are moments of connection among worlds. One of them is the performance of the *arrullos*, when "Saints descend from heaven, opening a 'window' that connects heaven and earth" (Quiroga 1994: 71). Similarly, the ritual gaiety of the *chigualo*, or ritual for dead infants, permits the ascent to heaven of the child as an angelito; local musical instruments such as the marimba, drums, and other percussion instruments open up the channels between this world and the worlds beyond. Within the Embera conceptualization of the Chocó, certain places such as caves and river sources are also locations of communication between worlds. The *jaibana* (local healer) establishes the relations and balance among worlds, also regulating the abundance or scarcity of natural species (Ulloa, Rubio, and Campos 1996). There are also other worlds that lie below this world. They are inhabited by *visiones* (visions), other beings, and certain animals; under certain circumstances humans might be kidnapped into one of them, never to return. The world immediately below is an inversion of this world (e.g., M. Escobar 1990: 41). It is inhabited by beings that resemble people but who feed on the smell of food and consequently have no ass; they are called *sinculo* (assless). Some animals, like crabs, are able to circulate between this world and that of the sinculo.

Also underneath is the *infierno* (hell), and certain music types and dances are associated with the invocation of the devil, much as the arrullos establish a connection with the *mundos de arriba* (upper worlds). Another world located underneath this world is that of the *sirenas* (mermaids) and *encantos* (enchanted beings), found in rivers and estuaries. Visiones, for their part, travel in specific places and are endowed with particular abilities. The *riviel* (a human ghost) and the *maravelí* (a ghost ship) are dreaded beings that circulate in the estuaries and the open sea. Along with the *madre de agua* and *indios de agua*, these visions enter this world and exert their influence through the mediation of water. Other visions are associated with the forest and the mangroves. The most well known is the *tunda*, who usually takes the form of a woman familiar to the person who encounters her—most commonly children wandering away from home into the forest or adult men walking alone in the forest at night. Those bewitched by the tunda are rarely fully brought back to this world, although the tunda is also thought to be friendly to humans at times, for instance, by giving men the ability to escape from prison. Again, these visions are an integral part of social life. Although few might have encountered the tunda, for instance, many claim to have seen her characteristic footprints; in this way, visions have their particular habitat and are not part of an abstract supernatural world. Moreover, the boundaries between the many worlds are not rigid and distinct. More than a radical separation, there is continuity between what moderns categorize as the biophysical, human, and supernatural worlds.[5]

The natural environment—the rivers, sea, and forest—has sustained the black groups of the Pacific for several centuries; the natural world thus has an intimate presence in the cultural imaginary of these groups. For some, this presence is elaborated into models and narratives that have borrowed from African, indigenous, and Catholic traditions. Oral literature bears witness to the integration of the natural and the human. Some see in the narrative of worlds and visions elements of an ecological ethic of mutuality and conservation that consists in warning humans not to overuse nature (e.g., Pedrosa and Vanín 1994: 75). The riviel, for instance, is seen as encoding a warning to fishermen to catch whatever they need and return home instead of staying alone at sea (39); some of the forest visions are meant to scare away the human colonizers and predators of the forest. In some tales, hunters end up at the opposite end of the hunt, with the animal (e.g., a *tatabra*, *Tayassu tajacu*) pointing the gun at him (35). Dozens of oral narratives, whether in verse, prose, proverb, or jokes, revolve around the relations between people, spirits, and the natural world and are told and retold with gusto (see, e.g., Friedmann 1989; and Friedmann and Vanín 1991).[6] The narratives cannot be taken in isolation as proofs of what I have called here, with Milton (1996), primitive environmental wisdom. Seen in the context of the complex models of the world, however, they are expressions of a cultural and ecological logic that I intend to highlight in this book. As in the case of most human communities, these "local models" are an expression of the systematic knowledge people develop about their environments (Haila and Dyke, eds. 2006).

Three additional features characterize the local model of nature of the black groups of the Pacific. The first entails a distinction among use spaces and is organized around the horizontal axis inside–outside, or *adentro–afuera*. The second is the existence of multiple worlds, structured around a vertical axis above–below, as discussed above. The third feature is a system of classification and ordering of beings that operates in terms of several important distinctions (see Restrepo 1996a, 1996b, 1996c). Some of these distinctions are based on categories that resemble occidental taxonomies but do not correspond to them. For instance, the *seres de este mundo* (beings of this world) are divided into the following groups: *animales, pájaros, mariscos y avichuchos* (animals, birds, mollusks and seafood, and avichuchos, certain weird and dangerous beings); *palmas, palos, bejucos, yerbas y matas* (palm trees, sticks or trees, vines, herbs, and plants); *cosas de la tierra* (things of the soil). The group containing animales does not correspond to the occidental category of animals since it neither opposes the plant world in its totality nor includes, say, birds and mollusks. Animals are *seres de servicio* (beings of service). They reproduce *por naturaleza* (by nature, that is, sexually) and cannot fly, although they can move. Avichuchos share some of the properties of animals but are not de servicio. Rather, they are usually prejudicial to people, often sting, and can never constitute food (e.g., scorpions). Mariscos are aquatic entities, usually de servicio, but not always. Beings in the second group, palos, yerbas, matas, bejucos y palmas, are seen as capable of movement but only as a whole. Yerbas are usually curative, matas have simple leaves and fleshy trunks, palmas are characterized by complex leaves and peculiar trunks, and bejucos have an elasticity and continuity that is opposed to the rigid, wooden structure of the palos. Finally, cosas de la tierra encompass a variety of entities that from an occidental perspective include mushrooms and small insects, thus constituting an intermediate realm between the other two groups.

These orders are crossed by three semantic axes: *manso–arisco* (tame–wild), *de lo alto–de lo bajo* (belonging to above or below), and *producido por el hombre–producido por la tierra* (produced by humans or by the earth or forest). According to the first axis, there are animals, yerbas, palos, pajaros, and so on that are tame and others that are wild. The second axis refers to an imaginary line in relation to the elevation from the ground; thus, deer and rabbits are de lo bajo since their food and territorial habits never rise above the ground; animals that have an arboreal life, such as the squirrel and the *perico* (parrot), are classified as de lo alto. This distinction also applies to other animals and plants: eagles and tall tress are de lo alto, whereas jaguars, quails, turkeys, and bushes are de lo bajo. These axes structure the concrete position of every individual entity belonging to any of the three orders of seres de este mundo. Finally, the category of *renacientes* (literally, reborn), very well documented in the Pacific (e.g., Friedemann 1974; Restrepo and del Valle, eds. 1996), is a central feature of the model. All the seres de este mundo are renacientes; they are located in a successive chain of generations stretching from the mythic foundation of the world to the future of times. For this reason there is never talk of extinction; of animals it is said, for instance, that they have gone farther away (*se han alejado*). Humans are part of the same order; some movement activists, indeed, prefer to use *comunidades renacientes*, rather than Afro-Colombians or *comunidades negras*, to refer to black people in the country (e.g., Cassiani, Achipiz, and Umaña 2002).

In conclusion, the construction of the natural world established by black groups of the Pacific, as shown by ethnographers, may be seen as constituting a complex grammar of the environment or local model of nature. This grammar includes ritual practices such as the ombligada, structured use of spaces, an ordering of the universe in terms of worlds and levels, and systems of classification and categorization of entities. The model constitutes a cultural code for the appropriation of the territory; this appropriation entails elaborate forms of knowledge and cultural representations in what amounts to "an original cognitive universe" (Losonczy) or "a dense universe of collective representations" (Restrepo) that is seen by some as an adjustment to rapid social and economic transformation (Whitten 1974 [1986]; Quiroga 1994; Arocha 1991). Accordingly, the environment is a cultural and symbolic construction, and the way in which it is constructed has implications for how it is used and managed.[7] The existence of diverse worlds and the continuity among them, in particular, are not irrelevant to the use of natural resources. How would sustainability and conservation look if approached from the perspective of the world construction of the black groups of the Pacific? This has remained an intractable question, as I discussed theoretically at the end of last chapter in reference to Leff's proposal for rethinking production. I will return to this issue below in the discussion of the traditional production systems of the Pacific.

II ◈ Nature Epistemologies and the Coloniality of Nature: Varieties of Realism and Constructivism

The knowledge of nature, as one should admit intuitively after the previous discussion, is not a simple question of science, empirical observation, or even cultural interpretation. To the extent that this question is a central aspect of how one thinks about the present environmental crisis, it is important to have a view of the range of positions on the issue. To provide such a view, even in a very schematic form, is not a simple endeavor, for what lies in the background of this question—besides political and economic stakes—are contrasting epistemologies and, in the last instance, foundational myths and ontological assumptions about the world. The brief panorama of positions presented below is restricted to the modern social and natural sciences.

I begin by introducing a general concept, that of the coloniality of nature. The concept of coloniality that has been applied to knowledge and power (see the introduction and chapter 4) also applies to nature. It is in the nature of coloniality to enact a coloniality of nature (e.g., Escobar 2003b; Walsh 2007). Martínez applies this notion in the case of biosphere reserves in the Zona Maya of Yucatán, in which "an essentialized notion of nature as wilderness and that is outside of the human domain becomes subject to a new kind of domination" (2004: 74). González (2004) and Apffel-Marglin (1998), among others, relate it to Western mechanistic views of nature and propose strategies of decolonizing knowledge as ways to decolonize nature and the land (see also Vaneigem 1994). The growing managerial rationalization of the environment has been usefully seen in terms of the Foucaultian notion of governmentality (e.g., Gupta 1998; Escobar 1999a; Luke 1999; Agrawal 2005), and it can be linked to a colonial sort of governmentality. Very schematically, the main features of the coloniality of nature, as established by myriad discourses and practices in post-Renaissance Europe and beyond, include classification into hierarchies ("ethnological reason"), with nonmoderns, primitives, and nature at the bottom of the scale; essentialized views of nature as being outside the human domain; the subordination of the body and nature to the mind (Judeo-Christian traditions; mechanistic science; modern phallogocentrism); seeing the products of the earth as the products of labor only, hence subordinating nature to human-driven markets; locating certain natures (colonial and third world natures, women's bodies, dark bodies) outside of the totality of the male Eurocentric world; the subalternization of all other articulations of biology and history to modern regimes, particularly those that enact a continuity between the natural, human, and supernatural worlds—or between being, knowing, and doing.

As are questions of identity (see chapter 5), nature epistemologies tend to be organized around the essentialist or constructivist divide. Essentialism and constructivism take contrasting positions on the relation between knowledge and reality, thought and the real. Essentialism is the belief that things possess an unchanging core, independent of context and interaction with other things, that knowledge can progressively know.[8] Concrete beings develop out of this core, which will eventually find an accurate reflection in thought, for example, through the study of the thing's attributes in order to uncover its essence. The world, in other words, is always predetermined from the real. Constructivism, by contrast, accepts the ineluctable connectedness between subject and object of knowledge and consequently the problematic relation between thought and the real; epistemological constructivism thus entails much more than the assertion that reality is socially constructed. The character of this relation yields varieties of constructivism. Much scientific research has tended to remain within an essentialist conception, although this predilection has changed dramatically over the past few decades.

Paradoxically, groups like those of the Pacific, who could be thought to be essentialist at heart, evidence more constructivist features, given the transformational character of their model.

There is an array of epistemological positions along the essentialist–constructivist axis, from the most established positivism to the most recent forms of constructivism, each with its respective philosophical commitments and political attachments.

1. *Epistemological realism*, which has two distinct varieties:

a. *Positivist science perspective*, the predominant approach to the relation between knowledge, thought, and the real. It assumes, first, the existence of nature as a distinct ontological domain and, second, a correspondence between knowledge and reality. This position upholds the distinction between the constructed and the naturally given, between knowing subject and known object, observer and observed, representation and the real. This epistemology reigns in most of everyday normal science, including those social sciences largely unchanged by poststructuralism (e.g., economics and much of political science). Ecology and biology function largely within this tradition, including mainstream cognitive science, with its notion that cognition is the process of representation of the world by a pregiven mind, external to that world (see Varela, Thomson and Rosch, 1991, for a critique). The broad philosophical framework in which this tradition fits is well known. It includes the project, dating from the seventeenth century, of emancipating human beings from need through a knowledge that would bring them ever closer to truth. The great innovation of this period was the use of knowledge to intervene powerfully upon the real; the knowledge of nature entailed the domination of nature through technology. This tradition of rationalism has always had its critics; it entered into crisis closer to modern times with the critique of metaphysics as the pursuit of logical truth and the use of reason as the only valid basis for knowledge and of humanism, that is, a perspective that places humans at the center of the universe and the condition of possibility of all knowledge (e.g., Vattimo 1991; Foucault 1973). In ecology, Merchant (1980) and Shiva (1993) have made a connection between positivist science, patriarchal structures, and environmental destruction that is well known. The sciences have also seen challenges to the dominant position of epistemological realism from within their own ranks, from Werner Heisenberg's uncertainty principle to theories of irreversibility and nonlinearity, self-organization, and complexity. In anthropology, the breakdown of epistemological realism started with interpretive anthropology and deepened with the so-called postmodern anthropology.

b. *Systems science perspective*. The development of systems approaches out of the work of von Bertalanffy and others (e.g., Bertalanffy 1975; Churchman

1968; Lazlo 1972; von Foerster 1981; see chapter 6 for further discussion) provided a valuable critique of reductionist forms of science. Systems theory, a tremendously important movement, enabled resistance to the fragmentation of the real, and it kept alive a way of looking at the potentiality of the real other than that afforded by reducing it to its parts. Systems thinking influenced salient trends in the 1970s, such as the early theories of self-organization (e.g., Jantsch 1980; Jantsch and Waddington, ed. 1976). With the development of systems analysis in business, the military, and ecology this approach tended to reproduce the basic tenets of realism at a higher level—namely, identifying truth as the correspondence of holistic knowledge with a total, albeit complex, reality. Nevertheless, to the extent that systems approaches emphasized the whole over the parts; relations, feedback, and interactions over independent variables; organization, process, and structure over the study of particular properties and the function of distinct elements, they contributed to unsettling the epistemology of mainstream science. Systems thinking informed the development of various ecological approaches in anthropology, from the early work on adaptation in the cultural ecology of the 1950s to ecosystems ecology approaches after the 1960s.[9]

2. *Epistemological constructivism*. The constructivist positions are more difficult to classify. The following are said to be the most salient ones in the nature-culture field; these are not distinct schools but partially overlapping positions. They do not necessarily constitute highly visible trends, and some are marginal or dissident within their fields, including biology. Finally, it is debatable whether all of them can be described in terms of a constructivist research program (e.g., phenomenology), although in these cases their effect vis-à-vis epistemological realism can be said to be similar to that of the constructivist proposals.

a. *Dialectical constructivism*. Marx can be said to have been among the first deconstructionists to the extent that his analysis of capitalism as a historical formation debunked the truth claims that described capitalism as a social order naturally functioning in term of individuals and markets. As the philosophical principle of historical materialism, dialectical materialism purported to transcend realism by proposing a different way of bringing together knowledge and the real; dialectical materialism sees social reality as undergoing constant transformation, the product of conflict and power, not as constituted by value-free knowledge. Dialectical materialism, however, did not give up its claim to being scientific and universal, and thus in many instances Marxism devolved into more conventional realist positions. Marxist dialectics continually questions the fragmentation of knowledge; in this way, and through its attempt at showing connectedness, relationality, and wholeness (totality), Marxist dialectics represents a strong constructionist program. Besides the transformation of historical materialism through ecology (see chapter 2)—the account of capital's restructuring of production conditions—the Marxist framework has produced the influential view of the dialectic of organism and environment, especially in the work of the biologists Levins and Lewontin (1985). In this view, there is a mutual process of construction between organism and the environment. By complicating the binarism between nature and culture, these biologists contributed to a rethinking of theories based on this cleavage, including evolution and the ontogeny–phylogeny relation, although the implications of their work for ecology have been less explored. A similar contribution, although from different sources, including theories of heterarchy, is from the field of historical ecology. This important field studies long-term processes in terms of changing landscapes, defined as the material—often dialectical—manifestation of the relation between humans and the environment (e.g., Crumley, ed. 1994).

An altogether different conception of the dialectical method has been developed by Murray Bookchin and the school of social ecology, building on socialist and anarchist critiques of capitalism, the state, and hierarchy. By weaving together the principles of social anarchism (e.g., balanced community, decentralized society, direct democracy, humanistic technology, a cooperative ethic, etc.) with what he sees as the natural dynamic that characterizes evolution itself, Bookchin developed a framework for a systemic analysis of the relation between natural and social practice. The dominant view of human nature as individualistic and competitive—which Bookchin interprets as suggesting that "man is undoing the work of organic evolution" (1986: 89)—led him to propose alternative principles for society, such as mutuality and cooperation. The cornerstone of his framework is the notion of dialectical naturalism, that is, the idea that nature presupposes a dialectical process of unfolding toward ever-greater levels of differentiation and consciousness. The same dialectic is found in the social order; indeed, social ecology poses a continuum between natural and social evolution (between first and second natures) and a general tendency toward development, complexification, and self-organization. Like the dialectical biologists, Bookchin rejects the idea of organisms adapting passively to a pregiven environment; from that he concludes that organisms have a self-directed behavior that makes up much of evolution. Extending Bookchin's insights, Heller (2000) identifies mutualism, differentiation, and development as key principles affecting the continuities between natural and social life, natural and social evolution. For social ecologists there is, then, an organic origin to all social orders, which is to say, natural history is a key to understanding social transformation.[10]

b. *Constructive interactionism*. This approach, proposed by Susan Oyama, deepens the insights of "the dialectical biologist" by infusing it with debates in constructivism in the social sciences and the humanities, including feminist critiques of science. Oyama's focus is on rethinking biological development and evolution, taking as a point of departure a critique of genecentric explanations in evolution—what Keller (1995) has called "the discourse of gene action" and what Oyama (2000, 2006) describes as "the central dogma" in the relation between ontogeny and phylogeny. Genecentrism, however, the tendency to endow genes with an all-determining power in explanations of biological processes such as evolution, is only the trigger for a substantial reexamination of a set of entrenched habits of thought in biology—such as the assumed divides between nature and culture, internal and external forces, chance and necessity, imputations of one-way causality, etc.—that she sees at play in models of natural selection, innateness, and heredity. Oyama's call is for a dynamic and holistic approach to biological processes, an approach she advances, in her own field, through the concept of "developmental system," defined as "a heterogeneous and causally complex mix of interacting entities and influences" that produces the developmental cycle of an organism (2000: 1). She also proposes a nondualist epistemology called constructive interactionism; this principle does not rely on a distinction between the constructed and the preprogramed ("reality") and upholds the idea that "our presence in our knowledge, however, is not *contamination*, as some fear, but the very *condition* for the generation of that knowledge" (150). What emerges from Oyama's work is a kind of biology that "recognizes our own part in our construction of internal and external natures, and appreciates particular perspectives for empathy, investigation and change" (149).[11]

c. *Phenomenological perspectives*. Tim Ingold has long argued against the Cartesian tradition in ecological anthropology and biology, which he sees as pervasive and deleterious, particularly the assumption of the divides between humanity and nature and living and nonliving things characteristic of most neo-Darwinist approaches (1992). His main source of inspiration for overcoming this dualism is phenomenology; with its emphasis on the embodied aspect of all knowledge and experience, this philosophical current enables him to propose an alternative view of the relation between humans and the environment and of how knowledge of reality is gained. Life happens in the engagement with the world in which humans dwell; prior to any objectification, humans perceive the world because they act in it, and, similarly, they discover meaningful objects in the environment by moving about in it. In this way, things are neither "naturally given" nor "culturally constructed" but the result of a process of coconstruction. In other words, humans really do not approach the environment primarily as a set of neutral objects waiting to be ordered in terms of a cultural project, although this certainly happens as well (what Heidegger, 1977b, called "enframing"); rather than this "designer operation," in much of everyday life "direct perception of the environment is a mode of engagement with the world, not a mode of [detached] construction of it" (Ingold 1992: 44). Knowledge of the world is obtained not so much through abstraction as through a process of "enskillment" that happens through the active encounter with things. These ideas dovetail with debates on local knowledge and local models of nature in anthropology, such as those of the Pacific discussed above.[12]

d. *Poststructuralist antiessentialism.* Donna Haraway's effort at mapping "the traffic across nature and culture" is the most sustained antiessentialist approach to nature. The notion of traffic speaks to some of the main features of antiessentialism, such as the complication of naturalized boundaries and the absence of neatly bounded identities, nature included. For Haraway, the world and the real do not determine knowledge, as the positivist view holds, but the other way around: knowledge contributes to making the world in profound ways. The disembodied epistemology of positivist science ("the god trick" of seeing everything from nowhere, as she descriptively put it [1988: 188]) is at the root of that particular identity construction, Man the Modern (white capitalist patriarchy), with its subordination of nature, women, and people of color. In her "generically heterogeneous" style (1997: 15), Haraway offers a profoundly historicized reading of the making of socionatural worlds, particularly by contemporary technoscience. Along the way, building upon other proposals for a feminist science, she articulates an alternative epistemology of knowledge as situated and partial but that nevertheless can yield consistent, valid accounts of the world (Haraway 1988, 1989, 1991, 1997).

A great deal of work being done today at the interface of nature and culture in anthropology, geography, and ecological feminism follows the strictures of antiessentialism, and it would be impossible to summarize them here.[13] Among the basic tenets of these works are, first, the idea that nature has to be studied in terms of the constitutive processes and relations—biological, social, cultural, political, discursive—that go into its making; second, and consequently, a resistance to reduce the natural world to a single overarching principle of determination (whether genes, capital, evolution, the laws of the ecosystem, discourse, or what have you). Researchers who follow these principles study the manifold, culturally mediated articulations of biology and history—how biophysical entities are brought into social history and vice versa (Escobar 1999a). Whether treating forests, biodiversity, or genetically modified organisms (GMOs), these analyses contain a lot of history, culture, politics, and some (not enough yet) biology. Third, these works show a fundamental concern with biological and cultural differences as historically produced. In this respect, there is an effort at seeing both from the center—looking at dominant processes of production of particular socionatural configurations—and from the margins of social and natural hierarchies, where stable categories might be put into question and where new views might arise (e.g., Cuomo 1998;

Rocheleau 1995a, 1995b; Rocheleau and Ross 1995); this resonates with the idea of looking at nature constructions from the perspective of the colonial difference. Fourth, the emphasis on connection is often couched in terms of networks. As Rocheleau (2000) puts it, researchers need to understand how living and nonliving beings create ways of being-in-place and being-in-networks, with all the tensions, power, and affinities that this hybridity entails. Finally, there is a reconstructive strain in many of these works that implies paying attention to particular situations and at times concrete ecosystems and to the social movements and political identities that emerge from a politics of difference and a concern for nature. The hope is that this concern will lead to envisioning actual or potential ecological communities in the midst of particular processes of cultural and political reappropriation of nature—what Rocheleau playfully calls instances of ecological viability.

3. *Epistemological neorealism*. While constructivism restored a radical openness to the world, for its critics the price is its incapacity to make strong truth claims about reality. To be sure, not all constructivists renounce the assumption of a world autonomous from human action (and, despite the excesses of some forms of constructivism, certainly none of them denies the existence of biophysical reality, as some ill-informed critics have claimed). In this respect, there is a spectrum of positions, from those who unveil objectively true mechanisms at play in nature (social ecology) or who posit nature as an independent ontological domain, even if not knowable in any direct manner (Leff), to those who would dispense with this idea altogether, such as the phenomenologists. But there is a growing set of epistemologies that could be called neorealist, including the following two positions:

a. *Deleuzian neorealism*. Manuel de Landa (2002) suggests that a nonessentialist and nonrationalist, yet realist, account of the world exists in the work of the philosopher Gilles Deleuze. Deleuze's starting point is that the world is always a becoming, not a static collection of beings that knowledge faithfully represents; the world is made up of differences, and it is the intensity of differences themselves—flows of matter and energy—that generates the variety of geological, biological, and cultural forms one encounters. In other words, Deleuze sees matter as possessing its own immanent resources for the generation of form. This difference-driven morphogenesis is linked to processes of self-organization that, in de Landa's view, are at the heart of the production of the real. Differentiation is ongoing, always subverting identity, even if giving rise to real biophysical and social forms, the result of forms of individuation that are relational and always changing. Instead of making the world depend on human interpretation, Deleuze thus achieves openness by making it into a creative and complexifying space of becoming.[14]

An additional point resides in de Landa's rendition of Deleuze. One of the problems with most epistemologies and ontologies of nature is that they are based entirely on the human experience; that is, they distinguish between the real and the nonreal according to what humans are able to observe. In the Deleuzian view, on the contrary, "we need to acknowledge that realism is about what is out there, irrespective of whether we see it or not" (2003: 11). One must overcome the "non-realist baggage" if one wants to arrive at a new ontological commitment to realism that allows one to make strong claims about, say, emergent wholes. "Deleuze is such a daring philosopher," de Landa concludes, "because he creates a non-essentialist realism. Once you divorce ontology from epistemology, you cannot be an essentialist" (2003: 11). In the end, de Landa advocates a new form of empiricism that allows one to follow the emergence of heterogeneous and multiple forms out of differences in intensity in the field of the virtual (2002, chap. 2; 2006). I shall return to this discussion in the chapter on networks, in which I situate the Deleuzian proposal within a larger trend toward "flat ontologies," theories of assemblages, complexity, and self-organization.

b. *Holistic realism* is a view that has been articulated most explicitly by the complexity theorist Brian Goodwin (2007). Goodwin builds his position from within biology, albeit as a critique of mechanistic approaches. His reading of research on emergence, networks, and self-organization leads him to conclude that meaning, language, feelings, and experience are not the prerogative of humans but are found in all living beings; creativity is an inherent aspect of all forms of life, and it is on this basis that coherence and wholeness is produced. His proposal is for a hermeneutic biology and a holistic realism that accept that nature exists and expresses itself in form in embodied reality, that opens up toward the epistemological role of feelings and emotions, and that, echoing Goethe and Spinoza, imagines a science of qualities that links empirical observation and theory building in novel ways. The implication is that scientists can become "co-creators of [the] world with beings that are much more like us cognitively and culturally that we have hitherto recognized. . . . We are within the history of that unfolding. . . . *The task before us now is to rethink our place in the stream of creative emergence on this planet in terms of the deeper understanding of the living process that is now taking form.* The life of form, of which we are a part, unfold[s] toward patterns of beauty and efficiency that satisfy both qualitative and quantitative needs in such a way as to maintain diversity of species, cultures, languages and styles of living" (100, 101, 110; emphasis added).

What, then, of the initial question, What is nature? Within a positivist epistemology nature exists, pregiven and prediscursive, and the natural sciences produce a reliable knowledge of its workings. For the constructivist interactionist, on the contrary, one needs to "question the idea that Nature has a unitary, eternal nature that is independent of our lives. . . .

Nature is multiple but not arbitrary" (Oyama 2000: 143; see also Haila and Dyke 2006). The positivist might respond that if this is the case, an invariant must remain, a central core of sorts that humans can know; the response misses the point in that for Oyama there cannot be one true account of nature's nature. For Leff (1986, 1993), while nature is a distinct ontological domain, it has become increasingly hybridized with culture and technology and increasingly produced by humans' knowledge. For Ingold (1992: 44), nature exists only as a construction by an observer; what matters to him is the environment, that is, the world as constituted in relation by the activities of all those organisms that contribute to its formation. While for social ecologists nature is real and knowable, such realism is not the same as that of the Cartesian subject, but is rather that of a knowing subject deeply implicated in the very process of world making. For the antiessentialists in the humanities and social sciences, biophysical reality certainly exists, but what counts most is the truth claims one makes in nature's name and how these truth claims authorize particular agendas that then shape social and biological being and becoming. Despite the neorealist approaches of complexity theory, finally, the continued dominance of epistemological realism must be acknowledged; it relies not only on its ability to muster credible forms of knowledge, but also on its many links to power: the link between science, production, and technology; the current emphasis on the production of life through the further development of biotechnical rationality; and in the last instance its ability to speak for Western logocentrism, with its dream of an ordered and rational society—now buttressed by genetically enhanced natures and humans—that most humans have learned to desire and depend upon.

Would it make sense to build bridges among realist, constructivist, and neorealist positions? To do so would require further inquiry into the actors, practices, forms of knowledge, institutions, and so forth that underlie each stance, which is beyond the scope of this work. Precisely because they are grounded in different epistemologies (and in some cases ontological assumptions), there are levels at which the positions are incommensurable. However, the knowledge provided by positivist science (e.g., conservation biology) could be relocated and utilized within a constructivist conception, and to some extent this is what has happened in the Pacific (see chapter 4). Conversely, biologists have attempted, with some epistemological consequences, to establish dialogues between science and local knowledge. Interesting overlaps exist: ecological economics, for instance, appeals to realism and at the same time is not oblivious to constructivism. But the divisions are real. I would like to take one more step by explaining briefly Leff's neorealist approach to the environment. I say neorealist because for Leff there is a real nature prior to culture and prior to knowledge, even if it is always known and appropriated (constructed) in culturally specific ways. Yet Leff also believes there has been a progressive "weakening" of nature with humans' intervention;[15] to be sure, such enervation has been taking place since the early days of human history, but it has adopted ever more ontologically transforming forms with the development of modern technologies, to the extent that today one may ponder whether in the face of the most recent molecular technologies humans have not reached a stage of "after nature" (see also Strathern 1992). Nature has become so inextricably hybridized with technology and culture that life itself becomes a hybrid of the biophysical and the technosocial. It becomes impossible to decide where biology ends and technology begins.

This leads Leff to insist that the most important theoretical need at present is to develop a concept of the environment that reflects this growing complexity. In this conceptualization, as for Ingold but in a much different manner, the environment is not nature; it is rather a concept that seeks to make visible the growing complexification of the real effected by the processes of hybridization just described. Included in the environment are thus also the effects of the hyper-marketization and hypertechnologization of life and the environmental problems they have created. For Leff, the existing compartmentalized sciences and epistemologies are utterly unable to describe this complexity. There is a tremendous paradox here: the modern exploitation of nature constituted a definitive intervention in the evolution of the ontological orders of nature and culture, hybridizing the real in ineluctable ways; each science (physics, biology, anthropology, etc.) was supposed to take charge of a given differentiated aspect of the hybrid entity, but these sciences are unable to offer a view of that reality that matches its complexity. This is the greatest paradox of modern knowledge; that is, given their focus on a particular object of knowledge, the sciences *cannot know* this complexity. For Leff, it is not simply a question of building interdisciplinary approaches. What is needed is an epistemological break with the sciences as we know them and the creation of an environmental knowledge (*saber ambiental*) that can account for the multiple determinations of the complex entity that has emerged as a result of the modern transformation. Leff calls the resulting entity "environmental complexity" (1998a, 1998b). The saber ambiental Leff advocates thus includes the real, which is now made up of the hybridized ontological orders of nature, culture, and technology; the effects on that real of the steady economization and technologization of life in general; and the sciences themselves, newly articulated into a new form of knowledge that goes beyond disciplines.

Leff's goal is to outline a new environmental rationality (*racionalidad ambiental*) that can orient reconstructive efforts. His rethinking of production as an integration of economic, ecological, and cultural productivities (see chapter 2) pointed in this direction. His conceptualization of saber ambiental goes even further. The environment is also *a potential* that emerges from the articulation of the real—the articulation of culture, environment, and technology—and that could lead the world to envision an as yet unknown ecotechnological productivity. In sum, Leff advances the construction of a substantive concept of the environment, but he does so through the characterization of various notions, including saber ambiental, environmental rationality, environmental complexity, and alternative productive rationality or ecotechnological productivity. These concepts see the environment as an always emerging complexity that results from the very intervention of knowledge onto the real that brings together the biophysical, the cultural, and the technological into what most people still refer to as nature. The environment implies both an epistemological concept and an orientation toward action, through the notion of environmental rationality. Leff is not talking about a single normative rationality, but about multiple rationalities according to the concrete and culturally specific processes of the appropriation of nature. In cases lying outside the mainstream of modernity, this rationality implies dealing with different cultural processes, nonscientific forms of knowledge, and the deconstruction and reconstruction of occidental knowledge and its possible hybridization with local knowledges.

The above discussion is couched exclusively in terms of occidental philosophical knowledge. In what ways does the discussion apply, or not, to non-Western peoples? Even if they might not have the category of nature, could one say they live outside of nature, that is, outside the ontological domain of the real that occidentals call nature? Leff and the neorealists might suggest that the answer to this question is no. Thus it is possible to speak of different cultural regimes for the appropriation of nature (e.g., capitalist regime, as in the plantations; organic regime, as in the local models of nature of the Pacific; and techno-natures, as in the recent biotechnologies, as I have proposed elsewhere [1999]). Positivists are good at providing scientific information about biophysical aspects of nature, yet they are unable to account for the differences among regimes because for them nature is one and the same for all peoples and situations; and these differences have biophysical implications that they either miss or are at pains to explain. Constructivists usually do a good job of ascertaining both the representations or meanings given to nature by various peoples and the consequences or impacts of those meanings in terms of what is actually done to nature (e.g., Slater, ed. 2003 in the case of rain forests). Yet they usually bypass altogether the question, central to neorealists, of the ontologically specific character of biophysical reality. Finally, it is hard to see how the neorealism derived from complexity might allow a different reading of the cultural dimension of the nature–culture regimes. Leff's is an initial attempt in this direction. Ingold (2000b, 2000c) also points in this direction with his insistence on the profoundly relational character of reality. Even if they are the result of processes of individuation, things do not exist in the real world independently of their relations. And knowledge is not merely applied but generated in the course of lived experience, including, of course, encounters with the environment. In sum, it is not easy to envision relations between the biophysical and the cultural, including knowledge, that avoid the pitfalls of constructivism and essentialism.

Ingold's assertions ring true when one looks at the local models of nature of the Pacific. This model is quite different from that of the capitalist nature of palmicultoras and camaroneras. It is also distinguishable, as we shall see in the fourth section of this chapter and in chapter 6, from the technoscientific model of the global biodiversity apparatus. The coexistence of these models has telling implications for how one thinks about nature, the environment, and modernity.

III ◈ The Traditional Production Systems of the Pacific

There is a close relationship between how people signify their natural environments and the way they relate to them, transforming them. There is no simple, direct connection between a given meaning and a strategy of resource use but a continuous embedding and reembedding of beings and things through significations and practices. In the next chapter, I will analyze the biodiversity conservation project for the Pacific, Proyecto Biopacífico (PBP), through which the journey of the local model of nature into modernity took a new turn. PBP was a deeply negotiated process between activists and project staff over a span of five years (1993–98). Out of this negotiation came the strong conviction, shared by activists and PBP staff alike, that the "communal production systems associated with the collective territories of the ethnic communities" should be seen as "the nucleus of the strategy for the conservation and sustainable use of biodiversity in the region-territory of the Pacific" (Sánchez 1998: 24). The traditional production systems (TPSs) of the indigenous and black communities came to be seen as deeply embedded in cultural and social systems, as having their own forms of knowledge and rationality, and as being the basis for food security and conservation. PBP staff located these TPSs in rapidly changing socioeconomic conditions and tried to ascertain how the systems themselves were changing. An entire conceptualization and characterization of these systems emerged from the work of PBP staff, communities, and activists, one that I can only touch on here.[16] As I argue below, the local models of nature and the TPSs combined subsume an entire strategy of appropriation of nature that constitutes a distinct nature regime.

For the "biocultural territory of the Pacific," a TPS was defined as "the complex ensemble of forms of knowledge and practices of gathering, production, transformation, and distribution of goods that are characteristic of ethnic groups and peasant communities; these forms are closely related to the availability of natural resources and to the dynamic and natural cycles of the ecosystems in which people live, and which constitute the productive basis of the said systems" (Sánchez 1998: 37). Traditional systems, in this view, are characterized by an integrated, ordered ensemble of practices, assessed in terms of their contribution to food security, on the one hand, and of the time devoted to cultivation, on the other. In most cases, agriculture occupies a prominent position, followed by fishing, collecting, and hunting. Together, these four activities constitute the basis of most systems. The next category is extractive activities, particularly mining and forest products, chiefly timber. Finally, there are some complementary activities, such as raising small animal species, handicrafts, and some services, such as river transportation and tourism. Following a cultural ecology framework and landscape ecology methodologies and influenced by agroecological studies in various parts of the world, PBP staff conducted detailed studies of crops produced, use of spaces and species, time allocation, valuation in terms of food security as well as subjective valuation, productivity, and so forth; these studies allowed PBP staff to investigate TPSs in a variety of ecological, social, and cultural conditions.

Generally speaking, TPSs are small in scale and geared primarily toward self-consumption; they do not obey a logic of accumulation but are driven by the principle of self-reproduction. They are organized ac-

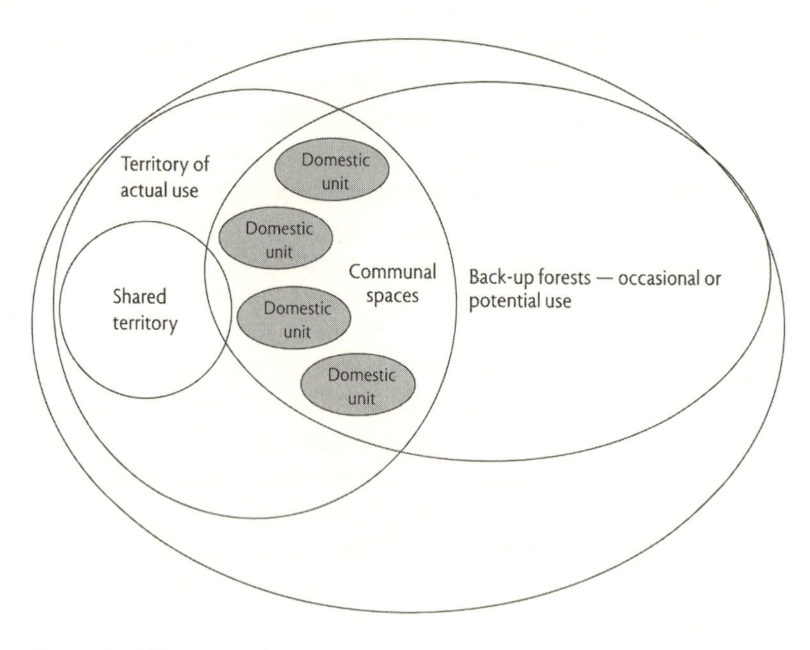

Figure 2. Spatial integration of the traditional territories of black communities of the Pacific region. *Source*: PBP (1998: 61)

cording to family-based and communal forms of territorial appropriation and often have a largely agricultural basis (corn, plantain, rice, fruit trees, cassava, depending on the place); fishing is also important in many cases. Besides family labor, there are reciprocal, communal, and kinship modes of labor; noncontractual agreements predominate, even if contractual forms are increasingly common. Afro-Pacific TPSs maintain a high degree of both productive and biological diversity, complemented by low-impact extractive activities that generate money income for particular purposes, from exceptional events and celebrations to calamities and the purchase of basic commodities (e.g., cooking oil, clothes, school supplies). The aim is for families to have a wide gamut of food sources throughout the year. In some regions, hunting has an important role in the diet and is a source of social role formation. These systems reduce people's dependence on outside inputs, use low-intensity technologies, and have a high ecological and production efficiency when they are able to maintain their diversity. As they become increasingly specialized, they lose their adaptive capacity, and their ecosystemic efficiency diminishes. As a general principle, the greater the local food dependence on these systems, the greater their conservation. For PBP staff, this meant it was necessary to recover and broaden the reliance on forest systems for food

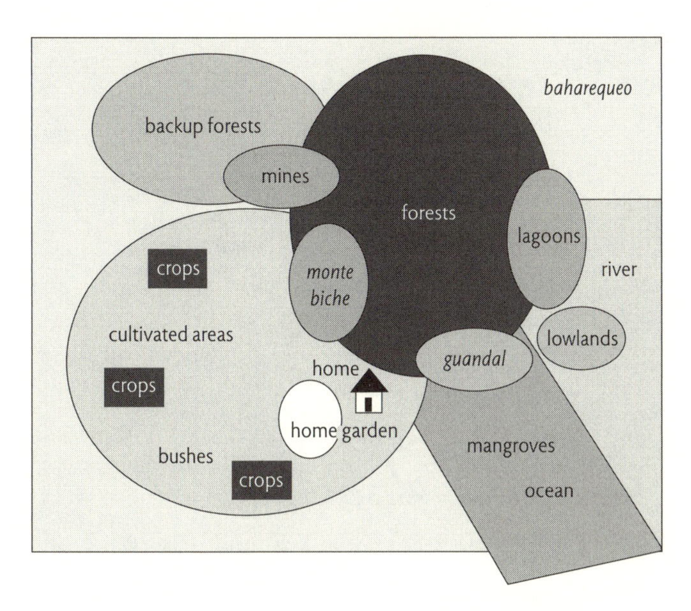

Figure 3. Spatial integration of the collective territories of the black communities of the Pacific region. *Source*: PBP (1998: 58)

in order to maintain their diversity and abet their conservation (see also Altieri 1995 [1987]; Rhoades and Nazarea 1999).

TPSs encompass family-based units and diffuse communal areas that are structurally integrated. Integration takes place according to the notion of territory of each ethnic group, which, for black communities, includes sea, mangrove swamps, forests, and the so-called support areas (*bosques de respaldo*), which are communally held and an essential part of actual or planned collective territories (figures 2, 3, above). Each system is characterized by a particular spatial distribution of crops and activities and by a complex set of flows that involve species, products, people, spaces, and money (figures 4, 5). The cultivation area of each family, variously referred to as *finca, cultivos, or colinos*, may correspond to one or several families, may be contiguous or not, and may have been obtained through a variety of means, from inheritance and exchange to purchase. Each system uses a combination of practices. Agriculture is characterized by a careful selection of seeds and plots, slash-and-mulch land clearing, differentiated use of spaces and time of the year according to the species, attention to the lunar cycle, association and rotation of crops, fallow periods, and family and reciprocal communal labor. Hence, for instance, in the alluvial planes coconut, rice, cocoa, *papachina* (a potatolike tuber, *Xanthosoma sagitteofolium*), beans, *yuca* (cassava, *Manihot esculenta*) maize, sugarcane, and plantain may

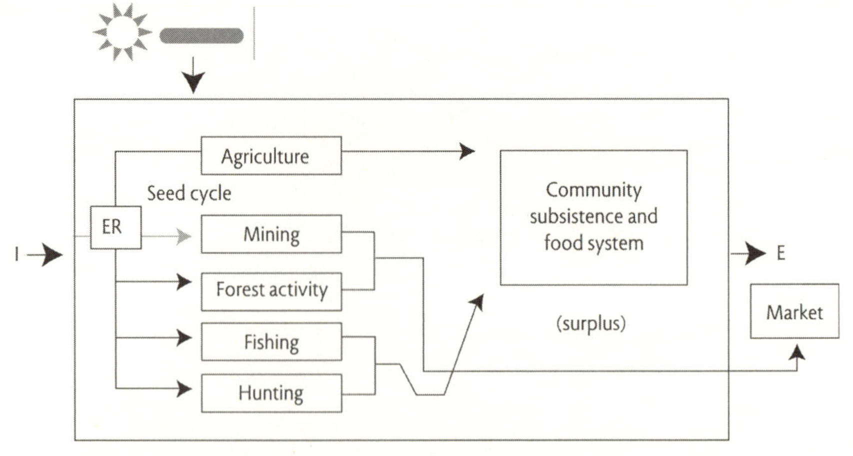

Figure 4. Traditional production systems of the black communities of the Pacific.

Source: PBP (1998: 47)

all be found in various combinations. Uphill, in the Andean slopes, there is less diversity of crops, but hunting and collecting may be more salient; there may also be some cattle in the sub-Andean areas, which in some regions threatens the existence of the cloud forests. Some sectors of the population might specialize in fishing, whereas many other communities raise pigs and chickens as food resources and a mode of saving; in some areas traditional mining of gold and platinum is important. *Zoteas*, rectangular wooden structures placed about two meters above the floor, are used for the cultivation especially of herbs and medicinal plants but also for some food crops, and are maintained by women. Zoteas are an important element of TPSs in the Pacific as they relate to local knowledge and agricultural biodiversity (Camacho 1998; see Arroyo, Camacho, Leyton, and González 2001 for an exhaustive study of zoteas in terms of species and genetic diversity, food security, and gender and ethnic dimensions).

The TPSs, in sum, have been characterized by high use of human energy, a marked sexual division of labor, some forms of reciprocal labor, collective distribution of the product by family and kinship group, a measurable use of barter (plantain is the main medium and measure), food security practices (e.g., salting of fish, pigs), and the fact that labor in not counted in market terms. To recall the discussion in chapter 2, many of these practices belong in the noncapitalist domains of the diverse economy. Yet the system is not a closed loop in that it interacts with the market. In the PBP conceptualization, such interaction is seen as taking two main forms: un-

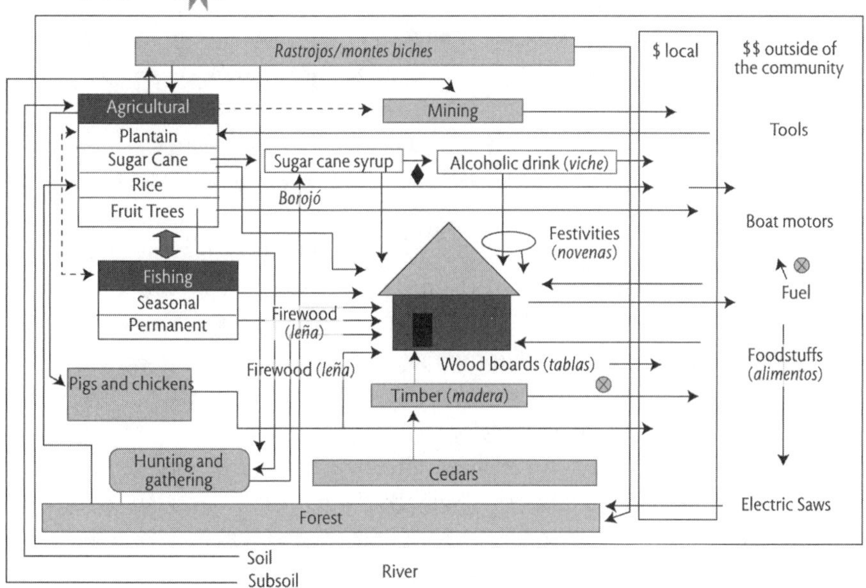

Figure 5. A traditional production system in the Chocó. *Source*: PBP (1998: 51)

equal exchange between the primary products of the TPSs and modern commodities; and *endeude* (indebtedness), characteristic of extractive activities, mainly timber extraction. The endeude depends on a host of social relations that make it difficult to characterize as capitalist. For PBP staff and activists alike, growing integration into market activities entails greater impoverishment, decline in food security, and loss of biological and product diversity. How, then, can these systems be assessed? A "subjective valuation" in terms of people's own priorities suggests that while traditional activities may be seen as less productive in terms of output and income, they are nevertheless highly valued by almost any other criterion, from food value to continuity and duration of the benefit. And although a quantitative evaluation of production shows lower yields as compared with the criteria of a market economy, it showed TPSs to be highly efficient in terms of food security and conservation of biodiversity.[17]

It is now easy to see the negotiated character of the PBP as far as the TPSs were concerned. First of all, TPSs were posited as constituting important elements of sustainability. This claim was substantiated by appealing to a quantitative and qualitative evaluation system that stressed a systemic ecological and cultural logic of use and management of resources. This evaluation made sense within this ecological conception, not outside of it. This does not mean it was totally relativistic, however. As the above epistemological discussion suggested, there is a degree of incommensurability among knowledge systems. PBP staff utilized the principle of multiple standards of valuation from ecological economics in an attempt to resignify TPSs as more conducive to sustainability. Second, PBP staff interpreted TPSs as a strategy for the in situ conservation of biodiversity—one that relies on local knowledge and models of nature, that is always producing innovation, and that is adapted to the difficult local conditions. This emphasis coincides with progressive arguments in the biodiversity field that emphasize in situ conservation as the chief strategy to be pursued (below).[18] To be sure, this strategy requires territorial control, a relative equilibrium in terms of intensity of use and ecosystem capacity, and an underlying network of social relations that makes the system workable. Macrodevelopment projects such as roads, the loss of knowledge, the extension of monocultures and illicit crops, and greater integration into the market, especially by young people, are seen as inimical to the success of a conservation strategy based on TPSs. As younger people intensify the extraction of forest product to maximize money income or simply lose interest in farming, they contribute to the rupture of the long-standing relationship with the territory, in terms of both cultural constructions and productive cycles (Camacho and Tapia 1997: 38).

The PBP staff engaged in the defense of a particular construction of nature by emphasizing traditional production systems; they also developed a sophisticated theory and management strategy for promoting sustainable development in the Pacific. They did so in the context of two main factors: the identity-centered black and indigenous movements and the space provided by the global discourse of biodiversity conservation. In doing so, PBP staff progressively deviated from the dominant biodiversity discourse; they endorsed a local cultural model of appropriation of nature—a local nature–culture regime. The dominant biodiversity discourse, on the contrary, emerged from an altogether different conceptualization of nature, as we shall see in the next section, before I examine in the concluding section the radicalization of the PBP approach in some social movement proposals and their implications for strategies of conservation and sustainability.

IV ◈ The Biodiversity Movement and the Rise of Technonature[19]

The concern with the loss of biological diversity in international development is new; the term itself was coined only in the mid-1980s and did not become prominent until later in the decade, especially following the Earth Summit of 1992 (UNCED, or United Nations Conference on Environment and Development, held in Rio de Janeiro), which created the Convention on Biological Diversity (CBD) as one of its main realizations.[20] Since then, concern with biodiversity has become an impressive science-cum-policy movement, resulting in a notable set of actors and interventions: a multiplicity of new institutional sites that speak about it, from international organizations to governments, NGOs, corporations, and grassroots groups; a host of strategies and interventions the world over, from basic taxonomic inventories to ambitious integrated conservation and development projects (ICDPs); and a growing array of expert discourses, from conservation biology and biodiversity planning to bioethics. In hardly a decade, the concern with biodiversity enabled the creation of a vast network for the production of nature and culture (see chapter 6).

The goal of this scientific and policy movement is to change people's attitude toward the natural world so that, as Edward O. Wilson, the single best-known spokesperson for the biodiversity movement, put it, we may "turn in our role from local conqueror to global steward" of nature (1995: xii). Most experts agree with Wilson that the twentieth was "a terrible century" in terms of species extinction. Under this realization, the conservation of biodiversity became a tireless task, a mission to be carried out in and on behalf of the magic, sacrosanct kingdom of wilderness, to continue with Wilson's idioms. By putting a scientific spin on the crisis, conservation biologists purported to become the authoritative spokespersons for an entire movement to save nature, having as its fundamental goal the "preservation of intact ecosystems and biotic processes" (Takacs 1996: 79). There is much to admire in the biodiversity movement, particularly the deep, genuine preoccupation with the fate of nature and the concrete efforts on behalf of particular species and ecosystems. But there is also much that is questionable from a number of perspectives, including the basic orientation of the concern because of its origins in particular scientific traditions; the limited analysis of the causes of biodiversity destruction; and, consequently, the built-in political choices in the proposed policy formulations.

First, as articulated in conservation biology and by mainstream environmental organizations, the biodiversity discourse, because of its reliance on evolutionary biology and ecology operating under Darwinian or neo-Darwinian paradigms, entailed the genecentric view of biological life. While scientific definitions of biodiversity emphasize the various levels of destruction—genetic, species, and ecosystems—in the long run the diversity of life is seen as being firmly anchored in genes as the fundamental level of diversity; biodiversity lies, in the last instance, in the natural stock of genetic material within an ecosystem. Diversity, to be sure, has been a long-standing preoccupation within biology; and systematics, the study of kinds and diversity of organisms, their characteristics and interrelations, has been of central import to this science (e.g., Mayr 1982). Nevertheless, after the "molecular wars" of the 1960s, which resulted in the dominance of molecular biology that has lasted to this day, and as various classical fields in biology accommodated themselves to the new situation, the use of molecular data became commonplace in many fields, including evolutionary biology and ecology (see, e.g., Wilson 1995: 231). Like so many other scientific fields and aspects of daily life, and despite the emphasis on systemic complexity by a number of authoritative proponents, biodiversity became a gene-centered enterprise.[21]

Little notice has been taken of the effect of genecentrism on the conceptualization of biodiversity. A parallel with health is instructive. Health, to be sure, is believed to depend on more than genes. With the promises offered by gene therapy, however, health is more than ever seen as lying within the narrow domain of the gene. And although biodiversity is seen as encompassing more than genes, the recognition of its genetic foundation suggests that it is in genes, not in the complex biological and cultural processes that account for particular biodiverse worlds, where ultimately "the key to the survival of life on earth" (a common phrase in the conservation trade) is supposed to reside. Genes are endowed with a determining power that, as the scientists trying to decenter genecentrism argue, they do not necessarily have. The consequences go beyond mere scientific debate. At issue are the ways in which genes become politically charged to enable particular truth claims. Hence the centrality, for instance in biodiversity discourses, of debates on genetic resources and intellectual property rights linked to them. And just as molecular genetics is summoned to tell humans the ultimate truth about themselves, so genetic resources—and their applications, such as GMOs—are seen as holding the key to progress, saving our planet, and even bringing about the end of hunger, as in some of the controversies surrounding bioprospecting ("gene rush") or, say, genetically modified ("golden") rice to solve vitamin A deficiency in Southeast Asia.[22]

A seemingly benign scientific discourse thus ends up as the basis of a complex system linking organisms and ecosystems, powerful tools, social institutions, private interests, and even the hopes and aspirations of millions. As many analysts have indicated, when linked to exclusionary property rights enforced by the World Trade Organization, the consequences of this tight system of truth telling linking science, policy, and economy can be devastating for the maintenance of local biodiversity. As a partial corrective, some biologists have constructively emphasized the need to take into account both the biological/genetic dimension and the cultural dimension of the informational content of biodiversity, arguing for a strategy that recognizes both genetic information and cultural knowledge (Nazarea 1998). The insistence on biodiversity conservation in situ, based on the idea of the coevolution of natural and cultural systems, also aims in this direction, especially when taking place under local control (e.g., Gari 2001, GRAIN 1995, 2000). Biodiversity frameworks are partly located at the intersection of genetic knowledge and globalization; this might result in strategies that are seen by local movements as an imposition of the global economy, as is the case with GMOs and transgenic agriculture (Heller 2002) or with bioprospecting and international negotiations on genetic resources that decontextualize biodiversity from its embeddedness in coevolutionary natural and cultural systems. In these cases, gene technology and patents are used to consolidate power over food and nature.[23]

Second, the diagnosis of the causes of biodiversity loss by the dominant discourses is largely conducted in a positivist tone that leaves many relevant issues and perspective out of view. There is widespread agreement in the conservation establishment that the main immediate threats to biodiversity are loss of habitats, introduction of nonnative species, and ecosystem fragmentation due to habitat destruction—all of which, again, are seen as eroding genetic diversity.[24] This diagnosis is backed by eloquent figures (e.g., rates of annual rain forest destruction), elegant charts (showing, for instance, loss of distribution by world region, ecosystem, or activity), powerful pictures (say, of key endangered species), and authoritative statements by scientists (usually first world biologists).[25] Representations of third world peoples, almost invariably showing them either as happy stewards of nature or as faceless hordes bent on destroying it out of poverty, are summoned to back up the diagnoses. The analysis of secondary causes rarely goes deeper than pointing at population and economic activity in an abstract sense. There is an underlying neo-Malthusian current in the mainstream analysis; for instance, in one of Wilson's products, an interactive CD-ROM program (Wilson and Perlman 2000), population growth emerges as one of the most preoccupying factors in relation to the fate of biodiversity; the program allows one to superimpose maps of diversity with maps of population growth. The Hall of Biodiversity at the American Museum of Natural History in New York similarly uses a simple interactive device: a world map sweeping through historical periods as an exponentially multiplying number of dots representing population change rapidly fill in world regions in a seemingly Malthusian nightmare. This facile conflation of population growth and loss of diversity, lacking any substantial analysis of underlying causes, is another troubling feature of the dominant biodiversity discourses. When reference is made to development, it is found to be necessary and sound in principle, even if the "conversion" of diverse-rich spaces into diverse-poor assets, such as agriculture, done in its name is regrettable (e.g., Swanson 1997). Rarely is mention made of capitalism, the endless resource need to satisfy the lifestyle of rich countries, or of the market framework.

As with the analysis of threats, so with that of potential solutions. The biologist Daniel Janzen (1992) summarized the three dimensions of a conservation strategy—knowledge, utilization, and management—in referring to biodiversity: "You've got to know it to use it, and you've got to use it to save it." First, an accurate and complete assessment of the biological diversity of the planet and its regions is seen as urgently needed. This includes basic taxonomy and systematics as well as the characterization of different habitats and ecosystems. Second, management strategies include international policies to encourage the conservation of biological resources (e.g., multilateral treaties such as the conventions on wetlands and on endangered species, CITES; the Global Environment Facility, GEF; international aid; debt-for-nature swaps), national actions (e.g., restricted trade in species, national biodiversity planning, protected area networks), and local level conservation programs or ICDPs. In situ conservation strategies such as rehabilitation of animal species in the wild and ecosystem protection and restoration are also important management strategies, as are ex situ facilities such as botanical gardens for plant species and seed and gene banks for genetic conservation.

Finally, the utilization of biodiversity is seen chiefly in terms of the principle of sustainability. Sustainable development was the concept introduced in the late 1980s to harmonize the goals of development with those of the preservation of the environment (Redclift 1987). This principle is also at the basis of conservation thinking. The idea is to bank on the biological, social, and cultural value of biodiversity. At the biological level, biodiversity is seen as the source of valuable genetic resources and as providing important ecosystems services (in relation to atmospheric processes such as gaseous composition and weather moderation, including carbon dioxide absorption; hydrological; biological and chemical, e.g., in terms of nutrient cycle and detoxification; agricultural; etc.). Making available local genetic resources for biotechnology use through biodiversity prospecting schemes with benefit sharing by local groups and national governments, and under the protection of appropriate intellectual property rights regimes, has been one of the most debated strategies for the sustainable use of biological resources. Payment for ecosystem services has been controversial but continues to be proposed as a conservation strategy.[26]

For some critics, the mainstream prescriptions amount to a "merchandising of biodiversity" (Martínez Alier 1996), "green developmentalism" (McAfee 1999), or a complex politics of cooptation (Asher 2000) that leaves intact the underlying framework of economics and the market that is inimical to nature in the first place. For others (e.g., Shiva 1993, 1997), the mainstream strategies uphold the logic of uniformity and the reductionistic approach of modernity, making possible a regime of biopiracy. (I will return to some of these critiques in chapter 6.) Is biodiversity as a discourse on nature primarily an elegant gimmick for the continued exploitation of nature by capital? I answered this question in chapter 2. Because of its contradictions, capital tends to restructure itself at the expense of production conditions. Some ecological economists see in this contradiction a tendency of capital to develop a conservationist bent that finds value in the protection of nature and resources rather than in its destruction as raw material (M. O'Connor 1993). Be that as it may, the question of whether the biodiversity movement contributes to transforming humans' understanding of nature in any significant way is an important one.

Biodiversity is arguably part of a larger set of processes that are altering humans' understanding and practices of the natural in very important ways. If one defines the *cultural regimes of appropriation of nature* as the concrete forms taken by the articulations between biology and history established by human action, it is possible to distinguish three situations. I introduced these regimes in passing earlier, and now, in light of this initial discussion of the discourse of biodiversity, I want to complete the idea. In the context of the Colombian Pacific, these regimes can be identified as having three sets of actors: black and indigenous groups, capitalist entrepreneurs, and biodiversity experts. The regimes embodied by each of these actors may be called organic, capitalist, and technonature. As we saw, the local models of nature of the black groups are not predicated on the strict separation between the biophysical, human, and supernatural worlds; these groups do not see themselves as manipulating nature out of culture, as moderns do (Strathern 1980). They certainly make classifications and distinctions, but these operate through different cognitive processes and tend to enact an integrated order (Descola 1996). It is precisely this integration of the natural, human, and supernatural worlds that modernity rejects, leading to a different cultural regime, that of capitalist nature. The result of complex historical developments associated with capitalism and modernity, capitalist nature, best represented by the plantation, entails the objectification of nature as external to humans and its subsequent treatment as commodity.[27]

With the development of molecular biology and genomics and their biotechnology applications, the organic and capitalist regimes are redefined. More clearly than ever before, nature becomes a matter of constant reinvention (Haraway 1991, 1997), and humans can now play with unprecedented combinations of the natural and the artificial. To the extent that each design is (or can be) specific, organic nature is always local, capitalist nature aspires to universality through coloniality, and technonature returns to a local logic—albeit these designs are usually couched in the anthropocentric universalizing logic of progress, development, and overcoming. Does technonature necessarily represent a new level in the domination of nature? Or does it open new, more ecological regimes? The answer depends on one's assessment of the new technologies themselves. On the utopian side, there are those who see in the new technologies the possibility of intelligent communities, including the design of sustainable socionatural worlds through technology (e.g., Lévy 1997). On a more pessimistic side (e.g., Virilio 1999), there are those who see in the new technologies a more radical delocalization, that is, a further, perhaps fatal erosion of the cultural logic of place and territory. For some visionary thinkers, the possibility of a radical renewal of socionatural worlds depends on an ecological consciousness—and ecosophy—that links the biological, the cultural, and the technological (Guattari 1990), or the world/cosmos, humans, and the supernatural/god, bringing about an unprecedented *cosmotheandric* experience (Panikkar 1993), or a relinking of nature, humans, and the spiritual world (Boff 2002, 2004).[28] For dystopian thinkers, the new technologies can result only in a perpetual logic of recombination fueled by informatics which would spell out the final decline of the organic. For Haraway (1997), the social and ecological character of the techno-bio-cultural assemblages will largely depend on the kinds of politics that various groups—artists, feminists, scientists, social movements—are able to bring to bear on the dominant tendencies.

Whatever the assessment, all individuals and groups today are confronted with aspects of the three regimes and their endless interrelations and overlaps. For social movements such as the black movements of the Colombian Pacific, the implication of this array of possible positions is the need to be mindful of and negotiate effectively among the three regimes. Biodiversity discourses and strategies constitute surfaces of engagement for the negotiation of regimes. Social movements are faced with the need to hold these three regimes in tension: the local regimes, which they want to defend and transform from a position of autonomy; the capitalist regime, the advancement of which they want to contain; and the techno regime, which, through processes of counterwork (see chapter 4) and politics of scale they want to utilize for the defense of identity, territory, and place. To conclude, let us see how these movements are seeking to redefine conservation from such a space of cultural intersection.

V ◈ Autonomy: A Social Movement Perspective on Conservation and Sustainability

There is a connection between loss of territory and cultural change that is experienced and expressed in strong terms by local leaders in the rivers of the Pacific. As we saw in chapter 1, locals place "loss of traditional values" (*pérdida de valores tradicionales*) at the top of the list of factors producing loss of territory (*pérdida de territorio*) and biodiversity. As a result of several years of debate and mobilization, many of the leaders have a complex understanding of the system that generates loss of territory and hence of biodiversity. The use of these categories might be tentative and uncertain, and they may be accompanied by local expressions and stories, but there is nevertheless a growing awareness of the interconnection between territory, culture, and political strategy.

How have activists come to see conservation and sustainability? Over the years, particularly in the period 1993–2000, the activists of PCN developed an entire political ecology framework to deal with these issues. The framework incorporates concepts of territory, biodiversity, life corridors, local economies, territorial governability, and alternative development. They have progressively articulated this framework in their interaction with community, state, NGOs, and academics. Briefly (see Escobar 1998 for a fuller analysis, and chapter 4), the territory is seen as a fundamental multidimensional space for the creation and re-creation of the ecological, economic, and cultural practices of the communities; it links past and present. In the past, communities maintained relative autonomy as well as forms of knowledge and ways of life conducive to certain uses of natural resources; in the present, there is a need to defend territory. The ensuing conception of territory highlights articulations between patterns of settlement, use of spaces, and practices of meanings-uses of resources, including traditional production systems. Activists are quite aware that the sustainability of these systems has been greatly compromised for most communities over the past several decades, becoming untenable in many areas.

For activists, biodiversity equals territory plus culture—there is no conservation without territorial control, and conservation cannot exist outside of a framework that incorporates local people and cultural practices. The Pacific is also thought about as "life corridors," veritable modes of articulation between sociocultural forms of use and the natural environment. For instance, there are life corridors linked to the mangrove ecosystems; to the foothills; to the middle part of the rivers, extending toward the interior of the forest; and those constructed by particular activities, such as traditional gold mining and women's shell collecting in the mangrove areas. Each of these corridors is marked by patterns of mobility, social relations (gender, kindred, ethnicity), use of the environment, and links to other corridors; each involves a particular use and management strategy of the territory.

More concretely, the territory is seen as the space of effective appropriation of the ecosystem, that is, as spaces used to satisfy community needs and to bring about social and cultural development. For a given river community, this appropriation has longitudinal and transversal dimensions, sometimes encompassing several river basins. Thus defined, the territory cuts across several landscape units; more important, it embodies a community's life project. The region-territory, on the contrary, is conceived of as a political construction for the defense of the territories and their sustainability. In this way, the region-territory is a strategy of sustainability and vice versa: sustainability is a strategy for the construction and defense of the region-territory. Activists are clear that sustainability cannot be conceived in terms of patches or singular activities or only on economic grounds. It must respond to the integral and multidimensional character of the practices of effective appropriation of ecosystems. The region-territory can thus be said to articulate the life project of the communities with the political project of the social movement. In sum, the political strategy of the region-territory is essential to strengthening specific territories in their cultural, economic, and ecological dimensions. Further, the region-territory can be said to be a management category of the ethnic groups, but it is more than that. It is a category of interethnic relations that points toward the construction of alternative life and society models. It entails an attempt to explain biological diversity from the endogenous perspective of the ecocultural logic of the Pacific. The demarcation of collective territories fits into this framework. Government dispositions violate this framework by dividing up the Pacific region among collective territories, natural parks, areas of utilization, and areas of sacrifice where megadevelopment projects are to be constructed.

Throughout the 1990s, PCN activists highlighted three interrelated guidelines for local actions: the life project of the communities, as described above; the territory as the space that sustains the life project and as a planning unit in the framework of Ley 70; and the organizational process centered on the appropriation and social control of the territory and as the basis for food security, self-subsistence, and autonomy. The focus on food security—or food autonomy, in the term preferred by activists—came about as a result of displacement in recent years, since food security and self-subsistence are seen as minimum guarantees to hold on to the territories. This has also been accompanied by a restatement that the *desarrollo integral* (integral, self-directed development) of the region must be based on the values and production practices of the communities. This implies a continued balance between activities geared toward subsistence and food security and income-generating activities for the market. In the last instance, the conservation of the diversity of life forms (natural and human) depends on a *proyecto de vida* (life project) based on the practices and values proper to the cultural vision of the communities.[29] This conceptualization is aptly summarized in figure 6 (from PCN 2000: 4).

This framework has been actively forged through practice and engagement in a wide range of situations. By most accounts, the PCN was the single most visible and important force engaged in the defense of the southern Pacific rain forest cultures and ecosystems in the 1990s, and it continues to be quite active. What made PCN important and in many ways unique was a combination of features that included a courageous, sustained political strategy vis-à- vis the state, NGOs, and local actors around territorial, cultural, and environmental problems and issues; the progressive elaboration of a sophisticated conceptual framework for

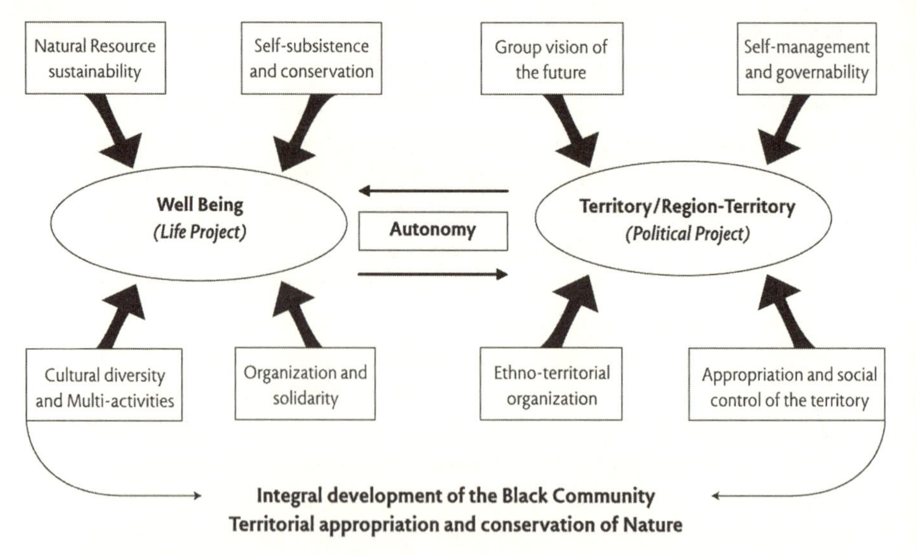

Figure 6. Basis for culturally and environmentally sustainable development. *Source:* PCN (2004: 4

problem analysis and alternative policy formulation regarding development, biodiversity conservation, and sustainability in the Pacific, just outlined (see also chapter 4); and a persistent engagement with concrete environmental conflicts and the search for solutions at local, national, and international levels, sometimes in collaboration with local or transnational NGOs (e.g., Fundación Habla/Scribe; World Wild Life Fund's Colombia Office, based in Cali; or even from state agencies, such as the National Natural Park System Office).

At this level PCN developed a coherent political ecology practice in terms of ecological distribution conflicts. Environmental conflicts have been particularly salient in four areas: (1) Particular cases of degradation and destruction of ecosystems, from gold mining (early to late 1990s) to the expansion of the African oil palm frontier (ongoing). In this regard, PCN activists have been instrumental in pushing the state to fulfill existing legislation (e.g., in the case of gold mining), leading the efforts at new legislation, particularly the cultural and territorial rights law, Ley 70, and originating international campaigns (gold mining, timber). (2) Deforestation caused by various practices, including the timber industry, agribusiness, and illicit crops (coca), heart of palm (*naidí*) canning, and cattle; this has led to PCN's participation in international fora, such as forest certification schemes. (3) Loss of biodiversity. PCN is actively involved in biodiversity conservation projects and international meetings in this area. As the armed conflict extended into the southern Pacific, this work became ever more difficult, particularly given the spread of illicit crops by armed groups. Despite these conditions, work has continued in this area. In July 2002, for instance, PCN was able to broker an agreement with the National Natural Park System Office of the Ministry of the Environment by which this office recognized the desirability of making the park system compatible with the collective territories, within a region-territory conception. The agreement also included the Park System Office's support for the communities' own conservation initiatives.[30]

Table 3 summarizes the main areas of concentration of this network of organizations in the environmental arena.

In some ways, PCN's conceptualization can be seen as a radicalization of the PBP strategy centered on the defense and required transformation of TPSs. It is also influenced by the emphasis of many movements on autonomy (the Zapatistas have been particularly articulate in this respect) and by the ethnicization of identities (see chapter 5). In the concepts of proyecto de vida and *vivir bien* (well-being and livelihood) lies a qualitative language of development that is culturally grounded. (One might find it hard to believe there are people in the world whose main aspiration in life is not ever-increasing consumption and material well-being, but this is indeed true of poor rural peoples of the world.) As in the case of TPSs, this social movement perspective can be seen as consistent with progressive thinking in ecology, such as that by Altieri, Leff, and Barkin, already mentioned for the Latin American context. Other approaches are worth highlighting from the perspective of PCN's conception. Writing about the indigenous people of Pastaza in the Ecuadorian Amazon, Gari (2001) argues that any conservation project should strengthen traditional ecological knowledge and practices, since these are essential to local (in situ) conservation and development. Many progressive national and transnational NGOs (e.g., Grupo Semillas in Colombia, Acción Ecológica in Ecuador, GRAIN and ETC worldwide) have developed political strategies to foster in situ conservation under local control, in great part because they have been developed in dialogue with social movements (see King 2000). In sum, like the activists, these thinkers and organizations see a tight connection between the conservation and use of biodiversity, the control of territories, and the construction of strategies of appropriation that emerge out of the meanings-uses that characterize the nature construction of local groups. I could buttress these arguments by appealing to the language of coevolution of natural, knowledge, and social systems (e.g., Norgaard 1995); or by mentioning studies about the greater productivity of small-scale agroecological systems; or by summoning the

Table 3. Ecological Distribution Conflicts in the Southern Pacific, 1992–2002

Environmental problems	Responses/Actions	Results
Pollution and destruction of ecosystems by industrial gold mining (several rivers, c. 1994–97)	• Conducted local, national, and international campaigns to stop industrial gold mining in several rivers of the central and southern Pacific. • Put pressure on the state to apply environmental legislation on mining. • Strengthened the organization of local communities to oppose the entrance of gold extraction equipment (*retroexcavadoras*).	• Slowed down industrial gold mining in some rivers. • Achieved expulsion of illegal miners in a few rivers.
Pollution and destruction of ecosystems by African oil palm processing plants and shrimp cultivation (Tumaco region, c. 1993–present)	• Put pressure on the state to utilize and create legislation to prevent further expansion of large-scale projects (African palm plantations, industrial shrimp), particularly Ley 70. • Contribute actively to the demarcation and titling of collective territories. • Empower local communities to deal with environmental problems and conflicts. • Act as the pivotal force for the creation of a novel environmental discourse in the Pacific.	Generated consciousness about the environmental and cultural impact of the oil palm and shrimp industries (even if without much significant effect on these industries).
Pollution and destruction of ecosystems by construction of oil pipeline through the rain forest (Buga-Bay of Malaga, 1995)	• Organized opposition to construction of pipeline. • Coordinated various groups to negotiate with government agencies.	Stopped construction of the intended pipeline.

(Table 3. continued)

Environmental problems	Responses/Actions	Results
Deforestation caused by timber industry, and by industrial licit and illicit crops (c. 1993–present)	• Promote legislation (i.e., Ley 70) to stop forest concessions. • Work with community councils to regulate and monitor timber concessions and activities. • Participate in international forums on forest issues (e.g., Global Action on Underlying Causes of Deforestation and Degradation). • Support the creation of community reserves (e.g., the Cortina Verde Community Reserve in the southern Pacific). • Persist in the denunciation of expansion of illicit crops by armed actors.	• Created greater involvement of local communities in the appropriation and management of territories and forest concessions. • Drew up local territorial action plans that involve alternatives to timber extraction in a number of river communities. • Made substantial contribution to the titling of a number of collective territories in the entire Pacific region (2,359,204 hectares). • Contributed to the management of a few of these territories on a sustainable basis.
Loss of biodiversity (1993–present)	Develop an alternative framework for biodiversity conservation (see narrative).	Brokered effective reorientation of biodiversity project for the Pacific.

vast repertoire of local knowledge that ethnoecologists consider of central importance for ecosystem management (e.g., Nazarea, ed. 1999); or, finally, by pointing out the fact that by holding the various nature regimes in tension social movement activists keep alive the possibility of different socionatural worlds.

Finally, the link activists make between biodiversity and food autonomy is increasingly recognized by experts observant of this relation on the ground. Rhoades's and Nazarea's (1999) pioneering work on traditional landraces (folk or indigenous plant varieties that are the result of many generations of farmer selection) shows how worldwide cultural knowledge has historically intertwined with biological elements in traditional (particularly marginal) agroecosystems, such as the TPSs of the Pacific, to maintain a dynamic ecological-evolutionary situation. Often, local agriculturalists and farmers have played an active role in maintaining diversity in situ by saving and managing local seeds and landraces in such a way that gene-rich pools of food crop species are maintained. The zoteas kept by women in the Pacific are a case in point. Landrace cultivators, of course, do not act solely for the sake of preserving seeds or improving diversity. They always operate with a complex system in mind that links seeds, food needs, ecological and human (e.g., labor, gender) requirements, the market, and a host of cultural aspects, from culinary considerations and the shape and color of landrace materials to ritual. The logic, if anything, is a fuzzy one, responding to the interconnectedness of plants, food, and culture. These practices are guided more by pragmatic concerns arising from a multiplicity of local demands than by the desire to come up with "best" or "high-yielding" varieties. Finally, there are clear socioeconomic and political implications of this conceptualization for in situ conservation. As Rhoades and Nazarea put it,

> Many indigenous communities with a firm sense of place are aware of the value and role of land and diverse crop inventories to their cultural survival, and communally strive to guard these resources. . . . In communities that have not yet been fully incorporated into commercial markets and still manage high levels of landrace biodiversity, future protection of such genetic resources requires values compatible with locally defined social and economic goals. Whether biodiversity is decreased or enhanced in the future may depend on the degree of self-determination of local communities to attain these locally defined goals. (1999: 225)

Here again one finds a high degree of consistency between the experts' view and that of social movements such as PCN. One also finds, of course, an implicit reference to the diverse economy. As Rhoades and Nazarea conclude, "Regardless of whether *in situ* conservation is an indigenous event or one stimulated by an outside-funded program, one principle must be preserved. That is, the degree of independence, even irreverence, necessary for the persistence of diversity is inversely proportional to the degree of integration into the market system and the degree of capture by the political vortex. . . . If the system of access and rewards can be restructured, we feel that local populations can use conventions that exist in the global marketplace to their benefit—and for the benefit of humankind—instead of being subjected to these conventions" (1999: 231). So do activists of food security and farmers' rights, and this determination shows at many levels, from their prioritizing of place-based systems to their staunch support of demonstrations against the World Trade Organization, as was amply shown in Cancún in September of 2003 by peasants from many parts of the world.

To sum up, biodiversity and sustainability have to be seen in their biophysical and cultural context and in relation to the empowerment of local groups, not in the commodified and decontextualized terms of much of the official discourse and the idioms of genetic resources and intellectual property rights. Failing to acknowledge this interconnection may well entail the failure of conservation. It may well be that biodiversity is the name we are giving to a complex, dynamic system that cannot be fragmented spatially or conceptually. Indeed, if one agrees with recent conceptions of rain forests as complex, self-organizing ecosystems, modern compartmentalized approaches to conservation become even more inappropriate. The logic of collective ownership of territory is more attuned in principle to this perspective than that of individual property, which tends to lead to greater fragmentation of habitats and communities. From an ethnoecological perspective, the multiuse strategies of local groups are more effective in maintaining habitat patchiness and heterogeneity and biological and genetic diversity. Given that the Pacific cannot be "conserved" without its people, it is imperative that future strategies heed these lessons.

Conclusion

Struggles over nature mediated by the construct of biodiversity continue to rage in many spaces. For instance, at the Eighth Conference of the Parties of the Convention of Biological Diversity held in Curitiba, Brazil, in March 2006, this struggle pitted social movements and radical NGOs against governments, corporations, and mainstream NGOs. While the former group issued a "Manifesto for Life: In Defense of Nature and Biological and Cultural Diversity" highlighting the need to control the commoditization of life, the advance of transgenic agriculture, and the destruction of ecosystem balance brought about by "neo-liberal colonization," the latter group met at their customarily luxurious settings to advance the conventional agendas of market-led conservation and sustainable development. Encounters of this type (discussed further in chapter 6) happen all the time at national and international meetings on biodiversity, biosafety, intellectual property right, forests, water, agriculture, trade, and so forth. In Latin America, many of these encounters were largely mediated by the decided opposition of social movements and many NGOs to so-called free trade agreements, resulting in the defeat of these agreements in many of the region's countries.[31]

The theoretical material of this chapter provides some food for thought for these mobilizations. Suffice it to say for now that the academic proposals of some political ecologists do find validation and resonance in the concrete experience of the Colombian Pacific. Conversely, and perhaps more important, the knowledge produced by activists is already finding its way into academic attempts at revisiting conservation and sustainability. This new circulation of knowledge, in its coproduction and overlaps but also in its specificities, is promising politically and environmentally. Throughout history, many place-based peasant and indigenous communities have internalized cultural and ecological conditions in the social relations and productive systems of their societies. Theoretically, one can say that in these cases the environment may be seen as a productive system based on the stability and productivity of the ecosystem and the cultural styles of the groups that live in it. These observations point at an alternative theory of production and subjectivity (following Gibson-Graham, 2006) that entails a triple articulation of ecological, cultural, and technological processes. Leff's notion of the environment is an attempt to account for this complexity; it is another way to pursue thinking about the economic, ecological, and cultural difference and struggles led by social movements.

My attention to questions of epistemology should make one attuned to the possibility that the proposals of social movements and ecologists reviewed here constitute a critique of the logocentric and Eurocentric conception and appropriation of nature that is characteristic of modernity—a strategy for decolonial relations to the natural. A decolonial view of nature and the environment calls for seeing the interrelatedness of ecological, economic, and cultural processes that come to produce what humans call nature. This implies the ability of constructing difference as the basis for both a critique of dominant nature–culture regimes and a guiding tool for efforts at reconstructing socionatural worlds. In other words, it is important not only to use theory but also to make visible options that might lie beyond theory (or beyond what theory is readily able to think). The practice of some social movements, as we saw in the case of PCN, suggests that a *cultural* redefinition of conservation, sustainability, and production is indeed possible. It remains to be seen whether this redefinition will flourish for more than a fleeting moment in a number of places throughout the world, as it did in the Pacific in the 1990s.

Because of their focus on difference, the political ecologies of social movements and critical intellectuals are ethical perspectives in that they entail an epistemic questioning of modernity and development that points at basic definitions of life. By privileging subaltern knowledges of the natural, these political ecologies articulate uniquely questions of diversity, difference, and interculturality—with nature as central agent. In Latin America, an emergent environmental thought builds on the struggles and knowledges of indigenous, peasant, ethnic, and other subaltern groups to envision the construction and reconstruction of local and regional worlds in more sustainable ways. Sustainability may thus become a decolonial project: thinking from existing forms of alterity toward worlds and knowledges otherwise. In the current landscape of a dominant coloniality of nature found in the capitalist and technonature regimes in the Pacific, this strategy will have to rely on the transformation of capital, development, and modernity effected by local groups and activists.

4 development

> *La afirmación del ser* [the affirmation of being] is an affirmation of the right to difference, of our cultures, our social mode of being and our view of life. This principle implies that the development plans for Afro-Colombian communities and regions cannot simply be made up of investment plans but should rather be channels to enable [*potenciar*] human development beyond the material solutions. This principle centers development on people, basing it on the decisions they take, thus contributing to their human dimension. Development plans inspired by this principle should result in the strengthening of people's capacity for decision making, creativity, solidarity, mutual respect, the valuation of their own culture, dignity, and consciousness of their rights and *saberes* [knowledges]. Cultural and ethnic identity and the feeling of ownership of the territory should emerge strengthened from this process, and there should also be the creation of broader horizons and spaces for people to be the actors of development.
>
> —PCN. 1994. "Principios para el Plan de Desarrollo de Comunidades Negras." Presented to the Technical Commission for the Formulation of the Development Plan for the Black Communities, Bogotá

Introduction: The Invention of the Pacific as a "Developmentalizable" Entity

In 1983, the first Plan for the Integral Development of the Pacific Coast (Plan de Desarrollo Integral de la Costa Pacífica, PLADEICOP), stated its call for development in the following way:

> This vast region harbors enormous forests, fishing, and mining resources that are required immediately by the nation; the region constitutes an area of fundamental geopolitical interest for the country. Hence the inevitabil-

ity of a state policy capable of understanding and assuming the integral development of the Pacific Littoral as a great national project. This project can no longer be postponed. (DNP/CVC/UNICEF 1983: 13)

Ten years later, the much more ambitious "Plan Pacífico: Una Estrategia para el Desarrollo Sostenible para la Costa Pacífica Colombiana" (DNP/CORPES 1992), funded by the World Bank and the Inter-American Development Bank, elevated PLADEICOP's goal to new levels by focusing chiefly on large-scale infrastructural investment (from electrification and basic services to transportation, ports, telecommunications, and so forth) in order to create the basis for regional capitalist development. Because of internal critiques and pressure from external funders, subsequent versions of the plan incorporated a more explicit environmental and conservation dimension; nevertheless, Plan Pacífico continued to have a largely economistic orientation, even if by the late 1990s it was couched in terms of "social capital," in which planners included institutional, infrastructural, and environmental capitals (see, e.g., DNP 1997).

At the same time Plan Pacífico was launched, the much more modest Proyecto Biopacífico (PBP) ($6 million for its first three-year phase, 1993–96, as opposed to $250 million for Plan Pacífico for the same period) started operations with the overall goal of the conservation of the region's biodiversity (GEF/PNUD 1993). In its early stages this initiative was a largely scientific and managerial endeavor for the conservation and use of biodiversity but had evolved, by the time of its dismantling in 1998, into a deeply negotiated project permeated by the language of ethnicity and territorial rights promoted by social movements. In February 2000, largely as a result of the learning process experienced around PBP, the recently created Pacific Institute for Environmental Research, along with the Ministry of the Environment (created in 1993) and Plan Pacífico, coauthored a policy document entitled *Agenda Pacífico XXI*, with the subtitle *Proposal for regional action for the biodiverse Pacific for the new millennium*. As the subtitle suggests, the document's scope was far reaching. The authors conceptualized their planning task as follows:

> In the last instance, the aim for the Pacific should be to develop leadership capacity for its own development in the midst of the advancing globalization of societies, taking its own diversity as a point of departure. . . . Because of its rich ethnic and environmental diversity, the *Pacífico Biogeográfico* is a unique and privileged region of the planet, with great possibilities for its inhabitants through an endogenously formulated sustainable develop-

ment strategy oriented toward the improvement of the well-being of the people and the preservation of its biological wealth. . . . To put it bluntly, the most important goal is the consolidation of the Region-Territory of the Pacific within the context of the Colombian nation. . . . The overall objective of the Agenda should be the formulation, in a concerted fashion and through a broad process of participation at the local level, of long-term policies, plans, and programs that respond to the ethnic, cultural, social, economic, and environmental reality of the Pacific, and that may guide its development and its articulation to the progress of the nation. (Instituto de Investigaciones del Pacífico 2000: 3, 14, 17, 18)

From PLADEICOP in the 1980s and the early Plan Pacífico to the *Agenda Pacífico XXI* of 2000, a great distance had seemingly been traversed in conceptual and political terms. To begin with, there was little in the first two plans of the language of cultural and biological diversity that would become prominent in the second half of the 1990s. There were no social movements to worry about, and, although "participation" was indeed considered, this was conceived as a bureaucratic problem for the institution to solve, not as a negotiation over knowledge and power, as it came to be with PBP.[1] PLADEICOP and Plan Pacífico belonged to an era of planning that saw no need for serious negotiation with local peoples. The *Agenda*, on the contrary, not only recognized that "indigenous and black community organizations have had a notable impact on the rethinking of planning and policy proposals for the region" (13), but also listed these organizing processes as one of the main potentialities for the region and one of its nine thematic axes. To be sure, this conceptualization was still immersed in the language of the development of the nation, the creation of infrastructure, and the use of natural resources. Yet the conceptualization of the Pacific developed throughout the 1990s, in great part under the political and intellectual thrust of social movements in a variety of spaces (PBP being one of the main ones), had been incorporated in the vision planners crafted at the dawning of the present decade.

How to assess this change in conceptualization and policy orientation? Opposite views tend to see it either as an intelligent co-optation by the state of the language of social movements that, to begin with, had been greatly enabled by the state itself after 1991 or, at the other extreme, as a result of the steady influence of the social movements on the state and its planning apparatus. The first explanation is persuasive up to a point; the emergence of the black and indigenous movements of the Pacific was indeed fostered by the economic and political reforms of the neoliberal state, particularly the Constitution of 1991. Yet to see these movements primarily as the result of these actions is to confer upon the state a coherence and prescience it rarely has, at least in the Latin American case. It is also to deny the vitality of the worldwide irruption of identities linked to ethnic and environmental concerns since the 1970s. In large measure, the answer to the question of the transformation described above depends upon the framework one employs and on the understanding of power such a framework implies. As will become clear throughout the chapter, this issue is central to how one interprets the entire experience of development in the Pacific in the 1990s, and how one visualizes the possible alternatives for the future.

With PLADEICOP, the Pacific was constructed for the first time as a "developmentalizable entity" (Escobar and Pedrosa, eds. 1996). The imaginary of progress had not been completely absent from the lands of the Pacific, to be sure. In 1921, Father Bernardo Merizalde, in his monograph on the Pacific written from the point of view of the Catholic missions, was already calling on the government to assist the region "by sending good functionaries, opening roads linking the littoral with the interior, protecting Tumaco by constructing a wharf and building a wall around it before it is swallowed by the sea, ending the isolation of Guapi, Timbiquí and El Charco by providing them with port facilities and a telegraph line, facilitating the development of mining and the exploitation of forest resources and, in sum, taking notice of this immense territory that because of its strategic location is called upon to be a reservoir of civilization and progress" (1921: 231). These goals were bound to be reached because, for Merizalde, the Costa "possesses wealth, beauty, and an exuberant tropical life" (231). His perspective, however, was more that of natural history than of rational state action. A state perspective was already emerging by the time of the publication of Sofonías Yacup's *Litoral Recóndito* (1934), to which I referred earlier. Yacup was born in Guapi in 1894, and by the time of his death in 1947 he had occupied important positions, including that of congressman of the republic. His conception of the Pacific has to be located within the post-1929 climate of nationalist liberal ideology that took hold in some Latin American countries, the kind that saw the rise to prominence of the famous socialist-populist leader Jorge Eliécer Gaitán in the 1940s. Yacup's reading of the littoral, while purged of the overtly racist and ethnocentric representations found in practically all previous works, from those of Caldas and Codazzi to Father Merizalde's, still portrayed the region as lethargic and recondite and saw progress in terms of the redemption of its people. In keeping with his liberal views, Yacup came close to endorsing the colonization of the Pacific by whites from Antioquia or by "healthy foreigners" because of their "sturdy individualism" (62). His book was a hodgepodge of reflections on liberalism, critiques of Congress, information on the Pacific, proposals for laws and decrees, and so forth, couched in a language that was closer to the nineteenth-century essay than to the scientific discourse that was to become prevalent in the development era.

There were other precursors of development and modernization in the Pacific. I have mentioned the cycles of boom and bust in the global economy; state territorial concessions to national and foreign investors, especially in forest and mining, played an important role in this regard: for instance, the gold-mining concession to a French company in the Timbiquí river at the beginning of the twentieth century, a veritable privatization of the river territory under semislavery conditions. These concessions "already introduced a modernizing bent to the region to the extent that the extractive technologies irrupted like powerful devastating forces on rivers, forests, and communities" (Pedrosa 1996: 76). Second, given that the Pacific was officially a "territory of [Catholic] missions," its education was entrusted to Catholic orders which, although not necessarily a modernizing element, helped prepare the ground for the seemingly perpetual delay in the modernization process. Third, plantations and colonization schemes, as well as fluvial and seaports, were important in fostering local and long-distance commercial activity and traffic, even if of limited scope. All of these antecedents, however, can be said to have taken place on the basis of a view of the Pacific as a region to be conquered and colonized, not developed. Introducing the imaginary of development required a further elaboration: it was necessary to create a collective consciousness of marginality and an awareness of exclusion, which would then open the way for intervention by the state and other actors, from entrepreneurs to international development agencies. Until the early 1990s, there was little in the official development plans for the nation that pertained specifically to the Pacific. Three notable exceptions were the Plan for the Development of the port city of Buenaventura in the late 1960s; the Cauca-Nariño Reconstruction Plan, drawn after the regional earthquake of 1979; and the Project for Integral Rural and Agricultural Development in Chocó's Atrato river, carried out between 1976 and 1984 with Dutch funding and technical assistance (Proyecto DIAR).

Not until PLADEICOP was the idea of a development plan for the entire Pacific articulated. Colombia's president at the time, Belisario Betancour, put it this way: "Colombia has always disregarded the Pacific ocean, leaving its littoral at its mercy. The government now wishes to devote the four years of its presidential period to setting down the foundations for a great policy for the Pacific and the elaboration of its first development plan" (DNP/CVC/UNICEF 1983: 13). The Pacific finally emerged fully as a developmentalizable entity, with dramatic consequences. First and foremost, the imperative of development was to replace the previous dynamic of Afro-indigenous colonization and limited, externally driven exploitation of resources (see chapter 1) with the modern imaginary of accelerated economic growth, intensification of natural resource extraction, development enclaves, urbanization, and so forth—in short, with the panoply of political technologies associated with modernity. From then on, a host of development strategies followed, including, besides PBP and Plan Pacífico, the Forest Action Plan for Colombia, the Action Plan for the Afro-Colombian and *Raizal* Populations,[2] and the Project for Ecological Zoning, among others (IGAC 2000). Linked to the development of the Pacific was a broader process of partial reversal of Colombia's Atlantic-oriented development (a spatial Euro-Andean occupation of the territory stemming from the European Atlantic and organized along the north-south axis defined by the Andes and the Magdalena and Cauca river basins) toward a "Pacific Era"; this reorientation was defined as the integration with the Pacific basin economies, with the concomitant need for macrodevelopment projects for the trans-Pacific integration of the country. This new orientation, celebrated with great fanfare in the early 1990s in national and international fora on the concept of a Pacific Era, entailed a spatial transversalization of Colombian development, spelled out in new plans for east-west road construction, ports, and even a new interoceanic canal (Pedrosa 1996). For the Pacific, the main consequence of this spatial shift was a reengineering of the development of western Colombia that disrupted significantly the biogeographic and cultural regime maintained by Afro-indigenous ecological and cultural practices. It is against this long historical backdrop that one can understand the actions of both development and social movement actors in and around this region.

The argument I want to make in this chapter is the following. Contemporary debates on development are subsidiary to discussions on globalization, on the one hand, and on modernity, on the other. The widely accepted thesis that globalization entails the universalization of modernity, where modernity is understood as a distinct mode of sociocultural organization that originated in Europe, leads to the conclusion that there is no outside to modernity, that from now on *it is modernity all the way*. This thesis takes various forms, each having different consequences for development. The dominant view is that Asia, Africa, and Latin America need to consolidate their modernity by pressing on with development through globalization. On the Left, the thesis that there is no outside to modernity suggests that modernity has to be overturned from the inside. A second perspective is contributed largely by anthropologists who advocate for the pluralization of modernity toward a conception of alternative modernities. These positions leave open two options for development: either its universalization as a cultural and economic logic, even if without the teleology that characterized it in the past; or development as a ceaseless negotiation of modernity in terms of the counterwork that local groups necessarily effect on the elements of development and modernity and toward more self-directed forms of modernity.

There are, however, a few sources that can be used as a point of departure to question the seemingly unanimous agreement on the universalization of modernity; such questioning results in a third scenario, that of alternatives to modernity. The main, but not the only, source of this option is a critical reinterpretation of modernity by a group of Latin American intellectuals from the perspective of coloniality. Modernity, according to these authors, cannot be understood without reference to the coloniality of power that accompanied it and that entailed the marginalization of the cultures and knowledge of subaltern groups. This conceptualization allows one to see how the local histories of European modernity have, since the conquest of America, produced global designs within which subaltern groups have had to live; it also makes understandable the emergence of subaltern knowledges and identities in the cracks of the modern colonial world system. These knowledges point at both a reappropriation of global designs by subaltern local histories and the possible reconstruction of local and regional worlds on different logics which, in their networked potential, may get to constitute narratives of alternatives to modernity.

The conclusion proposes that social movements (and, in different ways, policymakers and academics studying these actors) must hold in tension three coexisting processes and political projects: *alternative development*, focused on food security, the satisfaction of basic needs, and the well-being of the population; *alternative modernities*, building on the countertendencies effected on development interventions by local groups and toward the contestation of global designs; and *alternatives to modernity*, as a more radical and visionary project of redefining and reconstructing local and regional worlds from the perspective of practices of cultural, economic, and ecological difference, following a network logic and in contexts of power. That this framework is not entirely an intellectual fabrication but a plausible reading of the discourses and practices of communities, planners, and social movements is shown through an account of the avatars of three development projects in the southern Pacific in the 1990s: a coconut and cocoa commercialization cooperative, a cultural-cum-literacy project, and the biodiversity conservation project Proyecto Biopacífico. Part I of the chapter introduces briefly the framework of modernity/coloniality. It also summarizes recent debates on development, focusing on the notions of postdevelopment and counterwork. Part II presents the first two cases of counterwork, the cultural-literacy and cooperative projects. Part III provides an extended discussion of PBP, focusing on the project's transformation from a straight scientific and managerial approach to an ethnoterritorial perspective akin to that of *Agenda XXI*. The conclusion brings the argument full circle by restating the three dimensions of intellectual and political action regarding development today, namely, alternative development, alternative modernities, and alternatives to modernity.

I ◈ Globalization, Development, and Modernity

There are many ways to relate globalization, development, and modernity, according to the social processes that are highlighted and the theoretical frameworks used to interpret them. The review presented in this section does not pretend to be exhaustive. It is based on the most influential theories of modernity of recent decades. From the perspective of this chapter, the question, simply put, is, What is happening to development and modernity in times of globalization? Is development becoming naturalized, something that naturally will take place as part of globalization? Or is it rather recast as an explicit and still much-needed economic and cultural project? Is modernity finally becoming universalized? Or is it being left behind? The questions are all the more poignant because, from some perspectives, the present is a moment of transition: between a world defined in terms of modernity and its corollaries, development and modernization, and the certainty they instilled—a world that has operated largely under European hegemony over the past two hundred years if not more; and a new (global) reality which is still difficult to discern but which, at its opposite ends, can be seen either as a consolidating modernity the world over or, on the contrary, as a deeply negotiated reality that encompasses many heterogeneous cultural formations—and the many shades in between. The sense of transition is captured in Dirlik's (2001) question: Is globalization the last stage of capitalist modernity? Or the beginning of something new?

There are many ways to locate this transition. Some see the entire post–World War II period as a transition in the long history of the capitalist world system toward an as-yet-undetermined structure (e.g., Wallerstein 2000).[3] In political and economic terms I find it useful to locate its beginning in the late 1970s, with the inauguration of Thatcherism in England in 1979, followed by the Ronald Reagan and George H. W. Bush years and the definitive expansion of neoliberalism to most corners of the world. The first decade of this transition represented the apogee of financial capitalism, flexible accumulation, free market ideology, the fall of the Berlin Wall, the rise of the network society, and the so-called new world order. While this picture was complicated in the 1990s, neoliberal globalization still held sway. Landmarks such as the North American Free Trade Agreement (NAFTA), the World Trade Organization, Davos, Plan Puebla, and Plan Colombia were indications of the changing but persistent implantation of capitalist globalization. At the same time that a neoliberal global order was being instituted, however, signs of resistance appeared. In 1980, for instance, the foundation was laid for the creation of what became the United Nations Working Group on Indigenous Affairs, which was to be so influential in later years. In 1992, the United Nations Conference on Environment and Development (the Earth Summit in Rio de Janeiro) was an attempt to introduce an alternative imaginary to the rampant mercantilism then prevalent. For some, this meeting constituted the rite of passage to a transnation state, more negotiated and diverse than the proverbial "global village" under Western control (Ribeiro 2000).

From 1988, with the internationalized indigenous opposition to a large-scale dam on the Xingú river in Altamira, Brazil, the demonstrations against the General Agreement on Tariffs and Trade (GATT) in India in the early 1990s, the food riots in various Latin American capitals in the late 1980s and early 1990s, and the Zapatista uprising in Mexico to the large-scale demonstrations in Geneva, Seattle, Prague, Barcelona, Québec, and Genoa, the idea of a single, inevitable global order under the aegis of a capitalist modernity has been variously challenged. The result is that now—still within a transition period—it is possible to posit a different notion of globalization as made up of the actions and interests of many cultures, types of cultural politics, and imaginaries of social and natural life. The character of the transition was further muddled when, beginning with the first Gulf War but particularly after the attacks of September 11, 2001, and the invasion of Iraq of March 2003, there was a renewed attempt on the part of the U.S. elite to expand and consolidate its military and economic hegemony. This means that imperial globality requires an articulation of coloniality at a global level, even if taking specific forms for various groups and world regions.

Globalization and Modernity: Intra-European Perspectives ◈ The idea of a relatively single globalization process emanating from a few dominant centers remains prevalent. It is useful to review how this image arose in the most recent period and why it seems so difficult to dispel. At the root of the idea of an increasingly overpowering globalization is a view of modernity as essentially an European phenomenon. This assumption—found in thinkers like Habermas, Giddens, Taylor, Touraine, Melucci, Lyotard, and Beck as much as in Kant, Hegel, and the Frankfurt School philosophers before them—means that modernity can be fully explained by reference to factors internal to Europe. The views of Habermas and Giddens have been particularly influential. From this perspective, modernity may be characterized as follows:

1. *Historically*, modernity has identifiable origins: seventeenth-century northern Europe (especially France, Germany, England), around the processes of Reformation, the Enlightenment, and the French Revolution. These processes crystallized at the end of the eighteenth century.
2. *Sociologically*, modernity is characterized by institutions such as the nation-state and by some basic features, such as self-reflexivity, the continuous feedback of expert knowledge back into society, transforming it; the disembedding of social life from local context and its increasing determination by translocal forces; and space/time distantiation, or the separation of space and place, which means that relations between absent others become more important than face-to-face interaction (Giddens 1990). Modernity thus constitutes a new way of belonging in time and space, one that differentiates between past, present, and future (linear time and History), and that is tied to the spatiality of the nation-state above all.
3. *Culturally*, modernity is characterized by the increasing appropriation of previously taken for granted cultural backgrounds by forms of expert knowledge linked to capital and state administrative apparatuses (e.g., Habermas 1973, 1987). Habermas describes this process as the rationalization and colonization of the life-world. Modernity implies the creation of an order on the basis of the constructs of reason, secularization, the individual, expert knowledge, and administrative mechanisms–what Foucault (1991) called

governmentality and Vattimo (1991) the strong structures of modernity. Recently, Taylor (2004) has characterized modernity in terms of a social imaginary that privileges three social forms: the market economy, the public sphere, and the self-governing of people, with the forms of malaise they have brought about. Modernity also involves a way of being that, in contradictory fashion, highlights both perpetual change and the experience of the present—a dialectic of change and presence.

4. *Philosophically*, one may see modernity, on the one hand, in terms of the emergence of the notion of Man as the foundation of all knowledge and order of the world, separate from the natural and the divine (a pervasive anthropocentrism; Foucault 1973; Heidegger 1977; Panikkar 1993); on the other, in terms of the triumph of metaphysics and logocentrism, understood as a tendency—extending from Plato and some of the pre-Socratics to Descartes and the modern thinkers and criticized by Nietzsche and Heidegger, among others—that finds in logical truth the foundation for a rational theory of the world as made up of knowable and controllable things and beings (e.g., Vattimo 1991). For Vattimo, modernity is characterized by the idea of history and its corollary, progress and overcoming; Vattimo and Dussel point at the centrality of the logic of development to the modern order, including what Dussel (1996, 2000) calls "the developmentalist fallacy," or the idea that all countries have to move through the same stages and arrive at the same end state, by force if necessary.

On the critical side, the disembeddedness of modernity is seen to cause what Virilio (1999) calls global delocalization, including the marginalization of place (the here and now of social action) in the definition of social life. The underside of order and rationality implies the disenchantment that came about with secularization, the predominance of instrumental reason, and the normalization and disciplining of populations. As Foucault put it, "The Enlightenment, which discovered the liberties, also invented the disciplines" (1979: 222). Finally, modernity's anthropocentrism is seen by critics as connected with the phallogocentric cultural project of ordering the world according to rational principles from the perspective of a male Eurocentric consciousness. Although modernity did not get to constitute a total reality, it aimed at the purification of orders (separation between us and them, nature and culture), although inevitably producing its own hybrids of these poles and thus failing to reach its own goals (hence Latour's dictum "We have never been modern" [Latour 1993]).

Is there a logical necessity to conclude that the order sketched above—what Grossberg (2007) has called the West's commonsense view of modernity—is the only one capable of becoming global? For most theorists, there is. Giddens (1990) has made the argument most forcefully: globalization entails the radicalization and universalization of modernity. No longer purely an affair of the West, however, since modernity is everywhere, the triumph of the modern lies precisely in its having become universal. This may be called the Giddens effect; by this I mean a discursive effect that says, *From now on, it's modernity all the way, everywhere, until the end of time.* Not only is radical alterity expelled from the realm of possibilities, but all world cultures and societies are reduced to being a manifestation of European history. The Giddens effect seems to be at play in most works on globalization. No matter how variously qualified, a "global modernity" is here to stay. It might be seen as hybridized, contested, uneven, heterogeneous, even multiple, or as conversing with, engaging with, or processing modernity, but in the last instance these modernities end up being a reflection of a Eurocentered social order, under the assumption that modernity is now a ubiquitous, ineluctable social fact.[4]

Might it be, however, that the power of the dominant Euro-modernity—as a particular local history—lies in the fact that is has produced particular global designs in such a way that it has subordinated other local histories and their corresponding designs? Furthermore, as Mignolo (2000) suggests, might it be possible to think about, and to think differently from, an "exteriority" to the modern world system? Might it be possible, finally, to envision alternatives to the totality imputed to modernity and to posit a project that builds on but goes beyond the insight of alternative modernities to adumbrate the possibility of alternatives to modernity, not in the sense of a totality leading to different global designs, but as a network of glocal histories constructed from the perspective of a politically enriched alterity? This latter possibility, I argue, may be gleaned from the work of a group of Latin American theorists that, in refracting modernity through the lens of coloniality, unfreezes the radical potential for thinking from difference and toward the constitution of alternative local and regional worlds.[5]

II ◈ An Exteriority to the Global System? The Modernity/Coloniality/Decoloniality Research Program

The framework of modernity/coloniality/decoloniality (MCD) is grounded in a series of operations that distinguish it from established theories of modernity (see Escobar 2003 for a more complete account). These include the following: (1) locating the origins of modernity in the conquest of America and the control of the Atlantic after 1492, rather than in the commonly accepted landmarks, such as the Enlightenment or the end of the eighteenth century;[6] (2) a persistent attention to colonialism and the making of the capitalist world system as constitutive of modernity; (3) consequently, the adoption of a world perspective in the explanation of modernity, in lieu of a view of modernity as an intra-European phenomenon; (4) the identification of the subordination of the knowledge and cultures of groups outside the European core as a necessary dimension of modernity; (5) a conception of Eurocentrism as the knowledge form of modernity/coloniality—a hegemonic representation and mode of knowing that claims universality for itself and that relies on "a confusion between abstract universality and the concrete world hegemony derived from Europe's position as center" (Dussel 2000: 471; Quijano 2000: 549; Castro-Gómez 2005).

The main conclusions of this conceptualization are, first, that the proper unit for the analysis of modernity is modernity/coloniality—in sum, there is no modernity without coloniality, modernity/coloniality encompassing modern colonialism and colonial modernities (in Asia, Africa, Latin America, and the Caribbean). Second, the fact that "the colonial difference"—the differences suppressed by Eurocentrism that assert themselves today with social movements at the borders of European modernity—is a privileged epistemological and political space. The great majority of European theorists (particularly those "defenders of the European patent on modernity," as Quijano mockingly calls them [2000: 543]) have been blind to this colonial difference and the subordination of knowledge and cultures it entailed. Here is a further characterization of coloniality by Mignolo (quoted in Escobar 2004a: 218):[7]

> Since modernity is a project, the triumphal project of the Christian and secular west, coloniality is, on the one hand, what the project of modernity needs to rule out and roll over, in order to implant itself as modernity and, on the other, the site of enunciation where the blindness of the modern project is revealed, and concomitantly also the site where new projects begin to unfold. In other words, coloniality is the site of enunciation that reveals and denounces the blindness of the narrative of modernity from the perspective of modernity itself, and it is at the same time the platform of pluri-versality, of diverse projects coming from the experience of local histories touched by western expansion (as the Word Social Forum demonstrates).

The question of an exterior to the system is peculiar to this group and is easily misunderstood. In no way should this exteriority be thought of as a pure ontological outside, untouched by the modern. Exteriority refers to an outside that is precisely constituted as difference by a hegemonic discourse. This notion of exteriority arises by thinking about the Other from the ethical and epistemological perspective of a liberation philosophy framework: the Other as oppressed, as woman, as racially marked, as excluded, as poor, as nature. By appealing from the exteriority in which she or he is located, the Other becomes the source of an ethical discourse vis-á-vis a hegemonic totality. This interpellation of the Other comes from outside the system's institutional and normative frame, as an ethical challenge. This challenge comes from

> the interpellation which the majority of the population of the planet, located in the South, raises, demanding their right to live, their right to develop their own culture, economy, politics, etc. . . . There is no liberation without rationality; but there is no *critical* rationality without accepting the interpellation of the excluded, or this would inadvertently be only the rationality of *domination*. . . . From this negated Other departs the praxis of liberation as affirmation of the Exteriority and as origin of the movement of negation of the negation. (Dussel 1996: 31, 36, 54)[8]

This is precisely what most European and Euro-American theorists seem unable to envision, that is, the possibility of overcoming modernity by approaching it from the colonial difference. Critiques of modernity, in short, are blind to the epistemic difference that becomes the focus of the MCD framework. As Walsh put it, MCD "is useful, on the one hand, as a perspective to analyze the hegemonic processes, formations, and orders associated with the world system (at once modern and colonial), and, on the other, to make visible, from the colonial difference, the histories, subjectivities, forms of knowledge, and logics of thought and life that challenge this hegemony" (2007: 104).

For Dussel, the corollary is the possibility of "transmodernity," defined as a project for overcoming modernity not simply by negating it but by thinking about it from its underside, from the perspective of the excluded other. For Mignolo, the project has to do with the rearticulation of global designs by and from local histories. All of these ideas have led the MCD network to add "decoloniality" recently as a third leg to the collective project, making it into the MCD perspective. Many insist on the need to work with social movements operating from the colonial difference toward political projects grounded in a degree of autonomy on the social, cultural, economic, and epistemic spheres, including those movements that explicitly discuss a *pensamiento propio* (the movement's own thought), such as PCN and the Zapatista. These struggles are see as political and epistemic in character, as attempts at thinking other thoughts for other world constructions.

In short, the MCD perspective provides an alternative framework for debates on modernity, globalization, and development. MCD shows the shortcomings of the language of alternative modernities in that it incorporates everybody's project into a single, albeit diverse, project, thus missing the potential of the subaltern perspectives, for even in their hybridity subaltern perspectives are not necessarily about being modern but are heteroglossic and plural. In highlighting the developmentalist fallacy, MCD provides a context for interpreting the various challenges to development and modernity as so many projects that are potentially complementary and mutually reinforcing.

Development, Postdevelopment, and Beyond ◈ Development and modernization can be seen as the most powerful global designs that arose out of the local history of the modern West in the post–World War II period. In the process of becoming a global design, however, not infrequently it was transformed into local projects articulated from the perspective of subaltern histories and interests. The language of alternative modernities is an attempt to make visible this transformative potential of local histories on the developmental global design. It also points the way for a reconceptualization in terms of alternatives to modernity.

Over the past fifty years, the conceptualization of development in the social sciences has seen three main orientations: modernization theory in the 1950s and 1960s, with its allied theories of economic growth and development; dependency theory in the 1960s and 1970s (and related perspectives, such as Samir Amin's powerful framework of peripheral capitalism [1976]); and the poststructuralist critique of the development discourse. Modernization theory inaugurated a period of certainty in the minds of most world elites, validated by the promises of capital, science, and technology; this suffered its first blow with dependency theory, which argued that the roots of underdevelopment were to be found in the connection between external dependence and internal exploitation, not in any alleged lack of capital, technology, or appropriate cultural values. For dependency theorists, the problem was not so much with development as with capitalism. In the 1980s, poststructuralist critics questioned the very idea of development. They analyzed development as a discourse of Western origin that operated as an overarching mechanism for the cultural, social, and economic production of the third world. These three theoretical orientations may be classified according to the root paradigms from which they emerged: liberal, Marxist, and poststructuralist theories. Even if today these orientations overlap, they may be distinguished for purposes of analysis and in order to disentangle current debates (see table 4).

The deconstruction of development led poststructuralists to postulate the possibility of a postdevelopment era, defined as one in which development would no longer be the central organizing principle of social life. This notion could be restated today in terms of the construction of forms of globality that, while engaging with modernity, are not necessarily modernizing or developmentalist, precisely because they are built from the colonial difference. In the second half of the 1990s, the poststructuralist analyses became themselves the object of poignant rebuttals. This may be seen as a fourth moment in the historical sociology of development knowledge. Many of these works were directed against what was described as "the postdevelopment school." While these critiques do not constitute a unified body of work, it is possible to identify three main objections: (1) the postdevelopment critiques presented a homogeneous view of development and the development apparatus, while in reality there are vast differences among development strategies and institutions; (2) they romanticized local traditions and movements, ignoring the fact that the local is also embedded in power relations; and (3) they failed to notice the ongoing contestation of development on the ground. As I have indicated elsewhere (Escobar 2000, 2007), while most of these critiques are very pertinent, often the disagreements can be explained in light of contrasting paradigmatic orientations and by considering the changed context of knowledge production of the 1990s.[9]

A number of authors are taking a pragmatic approach by constructively incorporating aspects from the various trends and paradigms. Arce and Long (2000), for instance, have outlined a project of pluralizing modernity by focusing on the counterwork performed on development by local groups. Bebbington (2000, 2005; Bebbington and Batterbury 2001) has called for a notion of development that is at once alternative and developmentalist, critical and practicable, focused on the concept of livelihood, and understood in terms of place-based dynamics that occur within larger spatial processes, including the organizations making up aid chains. The gap between aid agencies and local organizational possibilities needs to be bridged to give appropriate support to livelihood issues. Grillo and Stirrat (1997) take their critique of postdevelopment as a point of departure for a constructive redefinition of development theory and practice. Fagan (1999) has suggested that the cultural politics of postdevelopment has to

Table 4. Development Theories According to Their Root Paradigms

Issue	Paradigm		
	Liberal theory	*Marxist theory*	*Poststructuralist theory*
Epistemology	Positivist	Realist/dialectical	Interpretivist/ constructivist
Pivotal concepts	• Individual • Market	• Production (e.g., mode of production) • Labor	• Language • Meaning (signification)
Objects of study	• "Society" • Market • Rights	• Social structures (social relations) • Ideologies	• Representation/ discourse • Knowledge-power
Relevant actors	• Individuals • Institutions • State	• Social classes (working classes, peasants) • Social movements (workers, peasants) • State (democratic)	• "Local communities" • NSMS, NGOS • *All* Knowledge producers (including individuals, state, SMS)
Question of development	How can societies develop / be developed through a combination of capital and technology and individual and state actions?	• How does development function as a dominant ideology? • How can development be delinked from capitalism?	How did Asia, Africa, and Latin America come to be represented as "underdeveloped"?
Criteria for change	• "Progress," growth • Growth plus distribution (1970s) • Adoption of markets	• Transformation of social relations • Development of the productive forces • Development of class consciousness	• Transformation of political economy of truth • New discourses and representation (plurality of discourses)

(Table 4. continued)

Issue	Paradigm		
	Liberal theory	Marxist theory	Poststructuralist theory
Mechanism for change	• Better theories and data • More carefully tailored interventionism	Social (class) struggle	Change *practices* of knowing and doing
Ethnography	• How culture mediates development and change • Adapt projects to local cultures	How local actors *resist* development interventions	How knowledge producers resist, adapt, subvert dominant knowledge and create their own.
Critical attitude concerning development and modernity	Promote more egalitarian development (deepen and complete Enlightenment Project of modernity)	Reorient development toward satisfying requirements for social justice and sustainability (critical modernism: delink capitalism and modernity)	Articulate ethics of expert knowledge as political practice (alternative modernities and alternatives to modernity; decolonial projects)

start with the everyday lives and struggles of concrete groups of people, particularly women, thus weaving together Marxist and poststructuralist proposals; Diawara (2000) implicitly makes a similar point by advocating a consideration of the varieties of local knowledge that are present in the development encounter. The relation between postdevelopment, feminism, and postcolonial theory has been another focus of fruitful discussion. Sylvester (1999) warns about the effect on our accounts of the world of our distance from those we write about; she advocates building connections between postcolonial theory and postdevelopment as a corrective to this problem and as beneficial to both. Other authors find in gender a privileged domain for weaving together elements of postdevelopment, postcolonial theory, and feminism into a new understanding of development (e.g., Parpart and Marchand 1995; Gardner and Lewis 1996;

Schech and Haggis 2000; Bhavnani, Foran, Priya Kurian, eds. 2003), with some maintaining an emphasis on ethnocentrism and the exclusion of women in developmentalist representations.

Echoing the ethnographies of modernity (see note 4, pp. 346–47), Arce and Long (2000) focus on the ways in which the ideas and practices of modernity are appropriated and reembedded in local life-worlds, resulting in multiple, local, or mutant modernities. Their key notion of counterwork involves a number of dimensions. First, a collective appropriation by social groups of the dominant practices of modernity into a shared cultural background that changes as a result.[10] This entails a dynamic repositioning of hegemonic elements within familiar contexts that involve the reassembly of elements, and a disembedding and reembedding of Western standards within local representations of social life (including what counts as modern). What the notion of counterwork adds to the ethnographies of modernity is the idea of an ongoing, endogenously generated task performed on incoming messages, elements, information, etc. that transforms what one thinks of as the modern and the traditional. Such continuous processing is no longer a hybridization of distinct cultural strands, but a series of self-organizing mutations driven by internal dynamics, even if often propelled by outside interventions. In the end, counterwork may result in new power claims in terms of deessentializing Western products of their superior power or the empowering of a group's self-definition, even as it changes.[11]

Recent approaches to the ethnography of development projects have similarly suggested that ethnographic research could lead to an understanding of both the social work of policy ideas and the transformation that locals effect on the projects and that this understanding could be utilized to link more effectively "the emancipatory intentions of policy and the aspirations and interests of the poor" (Mosse 2005: 240). This goal requires a detailed understanding of the relation between policy and practice as it is played out at many sites by a diversity of actors; interestingly, this understanding needs to be multipositioned in addition to multisited, the anthropologist acting as part insider and part outsider in several of these sites. The hope is that, given the reality of development, the critical ethnographer could illuminate the conditions for a more effective popular appropriation of the projects. Sinha (2006a, 2006b) has underscored that this process of appropriation also goes on at the national level, where political imperatives are crucial for negotiating development agendas; these agendas, indeed, have multiple lineages, some of which might even have little to do with Western intervention per se.

Closer attention to the interaction between state and civil society organizations, he adds, should give us a more nuanced account of the flows of power than in previous poststructuralist analyses, underscoring how development operates as a multiscale, hegemonic process that, as such, is transformed and contested all the time.[12]

Another meaningful word of caution has been provided by de Vries (2007). Most critiques of development, he argues, have overlooked the desire for development that many poor people have, precisely as a result of the same development apparatus dissected by poststructuralists; these desires exist in the gap between promises and meager realizations, "thus giving body to a desiring machine that also operates in between the generation and banalization of hope" (30). Treating development as an object of discourse, or chiefly in terms of governmentality, amounts to a disavowal of the subjectivity of people. Moreover, in de Vries's view, many critics fall into a position of cynicism, as if saying, "We know development is an apparatus of power, and we do not believe in it, and yet we do it anyway for pragmatic reasons." By acknowledging that development is a desiring machine and not only an apparatus of governmentality, he suggests, one might find new ways to radicalize development theory. This is so because "the desire for development both masks its impossibility and reveals a utopian dream" that goes against the historical project of capitalist modernization (37). It is thus important to see in people's desires a refusal to accept the banalization of development by the anti-politics machine, which implies "an ethics of sustaining the capacity to desire," to keep searching for what is in development more than itself, for demanding what development offers but cannot deliver (40). This insightful theoretical-political critique of Foucaultian approaches can be linked to Gibson-Graham's call for cultivating subjects of economic difference (see discussion in chapter 2). How could the very development apparatus be used to cultivate subjects of diverse developments and diverse modernities?

This question is adumbrated by Medeiros (2005) within the framework of coloniality. Local expressions of the desire for development, as she shows in her detailed analysis of rural development projects implemented in the Bolivian highlands with the support of the German Development Agency (GTZ), need to be seen against the background of the complex history of several hundred years of discrimination, including the experience of promises made and never fulfilled since independence. In the absence of this analysis, and despite their good intentions, development projects often end up reproducing old power and knowledge asymmetries. Even the most enlightened expert-driven participatory projects reenact the conditions of coloniality, including knowledge and ethnic hierarchies. As she also shows, indigenous peasants have their own situated understanding of development; it articulates their historical experience of modernity and coloniality. The local notion of development includes the acquisition of those tools of dominant knowledge systems that might empower them to envision and implement a viable future. However, when peasants speak of development as "awakening," this is not an expression of a demand for development; it is a subversive appropriation of a never-fulfilled promise. In other words, local talk about development is not so much about development per se as about history and culture—about the state, citizenship, difference, knowledge, and exploitation. It is about the colonial difference, namely, the communities' position within the modern colonial world system.[13]

If the Giddensian thesis suggests that globalization subsumes development—development becomes naturalized, even if without its teleology and without assuming any single model, precisely because all people come to desire the same things, namely, commodities and markets—Arce and Long imply that every act of development is at least potentially an act of counterdevelopment and that every act of counterdevelopment is potentially an alternative modernity—a modernity from below (2000: 21). If PLADEICOP and Plan Pacífico sought the implantation of a dominant Euro-Andean modernity in the Pacific, the Proyecto Biopacífico incorporated a logic of plural modernities. In a few of PLADEICOP's programs a logic of counterwork was even set in motion.

III ◈ Counterwork in the Pacific: The Dynamic of Cooperatives and Literacy Projects

Rural Development vs. Autonomous Cooperatives: Coagropacífico and the Struggle for Alternative Development ◈ The imaginary of cooperatives (*cooperativismo*) has had a powerful appeal in Latin American development for many decades, particularly in relation to rural development and peasant struggles.[14] The original cooperative model goes back to nineteenth-century England, and its mechanical application in Latin America was denounced in the 1960s as a form of intellectual colonialism (Fals Borda 1970). Promoted by the state as part and parcel of agrarian reform and integrated rural development programs, cooperatives became a tool of modernity—a form of rationalizing the production and commercialization of agricultural goods by small farmers. The cooperative movement arrived in the Pacific in the 1960s on the heels of the agrarian reform program. Early efforts at establishing cooperatives for the production and commercialization of coconut and African palm in the 1960s—for instance, in Guapi and Timbiquí—generated both strong interest and resistance on the part of local *agricultores*. The official demarcation of private holdings, the use of chemical inputs, seeds, and new methods of planting, and credit and accounting were all practices that were new to the farmers, and they were greatly resisted. Despite this fact and despite repeated failures, the imaginary of cooperatives caught on in the Pacific as a means of channeling resources to small producers and improving living standards (Grueso and Escobar 1996). Well-known cooperative institutes from the interior of the country (province of Santander) continued to train leaders, including those who were to work in the Pacific in the 1990s.

In 1986, a technical cooperation program between the Colombian and Dutch governments was started in the Tumaco area as a component of rebuilding efforts after the devastating earthquake of 1979. With the Autonomous Regional Corporation of the Cauca (CVC) as implementing agency, the resulting Convenio CVC–Hollanda identified a series of deficits in the agricultural development of the region, including lack of infrastructure and technical assistance, inadequate commercialization, and the lack of community participation in development. This diagnosis resulted in a project designed to promote the commercialization of coconut and cocoa among agriculturalists of the five rivers of the Tumaco region. From an initial group of 15 farmers, the project went up to 120 in a couple of years, at which time it was decided, with the participation of the farmers, that the project should adopt the cooperative model. The Cooperativa Multiactiva Agropecuaria del Pacífico, (Coagropacífico), was formally established in November 1989 with the aim of contributing to the well-being of the community through the rationalization of the production and commercialization of coco, cocoa, and plantain. Storage facilities were built on each of the five rivers, with a larger facility in Tumaco. Commercialization boards were created on each river, with a small number of local functionaries in Tumaco in charge of the overall management of the project. By 1993, Coagropacífico was selling 18,000 dozen coconuts per month and an unspecified amount of cacao, which it purchased from about 400 farmers. In 1996 and 1997, the amount of coconuts sold brought close to $250,000 per year.

From its inception, Coagropacífico visualized a three-step process that would end up in a member-controlled organization independent of external funding and expertise. An initial two-year period, in which the cooperative had high input of outside funding and expert advise, was to be followed by a period of comanagement, during which all activities would be decided upon collaboratively by the farmers, experts, and staff; this effort was to culminate in a stage of self-management in which local people would have been trained to fulfill all aspects of the management of the enterprise, from accounting to the links to outside markets. All development projects in the Pacific up to that point had failed to become self-sustaining, rapidly disappearing after the funding dried out and the state institution (or NGO) left. There was no precedent in the Pacific for what Coagropacífico staff referred to as *la transición*, which was to be completed by 1993 and would lead to local control of the project. This feature attracted the attention of local farmers and of the nascent group of activists working at the time on AT-55. Faced with the expansion of the African palm frontier, farmers found in Coagropacífico a way to hold on to their lands. The mostly urban-based Tumaco activists, on the other hand, found in the project a means to make inroads into the rural areas, the raison d'être of their strategy; some farmers and Coagropacífico leaders actually started to take part in the organizing process, particularly in PCN's *palenque* (regional organization) for Nariño. This climate pushed the CVC and local project staff alike to become invested in a successful transition, despite resistance by some at CVC and the doubts of many in the cooperative about their ability to keep the enterprise going on their own.

In this context, Coagropacífico leaders began to think about the need for communities to control external experts, train their own, and politicize their approach. At this level, a lot was achieved. An evaluation carried out in 2000 (Cordeagropaz 2000) highlighted the following achievements, besides the immediate positive effects on production and family incomes: increased leadership and organizational capacity; opening up of channels for participation in institutional spaces; awareness of family farming as a planned entrepreneurial process; training (*capacitación*) in a variety of skills, from pest control to farm management and community organization; and a sense of self-esteem among the agricultores and their families. On the negative side, the paternalism of the CVC, lack of diversification of products, failed credit schemes, and insufficient presence of the leaders in the communities, among others, were cited. The self-evaluation was contradictory. The project itself was contradictory in various ways. First, it pushed farmers to abandon their long-standing cultures of cultivation in favor of a new set of practices; these involved novel categories, such as *conocimiento técnico, planeación integral, capacitación, rentabilidad* (technical knowledge, comprehensive planning, training, profitability), and even labor and resources. For instance, farmers customarily cultivated several plots, often separated from each other; experts advised them to work on only one plot close to the household, on which all the technical interventions would be focused. While farmers followed these instructions for the most part, there was some resistance. In the Rosario river, for instance, one found farmers saying that they still visited their distant plots because "if we do not visit them they become sad and become attacked by pests" or that the land had to be "caressed," and "benefited" (*hacerle un beneficio*), thus evincing the existence of a local model of the land, much as Gudeman and Rivera (1990) demonstrated in other parts of Colombia. Another important contradiction was a dependence on the external institution despite the rhetoric and desire for full independence. By the time the CVC started its final withdrawal in 1995, the difficulties of independent life had become evident. Coupled with the importation of cheap coconut from Venezuela, this meant that the cooperative faced stringent financial troubles. Its survival for a number of years after that, however, reinforced the idea that it was possible to construct development under local control, if the communities indeed appropriated the projects as their own, if leaders put collective interests above personal ones, as was the case generally with Coagropacífico, and if locals finally "learn to put their trust in what is really ours."[15]

The Coagropacífico experience exemplifies the first type of project emerging from the relation between globalization, development, and modernity I wish to highlight, that of alternative development. Unlike mainstream development (e.g., macrodevelopment projects, African oil palm, or strategies such as Plan Pacífico), alternative development implies a level of contestation over the terms of development but without challenging its underlying cultural premises. While informed by idioms of progress and rational decision making, alternative development entails a struggle over the running of projects to reduce the control by experts and socioeconomic elites (whether it produces the same kind of experts and elites is another matter, but it need not). Alternative development projects might be most effective when focused on easily identifiable goals, such as income generation, production increases, or defense of particular ecosystems. A focus on livelihoods might be particularly conducive to alternative development (Bebbington 2000), especially in cases where it is linked to the social movement concept of *autonomía alimentaria* (food autonomy), and at this level it might enable a significant critique of mainstream development (e.g., as a strategy to protect the livelihood of the majority against the expansion of the African palm frontier). The contestation at stake, however, involves the counterwork visualized by Arce and Long only in a limited manner, which gives rise to the second project, that of alternative modernities. My second example illustrates this possibility.

Beyond Literacy: Popular Communications and the Early Pursuit of an Alternative Modernity in the Southern Pacific ◈ The Pacific of Colombia and Ecuador is a land of rich oral traditions and literature. There is not only a vast repertoire of oral forms, from poetry to storytelling, from songs to riddles, but also a great joy in inventing yet a new verse, in retelling a story in a way that makes it come alive, bringing awe and enjoyment to the listeners. Many of these forms and legends are based on Spanish models brought with colonization, but they have been adapted creatively by black people to their traditions and to local conditions; orality was also conditioned and influenced by the African heritage and by the lack of access to reading and writing during the period of slavery and most of the Republic. For the Timbiquí poet and expert in oral traditions Alfredo Vanín, in the oral world "the spoken language acquires a magical dimension that reaches unsuspected proportions. . . . The colonization through language engendered a kind of counter-language in the Pacific; the [Spanish] language was imposed; however, as a cultural reaction, other language was created in order to express the desire for liberty and belonging" (1996: 47). Oral tradition "involves norms of social and interpersonal behavior; it interprets the multiple and changing forms of life, and is ludic above all. It creates new premises for understanding the world, recreating the possibilities of existence . . . and it is in it that the archaic finds its utmost expression, and where modernity is assimilated, as indicated in a number of stories and *décimas*" (62).[16] These oral traditions have been increasingly permeated by modern forms, from the state and the economy to religion, education, migrants, and urbanization. Whereas oral literature is always the subject of change and innovation, the advent of modern media has posed new challenges and opportunities for this profoundly oral culture.

It was precisely to meet this challenge that the program discussed in this section, Gente Entintada y Parlante (GEP; literally, "people of the ink and the word"), was developed. Every effort at development and modernity involves literacy. For PLADEICOP, the high levels of illiteracy in the Pacific had to be corrected. In 1986 PLADEICOP approached a group of faculty at the Universidad del Valle in Cali to this end. The group deemed that a conventional literacy program was out of the question. In the 1980s, it was clear to many in Latin America that such programs uprooted people from their cultures in order to integrate them into an ideal project of modernization and that even in these terms conventional *alfabetización* (literacy) had largely failed. These were the post-Freire years, after almost two decades of a radical popular education movement, participatory action research, and intense politicization of the adult education movement in many parts of the world, particularly Latin America. A popular communications movement was also sweeping the country at that time, and together these two forces led to the formulation of an approach to literacy based on the appropriation of communications media by local groups. This was not to be a high-tech approach, however. Technologies had to be adapted to local conditions, with the overall goal of strengthening local cultures. The result was a powerful popular communications movement that engulfed many communities in the Pacific, especially the southern part, the effects of which can still be felt today, two decades later.[17]

What were the main tenets of the popular communications approach to literacy? How did this approach result in the GEP project? And how did GEP find its way into the Pacific? The starting point was the belief that "marginal and incipiently lettered cultures," such as the oral cultures of the Pacific, are not only active in enriching their traditions but capable of assuming control (*gestión*) over their literacy if given appropriate support. The approach had to be located within the daily lives of the communities so as to overcome the traditional subject-object relation peculiar to conventional literacy and development. At one level, what this support entailed was surprisingly simple: on the one hand, a ready-made supply of communications material, particularly paper, ink, cassette recorders, and simple rudimentary printing facilities; on the other, a basic information technology infrastructure for the planning and self-management (*auto-gestión*) of the circulation of the printed and recorded material on a self-sustaining basis, including paper supply, archives, filing systems, and so forth. The belief was that the very availability of materials in places where there was such a dearth of them would challenge popular actors, local artists, activists, and organizations to assume responsibility for their literacy and to produce their own communications materials. While the results were poor at first, the ensuing local editorial projects would become ever more legible and complex (Pedrosa 1989a).

The GEP project validated these basic principles. By the late 1980s, numerous groups had appeared in many parts of the southern Pacific, producing all kinds of materials of increasing sophistication and beauty. Narrators, singers, folklorists, writers, illustrators, poets, and printers started to establish communications relations with local farmers, store owners, artisans, fishermen, women shell collectors, miners, and small entrepreneurs and to depict their lives in color, ink, and printed and spoken words. This production was often guided by "thematic research" (originally a Freirean concept of investigating key local problem areas of reality around which a collective project of transformation could be elaborated). Among the first products to appear were drawings done on paper out of woodcuts carved in linoleum and color inks, simply but beautifully executed and often carved by illiterate people; they depicted everyday life in river, forest, and sea. These linoleum prints, of many sizes, circulated widely in GEP areas, attracting much attention; they were soon followed by loose sheets containing information about local problems, printed pages with local tales and traditions, and radio programs taped and broadcasted by low-power (e.g., 1 kilowatt) radio stations, also introduced by the GEP project. Over time, the movement promoted more ambitious programs, including video and an international festival devoted to the most important black musical form of the Pacific, the *currulao*.[18]

By "relocating the media laboratories and workshops within the very same context where marginality is lived" (Pedrosa 1989a: 51), then, the GEP made possible an active and participatory movement. Placing the workshop in the midst of communities (á la Latour in his analysis of Pasteur, 1988) enabled the production of literacy, understood as the construction of a literary culture specific to the Pacific and its "cultural reserves," to use Pedrosa's term. It activated a process, led by young people but with the incorporation of community members of all walks of life, by which people produced their own writings, ideas, and imaginations. In theory, a decentralized support system would ensure the functioning of the system. Neighborhood nuclei (say, in small river settlements) would be supported by zonal nuclei (e.g., in Tumaco or Guapi), and these by regional nuclei (Cali). In time, it was expected that the system would become self-sustaining as it marketed its editorial services and products. The result would be an entire "popular communications network" (*red de comunicación popular*) that would embody a complex sociolinguistic conception of literacy through a relatively simple architecture, enacting a geography of communications (Pedrosa 1989b). The system worked relatively well for some years. A number of popular communications foundations were established in some communities, local radio stations were created, and Fundación Habla/Scribe in Cali functioned as a regional support organization. All of this demanded great creativity in terms of the adaptation and utilization of limited resources for Pacific littoral conditions.[19] The result of this innovative conception of literacy was that "the frequent contact with multicolor inks and electronic sound, with paper, recorders and amplifiers turned the *entintados y parlantes* into important cultural actors in the Pacific, thus creating the means to construct and inhabit literacy" (Pedrosa 1989a: 65).

GEP was part of PLADEICOP's "basic social services" strategy, funded by UNICEF, the main objective of which was to support development. Popular communications were thus conceived as a development tool. In many ways GEP thus amounted to "popular communications without 'people'" (Rivas 2001). With rising ethnic and political mobilization (e.g., AT-55), a more dynamic appropriation of the popular communications movement started to take place. In part because of this, in 1991 Fundación Habla/Scribe was asked by CVC to relinquish control of the program. By then, however, the popular communications movement had spread well beyond the GEP itself, and the logic of *entintar* pervaded other domains.

Many of the GEP products played on a valorization of blackness, particularly its pictorial representation in ways that were seen as being more realistic portrayals of black people. As one of the foremost cultural activists participating in these approaches, the Tumaco artist and communicator Jaime Rivas said, until then, "representations of blacks repeated an iconography and humorist tradition that reduced black men and women to a black ink spot from which one could distinguish naïve and clumsy eyes, and the appearance of black people resembles more that of chimpanzees than human beings" (2001: 15). Despite resistance and controversy in the region over how to represent blackness (especially in relation to certain physiognomic features, such as the lips), locals took on the challenge of producing for the first time their own representation of themselves. At that point, the project became a space for the construction and strengthening of identity, promoting a more endogenous self-image, even if still in ambiguous terms vis-á-vis images of whiteness. In a similar vein, the strategic value of literacy was recognized; only by assuming literacy from an autonomous position, it was realized, only by asking oral tradition to tackle the question of literacy could such a tradition hope to survive and, indeed, thrive in its encounter with modernity and with a development project that was just beginning to make its presence felt.

In sum, while GEP generated communications for development, it also enabled a type of communications that questioned it. This counterwork on development originated in the cultural appropriation of the project by cultural activists and locals. Activists attempted to create an intercultural relation with the development project, one that passed through communication with themselves as different, as black. The movement of the entintados y parlantes also put pressure on the institutions to communicate differently with their target populations. In those cases in which the GEP was able to establish a strong basis in the communities, it became more than a development tool—it enabled expressions of cultural autonomy and an alternative vision of development. This was particularly true in the few cases in which the local group became associated with the ethnoterritorial organizing process that became prominent after 1991 (e.g., Fundación El Chigualo in Barbacoas, Nariño). In these cases, the project promoted a cultural and political approach to literacy centered on cultural rights and the territory. Celebrating its first ten years of existence in 1997, Fundación Habla/Scribe could characterize the communications production in the Pacific as taking place "in the framework of the Pacific's own cultures and problematic," thus breaking the long-standing invisibility through creative communications strategies (Ariza 1997). One of the twelve beautiful color posters that made up the printed collection on the rights of the black communities put it this way:

> There already exist in the Pacific multiple experiences with alternative communications that enable popular communicators to disseminate the utopia of life to which the communities themselves and the new generations aspire. The democratization of language [*la palabra*] and other means of expression has become a necessity so that the future is not imposed on us by centralizing institutions or by national or international interests that only seek to plunder our natural and cultural wealth, depriving the ancestral indigenous and Afro-Colombian populations of their rights over their territories and of everything they have constructed as a culture.[20]

In conclusion, the GEP project can be seen as an alternative modernity. In contradistinction with the alternative development case, GEP enacted a cultural challenge to the Euro-Andean vision of the Pacific and to the logocentric vision of development. The alternative modernity dimension of the relation between globalization, development, and modernity involves, in this way, a more significant contestation of the very aims and terms of development on the basis of an existing cultural difference and place-based subjectivities. An alternative modernity experience may or may not be explicitly linked to an ongoing political project or movement and may be enabled to a greater or lesser extent by a given political climate. As the GEP case illustrates, although the project was not directly linked to the ethnoterritorial political project of the social movements (with a few exceptions), the surrounding political character of the moment might have propelled GEP into a more openly cultural-political register. One may say of this experience, with Alfredo Vanín (1996: 65) in referring to modernity as a whole, that "the phenomena of modernity, the tastes of modernity, and the media of modernity seem only to navigate above guarded waters that hide old cultural contents that, not very clearly visible, are thus difficult to destroy." Or, as Gupta (1998: 9, 37, 104) says, that alternative modernities involve both the presence of development in the local imaginary and the fact that the "non-Western," far from being a vanishing tradition, is a constitutive feature of modern life. One does not have to rely on an idiom of a "primeval bed rock" (Vanín 1996: 63) that would still be alive in the Pacific to entertain the idea that a cultural background is at play here, enabling an endogenous process of counterwork that at the very least creates the possibility of alternative modernities. The biodiversity conservation project for the Pacific (1992–98) illustrates the trials and tribulations of a more explicit articulation of such a cultural background into an alternative conservation and development strategy. As we will see in the next section, while it did not succeed in claiming an alternative to modernity, it was moving in that direction.

IV ◈ Counterwork and Beyond: The Biodiversity Conservation Project for the Colombian Pacific

The biodiversity conservation project for the Pacific, Proyecto Biopacífico (PBP) was one of the first of its kind to be implemented in Latin America. Conceived within the guidelines of the Convention of Biological Diversity and the Global Environment Facility, it started out in 1992 with a conventional scientific framework. Although from the very beginning the project called for community participation and the consideration of local forms of knowledge and cultural practices, there was little of either over the first two years of the project, as an external evaluation mission determined in 1995. Coupled with mounting internal and external pressures, the evaluation triggered a reformulation of the project's conceptualization, the most salient feature of which was the ever-increasing role of black and indigenous activists at all levels of decision making and implementation. As a result, the project underwent a substantial mutation, from a conventional scientific and managerial conservation approach to a deeply negotiated project that crafted an alternative conceptualization of biodiversity. This conceptualization, as we began to see in chapters 1 and 3, emphasized cultural and territorial rights, on the one hand, and the project's location within an ethnoterritorial approach to social movement organizing, on the other. A closer ethnographic look is instructive of how the project was initially conceived (1992–95); of how the process of negotiation with social movements and the concomitant alternative conceptualization of biodiversity (1995–97) came about; and of how the project finally was eclipsed and transformed into more mainstream conservation policies and strategies after 1998.

A brief note about ethnographies of conservation: they have become a growing and important area in anthropology, geography, ecology, and other fields, often with a focus on integrated conservation and development projects (ICDPs). This trend addresses a "new politics of ecology" that brings together environmentalists, institutions, researchers, communities, and activists in complex and often unique ways at each site, all such activity against historical backgrounds that often include colonial histories of preservation, development projects, contrasting understandings of nature and landscape, and the like (Anderson and Berglund, eds. 2003). Attention has been directed, among other aspects, to the "moral geographies" introduced by conservation discourses (Bryant 2000); environmental NGOs' discursive and material practices in particular conservation projects, say, an ICDP in Papua New Guinea (West 2006); the crucial but often overlooked role of local workers in particular conservation projects (Sodikoff 2005 for a project in a biosphere reserve in Madagascar); the contradictions between development, ecotourism, conservation, and local landscapes and nature constructions (e.g., in the Mayan area of Yucatán, Martínez 2004); the difficulties in establishing pluralistic governing structures to mediate the tensions between conservation and development (e.g., Haenn 2005 for ICDPs around the Calakmul Biosphere reserves in southern Mexico); the relation between the extent of market integration by local communities and conservation possibilities and outcomes (e.g., Lu 2007, for indigenous communities in the Ecuadorian Amazon); and persistent Western biases in conservation vis-á-vis local emphases, cosmologies, and historical ecology (e.g., Velázquez 2005 for the case of the Wounaan of eastern Panama). In many of these works there is an effort at looking seriously at the encounter of rationalities, scientific and not, at epistemes or cosmologies, and, above all, at the practices of concrete projects.

The Proyecto Biopacífico ◈ One of the major realizations of the Earth Summit of 1992 was the Convention of Biological Diversity (CBD), ratified by most countries of the world, with a few exceptions (including the United States).[21] Another product of the Rio summit was the Global Environment Facility (GEF), set up by the United Nations Environment Program (UNEP), the United Nations Development Program (UNDP), and the World Bank as a multibillion dollar fund to be administered by the World Bank, biodiversity conservation projects being one of its primary obligations. One of GEF's initial operations was the Project for the Conservation of Biodiversity in the Colombian Pacific, or Proyecto Biopacífico (PBP), funded with nine million dollars from GEF and the Swiss government for the project's first three-year phase. Launched in August 1992, the project commenced activities in Bogotá in March of 1993. From its inception, in keeping with the then-prevailing understanding of conservation as defined by northern environmental NGOs, such as the World Resources Institute and the World Wildlife Fund (see, for instance, WRI/WWF/UNEP 1992), the objectives of PBP were geared toward ascertaining "the scientific, social, economic, and political elements necessary for a new strategy of biodiversity conservation and the sustainable utilization of the region's biological resources" (GEF/PNUD 1993: 3). In scientific terms, the point of departure was the need for a comprehensive mapping of the region's biodiversity, including its high levels of endemism. One of the anchoring points of the strategy was the existence of a National Park System that included thirty-three parks in the Pacific covering about one million hectares, most of which had various degrees of human intervention. Socially, the project focused on the black and indigenous communities, whose "cultural and socioeconomic isolation from the rest of the country" (10) produced endemic poverty and the lowest quality of life in the entire country. Despite this fact, and in the climate of the Constitution of 1991 and TA-55, PBP staff saw the participation of the communities as a great potential for conservation, even if in ways that had to be made compatible "with the social and economic development" of the country. PBP's first Operational Plan (1993–96) included a diagnosis of the destruction of biodiversity, highlighting deforestation, the destruction of particularly valuable or fragile ecosystems, for example, forest wetlands, and trends toward agroindustrial schemes and macrodevelopment projects, although without any degree of specificity.

PBP defined its overall objective as follows:

> The main goal of the *Proyecto Biopacífico* is to contribute to the consolidation of a new development strategy for the *Chocó Biogeográfico* based on the application of scientific knowledge and the identification of strategies for the management of biodiversity that guarantee its protection and sustainable utilization, and in a concerted fashion with local communities. (GEF/PNUD 1993: 16)

More than in the mainstream biodiversity approaches favored by northern NGOs, PBP registered from the start the correlation between biological and cultural diversity, and the importance of taking into account traditional knowledge and practices. Scientifically, however, the PBP conception did not stray far from the dominant vision, either in its definition of biodiversity in strict biological terms or in its overall view of conservation; this view favored an ecosystems approach that yielded a classification of subregions in straight biological terms, further differentiated only in terms of types of tenure (public versus private) and the degree of anthropic activity, or human intervention. Also in keeping with established approaches, PBP paid a great deal of attention to the socioeconomic benefits to be derived from what was imagined as the sustainable use of biodiverse resources. Consistent with these scientific, cultural, and economic analyses, PBP came up with an innovative operational strategy structured around four major areas of activity, defined as *conocer* ("to know," that is, to gather scientific and traditional knowledge); *valorar* ("to value," or the economic utilization of biodiversity); *movilizar* ("to mobilize," that is, to enlist the participation of the communities); and *formular-asignar* ("to formulate and allocate," which concerned policy formulation and allocation of resources). The four areas were defined as follows:

> *Conocer.* It seeks the recovery, synthesis, and generation of scientific, academic, and traditional knowledge about the region's biodiversity, as well as the accumulation of new knowledge. It aims at strengthening the national, regional, and local capacity to know biodiversity, and to use it appropriately.
> *Valorar.* This area has an economic character, and points at the generation of income and other benefits, with preference for local communities, through a process of valuing and strengthening those regional production systems that are compatible with the conservation of biological diversity.
> *Movilizar.* Of political character, this programmatic area seeks to improve the regional and local negotiation capacity concerning the conservation and sus-

tainable use of biodiversity, through processes of education, communications, and social organization.

Formular-Asignar. Intended to generate the political will and to create institutional spaces at various levels of government for the conservation and sustainable use of biodiversity, as well as to insure financial resources for participatory and concerted planning and management processes concerning the region's biodiversity. (GEF/PNUD 1993: 38)

The four areas were supposed to be integrated, but in practice they functioned largely independently of each other. There was also a significant gap between theory and practice during PBP's first two years. Whereas Conocer, for instance, was supposed to devote equal attention to scientific and traditional knowledge, in practice most subcontracts in this area went to academics and NGOs that couched their conservation projects in standard scientific and policy terms. In time, this tension generated the sense among activists and communities that PBP was going to exacerbate the extractive style of previous development activities in the Pacific. By 1995, in a climate of heated debates about genetic resources and intellectual property rights, including the drafting of the Andean Pact legislation on the subject, this gap became ever more noticeable, and the tensions affected the PBP technical team in Bogotá. Similarly, Valorar was meant to support local sustainable initiatives; in practice, this came about only toward the end of the project, following the adoption of a more participatory project structure after 1995. In its initial period, the utilization dimension of the project remained economistic, opened to productivist NGO initiatives, and guided by the need to ascertain the value of biodiversity and the opportunity costs of its conservation. Finally, Movilizar emphasized three strategies: education, communications, and organization; these tilted toward communications and educational projects that were functional to conservation and development at the expense of the alleged goal of strengthening the ethnoterritorial organizing processes. At this level, Movilizar subcontracted projects tended to favor practices of participation and rural animation that were well known in the development world since the late 1970s.[22]

PBP's first two years were thus riven by tensions, which were brought to the fore in an internal evaluation carried out in late 1994 and, more forcefully, by an external evaluation in the spring of 1995. Among the flaws found by the internal evaluation were lack of coordination among thematic areas, a need to come up with new mechanisms of participation, lack of coherence between a thematic focus and a territorial perspective, and a need to integrate more fully the regional offices into the decision-making process. These problems were taken as the basis of a revised planning process, undertaken in the second half of 1994. After conducting a series of workshops with black and indigenous organizations, the external evaluation team underscored lack of participation as the single most important obstacle to be overcome; indeed, this was one of the main criticisms leveled by the popular organizations against the PBP. Lack of participation had already been found to be pervasive in most GEF-affiliated projects (e.g., Wilshusen 2003; Guerra 1997) and was consistent with the kind of subcontracting done by PBP until then, which favored individuals (e.g., academics) and NGOs over political and territorial organizations. Coupled with strong pressures coming from ethnoterritorial organizations, the project evaluation led to drastic reform, which was to take place throughout 1995 and part of 1996. Reform had started slowly in 1994, as a result of consultation by PBP staff with black and indigenous organizations. However, its culminating point was not to come until June 1995, when a special task force or working group called the Equipo Ampliado (Expanded Team) was created. This resulted in a profound political and conceptual transformation that probably remains unparalleled in the experience of conservation projects in the world.[23]

The Equipo Ampliado included the main organizations of the region; in addition, it was conceived as the instrument for redesigning the entire project with the full participation of black and indigenous ethnoterritorial organizations, that is, not just any black and indigenous organizations but explicitly those that sought to vindicate cultural and territorial rights.[24] A series of national workshops had been held in 1994 and 1995 on Ley 70 and on the meaning of ethnic and territorial rights, culminating in the meeting at Perico Negro in June 1995, at which black and indigenous organizations of the Pacific laid down the basis for a shared conceptualization, including the concepts of territory and region-territory (see chapter 1). One of the central working principles throughout this process—a principle also at play in many other government strategies—was that of *concertación*. More than simply negotiation, concertación involved a sense of thorough discussion of clearly conflicting interests with the aim of reaching a jointly agreed upon formula. The term implied the sense of power differentials and, often, that of contrasting cultural backgrounds. The PBP was perhaps the first experiment in concertación of this type after the Constitution of 1991 and can be seen as a productive, if tension ridden, example of negotiation of ecological and cultural distribution conflicts.

These conflicts came sharply to the fore with the articulation of biodiversity worked out by the Equipo Ampliado and offered as the basis for the redesign of the project. Entrusted with the elaboration of the Operational Plan for 1995–97, the Expanded Team took it upon itself to initiate a substantial revision of PBP's framework and objectives through an active process of meetings and discussions, especially in the eight months following its creation; the effort began with an influential meeting in Piangua Grande (Buenaventura) in late August 1995, also attended by a large number of local organizations. It was positively stated that "the strategy of biodiversity conservation in the region-territory of the Colombian Pacific must have as its objective the strengthening of the capacity of the black and indigenous communities to appropriate, defend, and benefit from the territory as an *espacio de vida* (life space) (Procesos Organizativos de Comunidades Negras e Indígenas 1995: 10).[25] This view of biodiversity started by highlighting "the intrinsic relation between territory and culture" as determining for biodiversity and its utilization potential:

> For the black and indigenous communities the term "biodiversity" comes from the outside; it is not used by the people. The black communities have used the term *monte* [literally, "bush"] in more or less the same way as implied by the word biodiversity, since it is in the *monte* that diverse forms of life are found. The relation between nature (territory), culture, and everyday life is equal to biodiversity. The knowledges [*saberes*] of the black and indigenous communities have enabled a harmonious relation with the environment, thus favoring biological diversity. The landscapes of the Pacific would in fact be quite different if other peoples had inhabited it. This is why it must be recognized that biodiversity has two components: a tangible dimension—the manifestation of ecosystems, species, etc.; and the intangible, namely, those forms of knowledge—and the associated cultural innovations, uses, and practices—that allows biodiversity to be recognized today as a resource. . . . To speak of biodiversity from the perspective of the black and indigenous communities thus inevitably entails speaking of the ethnic, cultural, and social diversity of a people that has a particular relation to nature and the environment. (2)

The formula territory-culture-biodiversity thus became the central guiding principle for a new planning strategy. Far from the scientific and managerial vision of biodiversity upheld by the CBD, GEF, and the early PBP, the Expanded Team favored a culturally, socially, and politically rooted conceptualization. By emphasizing the de facto preference for autonomy that communities had maintained in their settlements, as well as the "cultural and territorial aggression" unleashed against them in the most recent decades, the ethnoterritorial representatives at the center of the Expanded Team situated biodiversity squarely in the domain of political contestation. Similarly, its vision attempted to challenge the economistic view of biodiversity by introducing a decommodified notion of nature and territory. The narrative of the organizations, as stated in Piangua Grande, did not stop there; they wanted an entire restructuring of the project: "Taking into account the conjunctural importance of biodiversity . . . we consider that the reformulation of PBP must depart from a reconceptualization of biodiversity and it must include a significant restructuring and redefinition of principles, policies, strategies, criteria, objectives, thematic axes, and specific projects" (4). As political guarantees for this process, they demanded equal decision-making power, the concertación and *socialización* of the project and its results through the *procesos organizativos*, and the appointment of technical support teams for the elaboration of the Operational Plan, to be chosen by the organizations.[26] The tone was decidedly political: "Based on these principles, the defense of the territory must be conceived of and assumed by communities and institutions from a perspective that links past, present, and future and that understands and develops the tradition and history of resistance of black and indigenous communities, as well as their desire to validate, maintain and develop a different and particular life project" (7).

The overall objective was to be looked at "on the basis of the social mobilization and the strengthening of ethnic groups' autonomy, the implantation of novel and effective spaces for participation, the socialization of scientific information, and the valuing of traditional environmental knowledge and practices," and the strategy was "the defense of the cultural and biological patrimony of the nation and the ethnic groups of the Pacific and the construction of alternative models of development" (Ministerio del Medio Ambiente/DNP/GEF 1996: 10). The project's general objective was "to improve the life conditions of the population and to guarantee the conservation of ecosystems and the sustainable utilization of biodiversity, by means of a strategy that promotes the local processes of appropriation, use, and conservation of the region-territory of the Colombian Pacific, and by guaranteeing the strengthening of the organizing processes of the indigenous and black peoples that inhabit it" (Ministerio del Medio Ambiente/DNP/GEF 1996, Anexo 2, p. 3). This goal would lead to particular objectives, such as the defense and promotion of traditional production systems, food security, territorial planning, valorization of culture, and so forth. In terms of research, the guidelines stated that there should be a process of consultation with communities and their organizations on the nature of the research, the incorporation of capable local researchers, the socialization of results, and the respect of intellectual rights and benefit sharing when applicable, in the framework of established legislation.

How to assess the transformation of this project? As an initial step, it is useful to characterize the transformation as a series of contrasts: from a scientific and managerial conservation project in 1992–94 to a cultural–territorial conceptualization; from a rational project of intervention by the state and the international conservation establishment to a conceptualization of planning as a political process; from marginal participation of political (social movement) organizations to a design in which the latter occupied a place of pride, in principle with equal decision-making power as the proper technical personnel; from a project conceptualization in terms of thematic areas to a territorial-based design that gave greater coherence to the thematic areas; from a largely biogeographical notion of subregions to a bio-cultural-political demarcation of "territorial action programs." This transformation was also visible at the level of knowledge: for instance, from an appeal to local knowledge in which this was seen as functional and complementary to scientific knowledge to a valuation of local knowledge as an important basis for conservation, with modern science and technology playing an important, yet supporting, role; from extractive research as the norm to a problematization of this practice and a set of alternative principles for research and its socialization. The view of planning changed from an expert practice effected on an external object (the biologists' Chocó Biogeográfico) to a negotiated but quasi-endogenous process from the perspective of the social movements' alternative notion of Pacífico Biogeográfico or region-territory of ethnic groups. The project's new functioning structure—with the Expanded Team at the center, articulating with ethnoterritorial organizations, the four thematic areas, the regional offices, and the National Office in Bogotá—was an integral component of this transformation (see Equipo Ampliado 1997a: 6 for the organizational chart).

This unprecedented process of concertación was neither easy nor completely successful, nor can it all be ascribed to the intelligence and determination of activists and PBP staff. By all accounts, the process was ridden with tensions, frustrations, mutual suspicions, and misplaced expectations.[27] Even though by the end of the negotiation and the progressive phasing out of PBP in 1998 most participants agreed it was a pioneering and largely positive experience, the difficulties were real.[28] The concertación faced difficulties at various levels: between ethnoterritorial representatives and the PBP technical staff; between regional coordinators and thematic area coordinators; between indigenous and black organizations; and within the organizations themselves. The organizations saw these difficulties largely as a learning experience, albeit an expensive one in terms of time commitments. The political character of the representatives of the organizing processes in the Expanded Team clashed at times with the more technical role of the Technical Team; this tension was felt particularly strongly by the regional coordinators, several of whom had a dual role as representatives of political organizations and of technical staff. On the more positive side, the participants emphasize the learning process and the growing acceptance of difference as well as the open acceptance of the eminently political character of the project. This emphasis meant that importance was given to process more than to results. The resulting Operational Plan of 1995–97, while not meeting all the new guidelines, reflected the process of concertación and enabled a final round of projects.[29]

The most ambitious goal of all, that of contributing to the reshaping of Colombia's environmental policy, remained elusive. The reason was simple: at the national level, the Pacific continued to be seen in terms of the expansion of the conventional development model. Building on the legal mechanisms of the 1991 Constitution and counting on worldwide trends that validated conservation and biotechnology, PBP staff bet on providing a countertendency to development strategies that not only destroyed the rain forest but did in benefit of private actors from outside the region. In the long run, conservation policies returned to a more conventional conception. By the time Agenda Pacífico XXI was enunciated, PBP had ceased to exist as part of a larger reorganization of Colombia's environmental sector; this reorganization, in gestation since the creation of the Ministry of the Environment in 1993, included the creation of five specialized research institutes. Two of these were the Instituto de Investigaciones Ambientales del Pacífico (based in Quibdó, the capital of Chocó province) and the Instituto de Investigación de Recursos Biológicos Alexander von Humboldt (based in Bogotá). As the entity responsible for implementing the dictates of the Biodiversity Convention, the Humboldt Institute was put in charge of a new National Biodiversity Action Plan. The plan fell back on a largely scientific, economistic conception of conservation, even though recognizing to some extent the value of cultural diversity and traditional knowledge (see Asher 2001 for an analysis of this plan). This change seemed to reconcile the contradiction between the prodevelopment government strategies (such as Plan Pacífico) and the late PBP's orientation, but only by enabling a reinsertion of biodiversity into the narratives of capital and technoscience.

The transformation of the PBP must also be seen in terms of larger contextual factors, including the expansion of biodiversity and social movement networks (see chapter 6). The external evaluation of 1995 triggered that transformation, and the intellectual and political community then growing around things Pacific fulfilled an important supporting function. Yet for a time it was the deeds of the Equipo Ampliado and the PBP staff that captured the imagination of all of these actors, despite tensions and disagreements. The PBP experience exemplifies the notion of counterwork not only because over time PBP came to see itself as a countertendency to development but in terms of the entire process of concertación and socialización that took place in the space of encounter between a conservation/development project and the procesos organizativos of the Pacific. It can reasonably be asserted that the post-1995 project design constituted an alternative conservation approach—an alternative political ecology framework (Escobar 1998; see figure 6). It also embodied an alternative modernity, to the extent that conceptions of conservation, development, nature, economy, and planning were all shifted to a significant extent in relation to the established discourses and practices of the state, market, and modern environmentalism.

Counterwork remained within the language and conceptual universe of modernity, however, and to this extent the social movement activists can be seen as agents of an alternative modernity. It can also be said, however, that the process allowed a glimpse of an alternative to modernity. Whether this third option or level of counterwork was actually made viable at any moment of the process remains an open question, given the discontinuation of the project. I suggest that in engaging fully with the apparatuses of modernity—such as planning, development, and technoscience—activists and planners sought to shelter place-based cultural constructions and practices (e.g., so-called traditional production systems; local models of nature; local knowledge; social, spatial, and territorial practices), thus keeping alive the more radical option of alternatives to modernity. Activists, that is, moved into the epistemic terrain and entertained the idea of a decolonial project.

I want to make two final comments in this regard. First, the potential for an alternative to modernity can be visualized from the angle of the colonial difference. The colonial difference in this case must be seen as a historical experience of many-layered subalternity and discrimination—the historical conquest of the region; its construction by ethnic groups under conditions of marginality; the brutality of the logocentric project of development. This difference, as I have repeatedly remarked, may be seen in terms of the persistence, reenactment, and transformation of practices of economic, ecological, and cultural difference. Second, the larger contradictions that pitted PBP against the developmentalist orientation of the state and the global economy can be worked out only up to a point through strategies of alternative development (e.g., Coagropacífico or the ethno- and ecodevelopment orientation of the *Plan de desarrollo para las comunidades afrocolombianas y raizales*), or alternative modernities (e.g., GEP or post-95 PBP). To deal with these larger contradictions, one must move beyond these levels toward a kind of thinking that intuits the possibility of alternatives to modernity. The PBP restructuring did not venture into this territory, for obvious reasons. (I will address further the theoretical possibility of alternatives to modernity when I discuss the concept of virtuality in chapter 6.)

Alternatives to modernity, then refers to that dimension of the relation between globalization, development, and modernity that imagines an explicit cultural-political project of transformation from the perspective of modernity/coloniality/decoloniality—more specifically, an alternative construction of the world from the perspective of the colonial difference. The dimension of alternatives to modernity contributes explicitly to a weakening of the strong structures of modernity—universality, unity, totality, scientific and instrumental rationality, etc. (e.g., Vattimo 1991; Dussel 1996, 2000) and the coloniality of nature, but from a different position. One should be clear also about what this concept is not: it has been said of the notion of postdevelopment that it pointed at a *real* pristine future in which development no longer exists. Nothing of the sort was intended. Rather, the notion intuited the possibility of *imagining* an era in which development ceases to be the central organizing principle of social life. The same with alternatives to modernity, as a moment when social life is no longer so thoroughly determined by the constructs of economy, individual, market, rationality, order, and so forth that are characteristic of the dominant Euro-modernity. Alternatives to modernity is the expression of a political desire, a desire of the critical utopian imagination, not a statement about the real in a strict sense, present or future. Operating in the cracks of modernity/coloniality, it gives content to the World Social Forum slogan, *Another world(s) is (are) possible* (Escobar 2004). But it should be kept in mind that alternatives to modernity are intended as a reformulation of the modern colonial world system but still operating within modern/decolonial critical languages, that is, without giving full weight to, much less bringing into life, those other worlds that nevertheless are part of what inspires the notion. The latter would constitute a different project (see Yehia 2006 for a discussion of this idea; Blaser forthcoming).

Conclusion

I want to offer some tentative conclusions for development studies and studies of conservation. In addressing activist trends in science and technology studies (STS), some authors have suggested a "turn towards the normative and the activist" that can be usefully applied to critical development studies (Woodhouse et al. 2002). The question is posed as follows: How should development be reconstructed to promote more democratic, environmentally sustainable, socially just, and culturally pluralistic societies? Anthropology's "extended reflexivity," its promise of "near-native competence" (its intimate knowledge of both development experts and local groups involved in a given development or conservation project), and its claims to superior knowledge based on ethnographic research could serve as the basis for such an articulation (Hess 2001). In the case of ethnographies of conservation, for instance, narratives of nature might provide a valuable standpoint for ethnographic studies of the coloniality of nature, particularly in the context of particular conservation-cum-development projects, and a basis for collaborative engagement with both local groups and environmental organizations and movements. Some questions to be asked include: What are the larger cultural worlds within which the narrative makes sense? How do communities, developers, and conservationists understand nature, themselves, and each other? How do development and environmental organizations mediate this understanding in contexts of power?

In an ideal scenario, the ethnographer would investigate in depth a particular ICDP with these questions in mind. Such multisited, distributed study would involve development of near-native competence with both environmentalists or developers and local groups and their organizations, including a complex interpretation of the self-understanding and actions of both sets of actors within historical and cultural backgrounds. There are many challenges to this kind of endeavor. Based on the enriched knowledge of the ethnographer, how can environmental organizations and communities be brought together into different dialogical situations and thus toward alternative collaborative approaches and projects? There are issues of positionality: the awareness of the political tasks the critical development and environmental practitioner takes on vis-á-vis both developers and communities. Equally important, ethnographer-activists might find themselves in the position of having to negotiate both the epistemological rift between anthropological interpretivism and development/conservation realism/positivism, on the one hand, and, on the other the epistemic rift between local groups' episteme and the modern organization's or ethnographer's worldview (toward *worlds and knowledges otherwise, diversality*). Finally, there might be pragmatic questions, such as how to implement the critique (including multiple translations and mediations across sites and actors) and, fundamentally, how to reconcile the modernist injunction to organize communities and manage the environment with nonmodernist (place-based, nonrationalistic) logics and life practices. These sets of concerns are what account for an ecology of difference approach.

Alternative development, alternative modernities, and alternatives to modernity are partially conflicting but potentially complementary projects. One may lead to creating conditions for the others. This is why I say that social movement activists, policymakers, and progressive intellectuals would benefit from holding in tension these three goals, conceptually and in terms of their particular strategies. This is, to some extent, the lesson of Proyecto Biopacífico. While the utopian goal of some of the region's social movements implied a desire to move toward alternatives to modernity, under today's difficult conditions they were able to move only a few inches along the paths of the first two dimensions. Coagropacífico, Gente Entintada y Parlante, and Proyecto Biopacífico found a meeting ground in the rivers, mangroves, forests, and plantations of the Tumaco region. Their advocates certainly did not walk in unison to the tune of alternative development or modernity, nor did they entertain a more radical movement. Yet the dominant project of development found in these expressions a contender that did not easily go away. Sadly, it is through the use of terror that the agents of capital and development seek to impose their project today. This throws any thinking about alternatives into disarray; it does not mean, however, that the question is irrelevant. On the contrary, in some ways it becomes even more poignant.

In other words, strategies should variously foster alternative development for livelihood and food autonomy (as minimum strategies of resisting, returning, and replacing, in cases where displacement or its threat is involved); alternative modernities that shelter the economic, ecological, and cultural difference that—even in the midst of a globalizing modernity—characterizes communities such as those of the Colombian Pacific; and alternatives to modernity and processes of decoloniality and interculturality predicated on imagining local and regional reconstructions based on such forms of difference. It is possible to differentiate among the following short-term scenarios for the Pacific: the victory of neoliberal designs, with the support of terror and coercion; reform within the boundaries of state and capital, in line with the goal of alternative modernities; and transformative projects involving explicit thinking from the colonial difference, including projects for a diverse economy. The balance among these projects will depend on factors such as the role adopted by the state, the strength of social movements, the path taken by the armed conflict, and the character of global networks that impinge visibly upon the region.

In the long run, only an effective interculturality, defined as a dialogue and mutual transformation ("impregnation," as Panikkar, 1993, calls it; see Escobar 2006 [1999]) of cultures in contexts of power, can prevent the further deepening of the triple economic, ecological, and cultural conquest currently under way in places such as the Colombian Pacific. I will return to these concerns in the conclusion. Next, however, I will consolidate my mapping of two of the main elements of the argument: the politicized black identities that emerged with tremendous force in the 1990s, constituting the most significant cultural-political fact of the decade in the Pacific; and the complex network dynamic in which these identities, as much as capital and technoscience, are immersed, making any straightforward reading of global and local processes impossible.

5 identity

It is not us who will save culture, it is culture who will save us.—PCN activist, Tumaco, 1998

¿Para qué nos sirven las identidades? Más aún, ¿para qué nos sirven las diferencias?

[What good are identities to us? What's more, what good are differences to us?]

—Betty Ruth Lozano, Afro-Colombian sociologist, at the Conference "Afro-Reparations: Memories of Slavery and Contemporary Social Justice," Cartagena, October 19–21, 2005

Introduction

Transitory Article 55 (AT-55) constituted a watershed in the history of the Colombian Pacific. As we have seen, the pace of activities regarding things Pacific quickened in the 1980s. Capitalists, state planners, regional development corporations, experts of all kinds descended on the Pacific as if it were a promised land—or "parachuted in," as the locals soon started to say mockingly in reference to the unduly short visits of state officials and development experts. In many places there was a ferment of cultural and organizational activities fostered by a progressive church, peasant unions, cultural activists, and the like. But it was only with the intense process of organizing around AT-55 and the subsequent Ley 70 that what was to become one of the defining features of the decade for the region as a whole came to the fore, namely, the emergence of unprecedented forms of black identity. There was nothing in the cultural, social, and political makeup of the Pacific that made this development necessary. To begin with, the long-standing forms of self-reference and belonging among the black groups of the Pacific had little to do with what after the Constitution of 1991 came to be referred to as "black communities" or, even more commonly, "black ethnicity." Gatherings of activists, experts, members of NGOs, local leaders, women's groups—all of them discussing the character and status of the newly discovered black communities—became a ubiquitous sight in the Pacific, from its river hamlets to its small towns and cities. A parallel flurry of activity could be seen at many national and regional state planning and development offices in Bogotá and in the main departmental capitals with territories in the Pacific, Cali, Quibdó, Popayán, and Pasto.

How can one understand this transformation? In the landscape of contemporary social theory, there are a number of pertinent concepts. Should one see the emergence of black ethnicity in the Pacific as an instance of the much-discussed class of "imagined communities" or "invented traditions"? Should one draw on the burgeoning literature on new ethnicities, which, along with fundamentalisms, are often cited as the single most important proof of the forceful return of forms of identification thought long ago left behind by the overpowering march of secular modernity? Or should one appeal to the impressive literature on identity growing out of many fields, from cultural studies and literary, feminist, queer, and critical race theories to social psychology, sociology, and anthropology? One could also examine how ethnicity was put into discourse, following Foucault, or appeal to identity-centered theories in social movements research. Any of these approaches–some of which are being applied to the Pacific—would yield rich results. The spectrum of theories that today seek to account for what is stated as identity is thus vast, a fact that itself calls for reflection. Why so much concern with identity in recent times? Who, indeed, needs identity, as Hall (1996) asks?

Broadly speaking, what has happened in the Pacific can be seen as a "relocation of 'blackness' in structures of alterity" and as the instauration of a novel order of alterization, as two of the most astute analysts of these changes have remarked (Wade 1997: 36; Restrepo 2002). The 1990s saw an unprecedented construction of difference through a series of practices that can be studied ethnographically. These practices involved issues as varied as memory, environment, culture, rights, the state, and production. Above all, they concerned the politicization of difference and the construction of a new political subject, black communities. If in the 1970s and 1980s anthropologists could still denounce the invisibility of black cultures in expert knowledge and state strategies (e.g., Friedman 1984), in the 1990s this feature was radically reversed. Suddenly there was a tremendous interest in things black, particularly as far as the Pacific was concerned. We have seen that, in broad terms, this was owing to the double conjuncture of the *apertura política* (political opening up) fostered by the Constitution of 1991 and the irruption of the biological as a global social fact and also related to the neoliberal policies adopted after 1990. What has to be explained is why these narratives took the form they did, particularly in terms of ethnic identity, cultural rights, difference, and black communities. And, very important for my purposes, how did social movement activists participate in this process? What knowledge and discourses about "ethnic identity" and the like did they produce? And how were they related to the discourses of other actors?

Should one see in the sudden appearance of black ethnicity in the 1990s chiefly a reflection of the power of the neoliberal state to create the conditions, even the terms, within which subaltern groups must couch their actions? Or, conversely, should one read in the emergent black identities a sign of the agency of this subaltern group, finally finding a workable formula for collective expression after decades of being silenced? As we shall see, while some analysts gravitate toward one or the other of these clear-cut positions, the answer lies somewhere in between. This in itself says little. The rest of the story will largely depend on whose voices and perspectives one privileges, one's space of enunciation, and the framework used to examine the encounter between the various actors in the play of identity.

This chapter is a partial statement on the play of identities in the Pacific since the early 1990s. I say *partial* because there is much that will be left out, in terms of both theory and ethnography, although it will be sufficient to provide a broad view of the process. Part I introduces some scholarly trends in the study of identity, including the necessary debate on essentialism versus constructivism and universalism versus particularism. Here I also summarize the main accounts by social scientists of the making of identities in the Pacific to date. The emphasis here is on those works which address the emergence of ethnic discourses after AT-55. Part II examines the articulation of a collective identity by the Process of Black Communities (PCN). Here, I start by introducing a theory of identity as a dialogical process that simultaneously highlights history and struggle, agency and determination. Keeping this framework in mind, and based on ethnographic research and collaboration with activists, this section goes on to recount the development of the social movement of black communities from the early 1990s until the present. Part III develops yet another approach to activists' identities, this time more phenomenological, based on the kinds of engagement and commitments activists enact at both individual and collective levels. The fourth part examines the slow and difficult articulation of the questions of women and gender within the movement. After a de facto decision to subordinate gender to ethnicity in the early 1990s, a politicized black women identity regime started to appear among various social movements in the second half of the 1990s.The concluding section asks, How does one assess the productivity of these movements and identities? Here I return to the modernity/coloniality/decoloniality framework and ask whether the collective mobilizations of the 1990s can in any way be seen in decolonial terms of alternative modernities and alternatives to modernity.

The following summarizes the overall arguments about identity made in this chapter:

1. Identity is a particular articulation of difference. Identities are the product of discourses and practices that are profoundly historical and thus always reside within an economy of power. Whether identity can ever escape this predicament is unclear.
2. Identities are constructed through everyday practices at many levels. From the realm of daily tasks and activities, which create microworlds, to the construction of more stable, albeit always changing, figured worlds, identity construction operates through an active engagement with the world. There is a constant tacking between identity, local contentious practice, and historical struggles that confer upon identity construction a dynamic character. This processual character of identities can be gleaned particularly well from activists' political strategies.
3. Identities are dialogic and relational; they arise from but cannot be reduced to the articulation of difference through encounters with others; they involve the drawing of boundaries, the selective incorporation of some elements, and the concomitant exclusion or marginalization of others. Gender, for instance, was often marginalized in the initial phases of activists' conversations on identity.
4. In political situations, identity involves ethical commitments by activists. Such commitment operates through a practice phenomenologists call skillful disclosing, that is, the creation of spaces in which new ways of knowing, being, and doing might emerge as historical possibilities out of given problematic situations (such as a form of oppression). Skillful disclosing involves the place-based articulation of the concerns of a collectivity in a way that shifts cultural understandings. Activists do so not as rational decision makers but as participants who are deeply committed to changing a particular situation through the production of alternative figured worlds. The practice of many of the PCN leaders can fruitfully be seen from this vantage point.

I ◈ The Emergence of Black Ethnicity in the Colombian Pacific in the 1990s

It is not difficult to appreciate the magnitude of the cultural and political transformation that took place all over the Pacific in the 1990s if one sees it in terms of identity. An entirely new identity regime emerged and took shape between 1990 and 1998. This regime was couched in terms of ethnicity and largely articulated around the concept of *comunidades negras* (black communities). The ethnicization of black identity can thus be understood ethnographically by focusing on the discursive and institutional practices associated with the emergence and dispersion of the construct of black communities. After making an initial inroad into approaches to identity to get a better sense of the questions that need to be asked, this section moves on to review succinctly the debates on these issues in the Pacific with the aim of providing an adequate background to the detailed examination of a particular social movement in the second half of the chapter. I start with a discussion of pre-AT 55 black identities, go on to present the analyses of the ethnicization of identity in the 1990s proposed to date, and end with a brief discussion of the various assessments of this process, particularly the question of the relation between identity, social movements, and the state. First, however, some general issues pertaining to identity and its politics.

Modernity, Identity, and the Politics of Theory ◈ As an explicit problematic of social life, identity is said to be thoroughly modern. Most authors have no qualms about stating that "identity is a modern invention" (Bauman 1996: 18). This is so, first and foremost, because identity "entered modern mind and practice dressed from the start as an individual task" (19); identity thus supposes the construct of the modern individual, full in its autonomy and free will, endowed with rights, and subject of his or her own knowledge. Identity is also seen as predicated on a modern logic of difference in which the subordinate term is seen as constitutive of and necessary to the dominant (e.g., Grossberg 1996). There is, finally, an idea recently put forth in the field of minority studies in the United States: that identity is "an epistemically salient and ontological real entity" (Martín Alcoff 2006: 5); identities are resources of knowledge for social change, particularly by oppressed groups (hence engaged in "identity politics"; see Martín Alcoff, Hames-García, Mohanty, and Moya, eds. 2006).

I will accept these diagnoses here and bracket the anthropological debates on whether identity, in the form of a definition of self, is a modern Western construct that is absent in many other cultures or whether there can be forms of belonging without identity. Suffice it to say that many anthropologists argue that the modern notion of the self—at least in the quintessential mode of the possessive individual of liberal theory—does not have a correlate among many non-Western or non-modern peoples.

There are certainly other notions of personhood in these cases, but not of "the individual" in the modern sense.[1] Be that as it may, the fact remains that there is a great deal of confusion about the meaning of terms that are often taken as being equivalent, including self, individual, identity, identification, person, subject and subjectivity. For now, the historical anchoring of identity in modernity is an important referent for the discussion of the Pacific case. If the question of identity has become pressing in modernity, can one infer from it that the problems to which it seemingly responds in the Pacific--and, consequently, the kinds of agency and political rationale that accompany it—can be fully explainable by a modern logic?

It is a commonplace to refer discussions of identity to the contrasting positions of essentialism and constructivism. According to essentialist theories, identity develops out of an unchanging core (recall also Oyama's discussion in chapter 3). This notion has resulted in unitary notions of ethnic, racial, and national identities, in which ontological identity is seen in terms of primordial group ties anchored in a more or less self-contained shared culture. Essentialist notions of identity, while still prominent in the popular imagination and in some scholarly works (often rekindled through self-serving reportage on so-called ethnic wars, clash of civilizations, balkanization, and so forth), are largely considered inadequate, if not outright passé. Most scholars and social movement activists today consider all identities to be the product of history and thus constructed. Agreement, however, ceases with this statement, to some extent because the various radical critiques of essentialism refer to different kinds of subjects (the bourgeois subject; the Cartesian, detached observer; the gendered subject; the agent of subject-centered reason, and so forth, in Marxist, poststructuralist, feminist, and other critical approaches), but also, importantly, because they have different political orientations and hence draw diverging political conclusions.

Poststructuralism has given great attention to conceptualizing identity, and Foucault has been the single most influential author in this area. Foucault's work underlined the production of subjects through discourses and practices linked to the exercise of power—practices through which the subject is objectified in various ways, for instance, through mechanisms of discipline and normalization, as much as practices of subjectification that the subject performs on himself or herself. For some, a theory of identity cannot be complete without an account of the subject's active self-constitution, a point Foucault left unfinished and which others have continued (e.g., Hall 1996: 15).[2] The contributions of

Butler and Laclau and Mouffe are among the most well known in this regard. Taking as a point of departure the contradictions and limitations of a representational politics within feminism—the fact that within this politics the category of women is produced and constrained by the very structures of power from which it seeks emancipation—Butler (1990, 1993) shows the multifaceted character of identity and the limitations of any attempt at constructing a stable subject. Her alternative is a constant questioning of representational politics through a critical genealogy of its practices. In the case of feminist politics, this genealogy reveals an underlying heterosexual matrix and an insufficiently scrutinized binary relation between sex and gender—precisely the constructs that ground the regulatory power of patriarchy. The problem is even more complicated, as Butler sees it, since there is no recourse to a utopian notion of a liberated identity outside the matrix of power defined by sex, gender, desire, and the body. Politics thus becomes a constant effort at displacing the naturalized notions that support masculinist and heterosexist hegemony; it becomes a question of making "gender trouble" by continually mobilizing and subverting the very categories that serve as the "foundational illusions of identity" (1990: 34).[3]

It can be said that for Laclau and Mouffe (1985; Mouffe 1993; Laclau 1996) all identities are troubled, certainly the collective identities associated with social movements. In these cases, unfixity has become the rule; all identities are relational and overdetermined by others, leaving a logic of articulation as the only possibility for the political construction of identity. This logic proceeds by the construction of nodal points around which meaning and identities can be partially fixed. We shall see the extent to which this type of articulatory logic was in effect in the Pacific and enabled the identity of black communities to emerge as a political subject. For this to happen, the existing relations of subordination (e.g., in cultural, ethnic, gender, or ecological terms) must come to be seen as relations of oppression, and this can happen only under certain discursive conditions that make possible collective action. In the case of the Pacific, discourses of ethnicity, cultural rights, and biodiversity played this role and made possible the interruption of subordination as usual, so to speak, and its articulation as domination. These discourses allowed activists to construct a novel narrative of the situation of the Pacific in terms of external impositions by the state, expert knowledge, and the global economy. At this level, activists appealed to the modern democratic imaginary, especially the idiom of rights, although, as we will see, their actions were not restricted to this imaginary. The articulatory model (or, as Laclau and Mouffe call it, the hegemonic form of politics) results in novel divisions of the social field on the basis of deeply constructed identities which are partially autonomous, even if their character is never given in advance but depends on the discursive articulations they establish. This raises the ultimate question of whether a strategy of opposition can lead to a strategy of construction of a new order and to a positive reorganization of the social.

The seeming fixation on an essentialist/constructivist divide has been strongly criticized by Comaroff (1996; see also Martín Alcoff 2006). Realist contructivists argue that behind identities lie particular sets of interests, thus falling back into an instrumentalist position; cultural constructivists, on the other hand, accept that identities are the product of shared signifying practices but tend to treat culture itself as a given; a third perspective, political constructionism, simplistically singles out the imposition of ideologies—e.g., by the nation-state—as the source of identities; finally, radical historicism follows Marxist insights in its belief that social identities are the result of the working out of inequalities at the level of consciousness and culture. Contrary to these positions, Comaroff sees identities as relations that are given content according to their ceaseless historical construction; when applied to ethnicity, he privileges relations of inequality: "ethnic identities are always caught up in equations of power at once material, political, symbolic; more often their construction involves struggle, contestation, and, often times, failure" (1996: 166). Once constructed, "ethnic identities may take on a powerful salience in the experience of those who bear them, often to the extent of appearing to be natural, essential, primordial" (166). Finally, the conditions that give rise to ethnic identity are likely to change, which means that those sustaining it are likely to be quite different (this became patently the case in the Pacific after 1998). Norval (1996) adds two important questions to this conceptualization: the construction of discursive horizons of meaning that go along with how communities interpret their belonging (e.g., by drawing boundaries), and the construction of political imaginaries.

The concern with power and politics can be resolved into theories of identity in various ways. For Grossberg (1996), the discourse of identity as the grounds for struggle, albeit important, is limiting since its account of its own location within modern forms of power is narrow and, as such, identity politics cannot provide an ample basis for new political communities. Identity's modernist bend, in Grossberg's view, relies on three logics: difference, individuality, and temporality; identity as difference tends to see the subordinate term as necessary for the dominant; this ignores the positivity of the subaltern "as the possessor of other knowledges and traditions" and imputes meaning to subaltern struggles before they even begin (1996: 92; see also Guha 1988). More generally, theories of identity "are ultimately unable to contest the formations of modern power at their deepest level because they remain within the strategic forms of modern logic" (93). To the logic that transforms identity into relations of difference, he opposes a logic of otherness, productivity, and spatiality. A perspective of otherness, first of all, enables an examination of identity-as-difference as itself the product of modern power. Second, while theories of otherness see both difference and identity as effects of power, they do not reduce the question of the other to being merely constitutive or relational; in other words, otherness is granted its own positivity. In the case of the Pacific, the positivity of black identities cannot be reduced to an articulation of difference dictated by the dominant Euro-Andean order. To do so would not only reinscribe them into modern power (a form of coloniality), but also entail denying their otherness as positivity and exteriority.[4]

Debates about the relation between identity and politics have raged in many countries where multiculturalism has become an important project, chiefly in the United States but also in some Latin American countries. Attacks on identity politics in the United States have led some researchers and activists to formulate a realist view in which identities are defined as "socially significant and context-specific ideological constructs that nevertheless refer in non-arbitrary (if partial) ways to verifiable aspects of the social world." Identities are thus "markers for history, social location, and positionality. They are always subject to an individual's interpretation of their meaningfulness and salience in her or his own life, and thus, their political implications are not transparent or fixed." In terms of identity politics, this is "in itself neither positive nor negative. At its minimum, it is a claim that identities are politically relevant, an irrefutable fact. Identities are the locus and nodal point by which political structures are played out, mobilized, reinforced, and sometimes challenged" (Martín Alcoff, Mohanty, Hamas-Garcia, and Moya 2006: 6, 7). Extended to groups, this concept is useful for looking at the situation of black identities in the Pacific, with the caveats about individualist and modernist biases already signaled. In this chapter, I will consider identity as a particular articulation of difference, and the politics of identity as an instance of the politics of difference that operates largely in the cultural register, although implicating ecological and economic difference in some fashion.

The Ethnicization of Black Identity in the Southern Pacific in the 1990s ◈ The significance of the transformation in the identity regime that took place in the Pacific in the 1990s can be gauged by considering how identities were previously constructed.[5] We already have a general idea of black cultural practices in the Pacific from previous chapters. For instance, we encountered Losonczy's account of the *ombligada* and her claim about the existence of a black cognitive universe that, while constituting a coherent whole with a logic of its own, is always changing and flexible. Critical of approaches that emphasize either a protoidentity based on the survival of African traits or the inability of black groups to articulate an identity given the harshness of their marginalization, Losonczy proposes a view of black identities in the Pacific as an "interstitial identity, the result of violent historical discontinuities." This identity was constructed "in terms of an underlying strategy that systematically reorganized exogenous cultural materials," resulting in "a cauldron of identities with open and fluid boundaries" (1999: 15, 16). This applies as much to the past as to the present. In the past, this strategy brought together elements of diverse provenance—Catholic, African, indigenous, modern—in ways that called for its own kind of collective memory. This memory is founded on two puzzling absences: about African origins and about slavery. This is not a total erasure, however, for while explicit memories of these events are nonexistent, they can be gleaned from a series of rituals and symbolic practices, such as mythic narratives and rituals of death and the saints, whose performance, iconography, and musical elements evidence the syncretism of African, indigenous, and Catholic forms (Losonczy 1997, 1999; see also Restrepo 2002).[6]

The anthropologist William Villa applies a similar argument to the most recent past. Speaking of the black society of the Chocó, he states that "identity is an artifice made of fine thread [*filigrana*], a tapestry woven from materials from diverse sources, caring not about their provenance nor about the final product" (2001: 207). If throughout most of the twentieth century this strategy entailed a relation to the state and established political parties and its incorporation into local identities, since the late 1980s this logic of recombination of fragments has compelled black peasants to discover ethnicity. Like most other commentators, Villa highlights the role of the progressive church, expert discourses, and some development projects as the source of the fragments for the new identity. For Villa, then, the new identities are part of a process that extends as far back as colonial society and finds in the current ethnicization its most recent phase; they are part, to use our terminology, of the working out of the colonial difference. For Losonczy, the recent appeal to the idiom of Afro (as in the label of Afro-Colombian) implies a return to the two mythic themes of origins and slavery. For her, however, the reinsertion of identity into this narrative takes place on the grounds of a modern, linear conception of history and thus runs counter to the previous dispersed regime of identity. It remains to be seen, she concludes, whether this process of bringing black ethnicity into a dialogue with modernity (through a sort of "neo-traditionalist memory") will succeed in forging a new identity.[7]

Losonzcy's argument that the ethnicization of identity in the 1990s represents a departure from long-standing logics of identity is shared by other students of this process. What was the previously existing identity regime? A common starting point is the assertion that pre–AT 55 identities were largely localized, fluid, and diverse. Identities in riverine settlements were (and continue to be to a greater or lesser extent) strongly place-based, anchored in the river or, more generally, "the aquatic space" (Oslender 1999; Hoffmann 1999; see chapter 2). To the question, Where are you from? the most common answer was, and continues to be in most cases, to refer to the river of habitation. Concepts of territoriality were intimately linked to kinship relations, labor practices, and, as we have seen, an entire grammar of the environment. Beyond that, some general ways of self-reference did exist, one of the most common being that of *libre* (free person). The origin of this term is surely to be found in the colonial racial taxonomy (Wade 1997); however, its usage is far from simple. Like the other common category, *renaciente*, *libre* "has a particular meaning in a complex articulation inside a deeply woven set of categories . . . thus the notion of *libre* is not just the local transcription of a racial category of 'black' as simply opposed to 'white' and 'indian' " (Restrepo 2002: 99). In fact, what obtains is a "polyphony of identities" that includes multiple notions, such as *cholos* / wild indians (*indio bravo*) / *indios* / *naturales* / *paisas* / serranos / gringos / *culimochos* / *libres* / morenos / negros (2002: 101). This fluid, mobile system of identities is further complicated by notions of belonging, occupation, and so forth. One is a worker, a peasant, a *leñetero* (firewood collector), a fisherman, a *conchera* (shell collector), or a *costeño* (coast dweller). These denominations made up the most common subject positions before the emergence of ethnicity.

It is this regime of identity which is seen as having been turned topsy-turvy by the arrival of black ethnicity.[8] To be sure, not all of the subject positions mentioned above disappeared overnight; they have been resituated and reconstituted. But the arrival of the 1990s signified a noticeable rupture with the existing articulations of identity. The overall goal was the relocation of blackness in the cultural and political imaginary of the nation—in short, a new politics of representation. In the northern Chocó the ACIA experience of the 1980s moved toward the articulation of an ethnic identity proper (Villa 2001), but in the south the single most important catalyzer of the process was AT-55. AT-55 inaugurated a series of institutional and political practices that resulted in a powerful discursive construction of ethnic identity in terms of the black communities.

First of all, AT-55 called for the creation of institutional mechanisms to move forward with the development of the law (what became Ley 70 in July 1993). These mechanisms, a national-level Special Commission for Black Communities, plus departmental-level Consultative Commissions, were mixed bodies that included state officials, experts, and representatives of black communities and organizations. This spurred tremendous activity. Novel organizations and ways of thinking emerged on the basis of categories that had little previous reference in the Pacific, particularly those we have encountered already operating in various guises, such as territory, culture, environment and, very important, the *comunidad negra*. Here is how Ley 70 defined *comunidad negra*: "Black community is the group of families of Afro-Colombian descent who possess their own culture, a shared history, and their own traditions and customs in the context of the town/country division, and who exhibit and preserve a consciousness of their identity that makes them distinct from other ethnic groups" (Ley 70, Article 2, paragraph 5).

Numerous observers have commented on how the entire law and particularly this definition, which was significantly shaped by anthropologists and other experts in the Special Commission, were based on the only known existing model of alterity, that of indigenous peoples. Be that as it may, the terms introduced by AT-55 and the subsequent law became the nodal points for the articulation for the first time of a politics of black ethnicity on a national scale. By the end of the 1980s, several conditions had prepared the ground for these categories to take root. In the southern Pacific, as we have seen, displacement from the land by *camaroneras* and *palmicultoras* was advancing rapidly; development projects such as PLADEICOP, biodiversity discourses, state decentralization requiring an active local subject, and some forms of organizing prompted particularly by the church and peasant unions—all of these factors meant that the concerns expressed in AT-55 found resonance among some local groups. As a prominent Tumaco activist put it, "With all of these changes, AT-55 appeared as a great possibility for self-defense. . . . that's how the Process of Black Communities was born, and that's how we started talking about the reconstruction of our identity as a black people" (Cortés 1999: 133). What ensued was a veritable pedagogy of alterity through which experts, missionaries, and progressive church people, state officials, development workers, and scores of activists put the novel language into circulation throughout the rivers, hamlets, and towns of the Pacific (Restrepo 2002). This was not, however, a top-down exercise, as the rearticulation of identity was produced in the very process of interaction between experts and ethnoterritorial organizations, experts and communities, activists and communities, advisers and local groups, etc. These interactions involved a constant negotiation of the terms and of the practices themselves.[9]

The interactions took place through a multiplicity of practices, including workshops and meetings in cities and river settlements, map making and census exercises in communities, the traveling of territories for collective titling, and so forth; most of these practices were new to the communities, and it was through them that ethnicity was put into discourse. The ethnography of this process shows that the putting into discourse of ethnicity relied on a series of operations that naturalized identity, portraying black groups as wise environmentalists; emphasized traditional production practices; represented communities as existing in discrete, bounded settlements; and highlighted specific economic rationalities, traditional forms of authority, and so forth. These operations located black groups in space and time in particular ways; they tended to objectify notions of territory and culture—again, from a fluid, dispersed regime to a more modern, normative one. The result was a significant rearticulation of local experience. As Restrepo concludes, "To think about the local population in terms of black community, with a territory, traditional production practices, an ethnic identity and a set of specific rights, was an exercise in the construction of difference that became possible in the Pacífico nariñense only in the decade of the nineties with the institutional and social positioning of this new regime of representation" (2002: 81, 82).

It would be a mistake, however, as Restrepo and other authors go on to say, to see this regime solely as an artifact of Ley 70. The emergence of ethnicity, however, poses the thorny issue of the assessment of the transformation described by anthropologists and others. In anthropological terms, the black community is not a self-evident empirical fact (as it might have been for many experts and activists) but a complex cultural and political construction. A main question in this regard is the relationship between this construction and the state and the economy. An earlier answer to this question saw the new paradigm of ethnicity as the result of a double trend—a sort of scissors effect—toward neoliberalism in the economy, on the one hand, and state political opening after the crisis of developmentalism, on the other. This transformation called for new subjects of state management and took the form of decentralization and pluriethnicity; the result was a state framing of black identity according to the indigenous model and in a largely essentialist formulation (Gros 1997; Wade 1995). Subsequent debates by these same authors and others have introduced complexity into this interpretation. Granting that what is at stake is an instrumentalization of identity, this process is no longer seen solely as a response to state manipulation. To be sure, there is connection between the 1990s identities and the neoliberal state; however, the identities can be seen only in part as a result of the state. As the earlier model of a populist and racially homogenous project of nation building entered into crisis, the state promoted the creation of new identities it was unable to control (e.g., Gros 2000; Pardo and Alvarez 2001). Coupled with greater attention to transnational factors emphasizing cultural diversity, environmental causes, and rights, the new interpretation complicates the more functionalist readings of a few years before.

Today's sophisticated ethnographies of the state focus on the partially overlapping and contradictory articulations between state and social movements. On the side of social movements, one may find forms of cultural politics that contest the state's institutional discursivity at the local level. On the state side, the analysis has focused on the local and everyday practices of state formation and functioning; this analysis shows that local pressures on the part of social movements do have an effect on the state, transforming it and influencing its practices. The analysis of Proyecto Biopacífico validates both of these claims. At the local level, state and social movements are engaged in a contest over the production of meaning; indeed, the very binarism between state and civil society tends to dissolve in places like the Pacific, so that the ethnicization of identity comes to be seen as a shared process. This is so because the state hardly functions as a coherent entity but is riddled with internal tensions and conflicts across levels, institutions, and programs. For instance, in some cases, local government officials and functionaries might seek to capture the boons of the "eco-ethno boom" in vogue to their own advantage, while in other cases they might become allies of local organizations who also operate as brokers between the state and local communities (Alvarez 2002). In yet other cases, access to national funds might be achieved through alliances at the local level among the various actors. Moreover, as we saw in the case of PBP, it could be that what one finds is that state offices are made up of largely mixed spaces that bring together civil servants, experts, and movement activists—which again makes it difficult to determine where the state begins and ends or whether it can even be identified clearly as a discrete entity. Clientilestic practices are not necessarily inimical to these mechanisms (Alvarez 2002; Pardo and Alvarez 2001).[10]

For Alvarez, the ethnographic analysis of the state "makes use of a less deterministic reading of power that considers the possibility of localized change within an all-pervasive neo-liberal predicament" (2002: 73). Despite the greater nuance and sophistication of the ethnographic analyses, however, a lingering state-centrism remains. For one thing, many analysts continue to see the play of truth and identities as taking place in strictly modernist terms; in this way, what social movements have sought and to some degree accomplished (especially if one stops counting in 1998) was the creation of a modern identity with greater capacity of instrumentalization as a means to more effective inclusion into the national project of modernization and development than that allowed by the older regime of *mestizaje* (e.g., Gros 2000). In some analyses, this option is seen as inevitable since traditional authorities and worldviews have already been largely dissolved by the state's institutional logic anyway (Villa 2001). The aim of the movements is thus seen as the consolidation of strong regional ethnic identities capable of negotiating with the state the terms of their insertion into society and the economy. With the implosion of the war in the Pacific after 1998–2000, this project largely collapsed. That the project ended, however, does not diminish the fact that some important gains at the levels of identity, territory, and culture (including collective territories) were accomplished in some cases, particularly at the local level; the gains, however, came at the price of having to adopt the institutional agendas set by the state, albeit through a process of *concertación*, as I showed in the case of the Proyecto Biopacífico. The availability of state monies for environmental projects over the five-year period 1993–98 brought state and social movements together into this logic. The much-desired regional force, however, never crystallized (e.g., Pardo and Alvarez 2001; Alvarez 2002).

Restrepo's (2002) poststructuralist analysis concludes by stating that representations of blackness in terms of ethnicity—or in whatever terms—constitute a field of discursive and hence political contestation; there is no necessary correspondence between a given social location (as blacks) and its representation. Ethnic identities may appear to be essential to some actors, as an instrumental imposition by the state to others (e.g., black elites), or as a space for maneuvering to still others (e.g., social movement activists). What has made the black community thinkable and material is precisely the dense interweaving of expert, state, place-based, and activist techniques with their corresponding mediations. The result has been a significant reconfiguration of modalities of power, which speaks of the profoundly political character of identity, an aspect that social movement activists know only too well, as we shall see in the next section.

To sum up, the experience of the 1990s in the southern Pacific exhibits some of the main features contemplated in contemporary theories of identity. To begin with, the historical character of black identities (before and after AT-55) has been firmly established. This means that identities are created by ensembles of discourses and practices. The main discourses of articulation in this case concerned nature (biodiversity) and cultural difference; other important discourses were alternative development and rights. These discourses centered on the notion of black communities, and it was largely this notion that allowed black identities to become part of the discourse. This discursive construction operated through a multiplicity of practices bringing together communities, activists, experts, state officials, academics, and NGOs in various combinations—workshops, special commissions, development and conservation projects, women's projects, legal procedures, and so forth. Largely modeled after the indigenous experience, black ethnic identities were relational and conceived of chiefly as distinct from a dominant Euro-Andean other (whites, or paisas, the whites from the Antioquia region, neighboring the Chocó). As a modernist tool, the construction of black ethnic identities may be seen as part of a process of negotiating a different mode of representation and insertion into national life with the state and society at large. What was at stake was a rearticulation of belonging—a new discursive horizon of meaning—that enabled the creation of an unprecedented political imaginary in terms of difference, autonomy, and cultural rights. Like all modernist identities, black ethnicity was enmeshed in a representational politics enabled by the very structures of power from which it sought to free itself. The extent to which black activists were able to "trouble" this identity along the way remains to be seen. The bottom line is that at least for some years black ethnicity introduced an unprecedented economy of power and visibility in the southern Pacific.

Yet this is not all. To anticipate a bit, the analysis in the following section raises some questions about both the limits of certain theories and the interpretations of the ethnicizing process presented above. Do the discourses and strategies of the social movements evidence only a conversation with modernity? Or do they intuit a decolonial project irreducible to modernity in some sense? Can the play of identities be explained solely in terms of the state and the economy? Or, conversely, could a different understanding of agency lead to a partially different reading? To what extent were activists aware or unaware of being engaged in a give and take with the state? Did that awareness make a difference? Do activists craft long-term visions—beyond and perhaps despite modernity—that could legitimize a different interpretation of their actions? Does the activists' frontal encounter with global coloniality not lead them to envision a politics of difference that at least at some level could be seen as defying the logic of capital and the state from another epistemic space, even if not altogether different? Can a politics of difference be based not only on the exteriorization of an Other but also on positive constructions of place and culture? If this is the case, what other kinds of conversations were activists engaged in besides and beyond ethnic talk? Were indigenous and white identities really the main reference points for black ethnic construction? Or how, for instance, does nature complicate this picture? Finally, what happens when one shifts the framework of interpretation? And how does one reconcile contrasting readings?

II ◈ The Social Movement of Black Communities of the Southern Pacific

On January 3, 1994, at the conclusion of our first year of research in the Pacific, our small research team held a daylong conversation in Buenaventura with a group of eight activists of what was then called Organización de Comunidades Negras (OCN). Based on a set of questions we had circulated in advance, the conversation was intended as a reflection on the organizing process to date. The discussion centered on achievements and needs, the scope of the movement, relations with the state and other black sectors of society, interethnic and gender relations, and so forth. Inevitably, the debate kept going back to persistent issues such as ethnicity and difference, environment and the need of the black sectors to have their own vision of development, and the heterogeneity of the black experience. Some unprecedented topics were also broached, chiefly gender. Twelve years later, the various OCNs had been incorporated into the network called Proceso de Comunidades Negras (PCN), the movement had grown, waned, and resurfaced, crafted a strong identity and seen that identity ebb, risen to national prominence and then scaled down owing to unfavorable conditions. By 2007 (the time of this writing) a core group of activists remained and a cadre of younger ones had joined the network, still focused on some of the key issues that moved them to organize from the beginning—difference, territorial autonomy, and identity—and some new ones, from economic and social rights to racism and reparations.

This section describes and analyzes the emergence and transformation of the black movement network organization PCN. As the most visible expression of a larger "social movement of black communities" of the southern Pacific beginning in the early 1990s, PCN may be seen in terms of the crafting of individual and collective identities in local contentious struggles. These struggles, of course, exist within larger contexts linking communities, the region, and the nation to longer networks and broader socioeconomic, cultural, and political histories. What links the various levels of identity are discourses of articulation. Rather than approaching the articulatory practices primarily from the side of the state and the economy or at an individual level only, I want to see what happens when one looks at the politics of articulation from the epistemic and strategic perspective of the collective agency of the activists.

Identity, History, and Agency ◈ My next rendering of theories of identity resonates with Grossberg's in its attempt to put identity in flux without referring it only to the logic of difference and similarity. Building on the work of two Russian scholars of the first part of the twentieth century, the psychologist L. S. Vygotsky and the literary theorist Mikhail Bakhtin, Holland, Lachicotte, Skinner and Cain (1998), and Holland and Lave (2001) develop a dialogic, practice-oriented, and processual understanding of identities. Identity in this view is a complex form of self-understanding improvised from the cultural resources at hand in a historical background. Their focus is on the intersection of person and society, the individual and the collective, and on how power and culture are negotiated at this intersection to produce particular identities in ways that evince the structured as well as the agential dimensions of the process. Against the most radical constructivist approaches, the dialogic focus develops a conceptualization of history-in-person that makes room for relatively stable identities. As ethnographic studies of the self show, discourses and practices are not only determinants of the self but also tools for identity construction—in short, "socially and historically positioned persons construct their subjectivities in practice" (Holland et al. 1998: 32). This conclusion is agreed upon by much poststructuralist theory; in addition, these authors lean on Bakhtin's idea of the ineluctably dialogic nature of human life to arrive at a notion of "codevelopment—the linked development of people, cultural forms, and social positions in particular historical worlds" (33).[11]

The production of identities in people's interactions with other people and with objects entails the construction of cultural worlds; this takes place through recursive improvisations in a sedimented historical background and also involves various kinds of mediations (e.g., symbolic, linguistic, and other "tools of agency"). In some cases, such as that of activists, these cultural worlds can be thought of as "figured worlds," defined as locally situated, culturally constructed, and socially organized worlds that make visible people's purposeful and reflective agency, that is, their capacity to remake the world in which they live. Although these worlds are subjected to continuous adjustment and reorganization, they may achieve some durability. They are worlds in motion; indeed, "this context of flux is the ground for identity development" and sets the conditions for "a space of authoring" (Holland et al. 1998: 63). Figured worlds of this sort are spaces in which cultural politics are enacted that result in particular personal and collective identities. One can think of activists as having certain competencies for, literally, figuring worlds through a variety of practices, articulations, and cultural artifacts. They do so through forms of "situated learning" in "communities of practice" (Holland et al 1998: 56, citing Lave and Wenger 1991). As in the case of PCN, "identities become important outcomes of participation in communities of practice" or, alternatively, "formed in the process of participating in activities organized by figured worlds" (57). PCN can be seen as an example of a cohesive community of practice in the process of constructing a relatively stable figured world or set of such worlds—including themselves as a group.

Next to the dialogic dimension of identity, as a second leg in the authoring perspective, is its historical dimension; this is fruitfully conceptualized by these authors through the notion of "history in person." At this level the emphasis is on mapping the relations between social, political, and economic conditions, on the one hand, and the identities-in-practice produced within such conditions, on the other. The concept of history in person invokes simultaneously the structuring effect of historical conditions and the actors' mediation of this process through the production of cultural forms that take the historical conditions themselves as resources for self-authoring. Holland and her coworkers introduce two useful notions in this regard: "local contentious practice," that is, the actors' situated participation in local conflicts that are generative of identity; and "enduring historical struggles," those larger processes that constitute the background within which the former category is located. It is not difficult to see how these two concepts might apply to the Pacific. Local contentious practices around specific territories and their biodiversity, for instance, are clearly linked to broader struggles concerning development, race, environment, the state, rights, and globalization. These enduring factors are deeply significant and may even determine the possibility of social existence. Ethnographically, this calls for documenting how "historically institutionalized struggles" linked to long-term conflicts may lead to sustained identities in contentious local practice.

To sum up, local contentious practice mediates between history in person and enduring struggles. Identities and struggles are always unfinished and in process—persons and institutions are never entirely made up previous to and independently of their encounter. Identities are formed in dialogue, if not struggles, across difference, which in turn involves the creation and at times dissolution of boundaries between self and others. In short,

> The dialogic selves formed in local contentious practice are selves engaged with others across practices and discourses inflected by power and privilege. . . . In the course of local struggles, marginalized groups create their own practices. These practices thus provide the means by which subjectivities in the margins of power thicken and become more developed and so more determinant in shaping local struggles. . . . Identities are formed in practice through the often collective work of evoking, improvising, appropriating, and refusing participation in practices that position self and other. They are durable not because individual persons have essential or primal identities but because the multiple contexts in which dialogical, intimate identities make sense and give meaning are re-created in contentious local practice (which is in part shaped and re-shaped by enduring struggle). All of the multiply authored and positioned selves, identities, cultural forms, and local and far-reaching struggles, given together in practice, are bound up in making "history in person." "History in person" thus indexes a world of identity, action, contentious practice, and long-term transformative struggles. (Holland and Lave 2001: 18, 19, 29, 30)

In authoring selves, individuals and collectivities also construct "stabilized social audiences," whether real or idealized; there are periods when identity becomes habituated and stabilized, so that one is no longer aware of its production since the orchestration of identity has somehow endured, even if for a time. This, however, does not necessarily mean that in the case of groups identities are uniform, as we will see in the description of the collective identities by social movement activists.

History and Agency in the Practice of the PCN ◈ The following statements, taken from the interview of January 3, 1994, with the OCN of Buenaventura (see OCN 1996), summarize some of the most important themes in the early development of the PCN, to be discussed in this section: organizational development (construir organización), the concept of black identity, and the activists' view of their relation to the state and the larger social and political context.[12]

> The recognition of rights is a strategy by government to institutionalize problems and make them manageable. It is true that the political opening might function to cushion the economic opening. For the government, Ley 70 is a negotiation tool because it knows that the goal of the organizing process goes well beyond the text itself. The goal of the process is organizational and meant to open up political space. . . . We have to negotiate to the fullest extent possible; it's a way of buying time; we cannot both construct the organization and fight on all fronts at the same time because we are weak. We need to construct the organization (*construir organización*), our own vision of development, and this requires time, energy, and perseverance. . . .
> What happens with "identities"? What is important is to give content to the concepts of black person and black community. Some groups have cultural specificities, but there are some elements in common. We have to engage in collective construction, taking into account what the different groups think, that is, visualize a project in which we can all fit and develop our potentialities and desires. Not because you are black are you part of the *comunidades negras*. One becomes part of the black community if one's lived experience finds expression through practices that reflect the cultural values of these communities. . . . The community does not emerge just because you put four black people together. If it did, Buenaventura would be the largest and most important black community in the country!
> As people become urbanized and "civilized," they lose a great deal of their identity. . . . I cannot assume I am a member of the black community if I am not aware of the fact that besides having a black skin there exists an entire conception of the world and of life that is different and that one starts to lose as one becomes civilized. *The social basis of the black communities lies with those who maintain certain rhythms and daily practices as black community and who do so as part of an historical process. It is those whom we need to reach. We need to ascertain what makes us different* (emphasis added).

The development of PCN's organizational structure and strategy has been a steady process. This process has been fed by the formulation of a set of organizational and political principles in which identity figures prominently. As I have suggested in previous chapters, the identity and organization of the collectivity called PCN has been developed through local contentious struggles—the encounter, primarily, with the state, but also with other groups, including experts, other black groups, and local elites—against the background of the larger theater of conflict and social struggles in Colombia and the world at large.[13]

Building Organization and Identity ◈ In its initial years (1991–94) and in the context of AT-55 and the struggle over the formulation of Ley 70, PCN gave preeminence to the social control of the territory and natural resources as a precondition for the survival, re-creation, and strengthening of culture. This emphasis was reflected in the geography of the movement's practices as much as in its organizational strategy. There was an initial focus on river communities, where activists geared their efforts toward advancing a pedagogical process with the communities on the meaning of the new constitution; reflecting on the fundamental concepts of territory, development, traditional production practices, and use of natural resources; and strengthening the organizational capacity of the communities. This sustained effort served as the basis, during the 1991–93 period, for the elaboration of Ley 70, on the one hand, and to firm up a series of politico-organizational principles, on the other (see below). It also helped PCN activists recognize the various tendencies, trajectories, and styles of work found among the array of black organizations involved with Ley 70.

The first Asamblea Nacional de Comunidades Negras (ANCN, National Conference of Black Communities) took place in July 1992 in Tumaco and was attended by representatives from all over the Pacific, the Caribbean, and the Norte del Cauca regions. Its principal conclusions were aimed at laying down a framework for the regulation of AT-55. At the Second National Conference in May 1993 the delegates revised and approved the text for the law negotiated by government and black community representatives in the ambit of the High Commission created for this purpose. The collective elaboration of the proposal for Ley 70 was a decisive space for the development of the movement. This process was advanced at two levels, one centered on the daily practices of the communities, the other on a political reflection by the activists. The first level—under the rubric of "the logic of the river"—sought the broad participation of local people in the articulation of their rights, aspirations, and dreams.

The second level, although having the river settlements as referent, sought to raise the question of black people as an ethnic group, beyond what could be granted by the law. This level saw the development of a conceptualization of the notions of territory, development, and the social relations of black communities with the rest of Colombian society. This conceptualization took place in a dialogical process with a host of actors, including traditional black politicians linked to the liberal party, who attempted to manipulate the process to gain electoral advantage, and, of course, the state. Government officials realized that the demands of the organizations went well beyond the desire for integration and racial equality as it had been maintained until then by other sectors of the black community. Besides, black organizations mounted a strategy of persuasion and consciousness raising among the delegates to the High Commission appointed by the government for the regulation of the AT-55. The process as a whole constituted a veritable social construction of protest (Klandermans 1992) that culminated in the approval by the Senate of the version of the law (Ley 70) negotiated with the communities.

The Third National Conference was convened in September of 1993 in Puerto Tejada, south of Cali, in the Norte del Cauca region, like Tumaco a predominantly black town. Attended by more than three hundred delegates, the conference debated the politico-organizational situation of the black communities. At the time of the conference, black sectors linked to the traditional political parties and eager to capitalize on the unprecedented legal mechanisms on behalf of the black communities, began to adopt a discourse of blackness that for the most part did not go beyond the question of skin color. Recognizing the existence of these sectors and the diversity of the social movement of black communities, the conference members proposed a self-definition and characterization of themselves as a sector of the social movement of black communities composed of people and organizations with diverse experiences and goals but united around a set of principles, criteria, and objectives that set them apart from other sectors of the movement. In the same vein, they represented a proposal to the entire black community of the country and aspired to construct a unified movement of black communities able to encompass their rights and aspirations.[14]

The objective of the organizing process was stated as "the consolidation of a social movement of black communities for the reconstruction and affirmation of cultural identity," leading to an autonomous organizing strategy "for the achievement of cultural, social, economic, political, and territorial rights and for the defense of natural resources and the environment." One of the central features of the conference was the adoption of a set of politico-organizational principles that, in the activists' view, encompassed the practice, life-world, and desires of the black communities. These principles concerned the key issues of identity, territory, autonomy, and development:

> 1. *The reaffirmation of identity* (the right to being black). In the first place, we conceive of being black from the perspective of our cultural logic and life-world (*cosmovisión*) in all of its social, economic, and political dimensions. This logic counters the logic of domination that intends to exploit and subject our people. . . . Second, our cultural affirmation entails an inner struggle with our consciousness; the affirmation of our being is not easy, since we are taught in many ways and through multiple media that we are all equal. This is the great lie of the logic of domination.
>
> This principle identifies culture and identity as organizing axes. As we shall see, despite its seemingly essentialist tone, it also partakes of a conception of identity as constructed.
>
> 2. *The right to the territory* (as the space for being). As a vital space, the territory is a necessary condition for the recreation and development of our cultural vision. We cannot be if we do not have a space for living in accordance with what we think and desire as a form of life. It follows that we see the territory as a habitat and space where black people develop their being in harmony with nature.
> 3. *Autonomy* (the right to the exercise of being/identity). We understand autonomy in relation to the dominant society, other ethnic groups, and political parties. It arises out of our cultural logic. Thus understood, we are autonomous internally in the political realm, and aspire to social and economic autonomy.
> 4. *Construction of an autonomous perspective of the future.* We intend to construct an autonomous vision of economic and social development based on our culture and traditional forms of production and social organization. The dominant society has systematically imposed on us a vision of development that responds to its own interests and worldview. We have the right to make known to others the vision of our world, as we want to construct it.
> 5. *Declaration of solidarity.* We are part of the struggle for rights of black people throughout the world. From our own particularity, the social movement of black communities shall contribute to the efforts of those who struggle for alternative life projects.[15]

This declaration of principles constituted a rupture with the political and developmentalist formulations of the Left, past black urban organizations, and traditional political sectors. The differences existed around four main issues: (a) the perception of history and identity; (b) the views and demands concerning natural resources, territory, and development; (c) the types of political representation and participation of the communities; and (d) the conception of organizational strategy and modes of construction of the movement. With this strategy, the PCN sought to become a source of power for the black communities vis-à-vis the state and other social actors and to contribute to the search for more just and viable societal options for the country as a whole. From then on, the PCN strategy and its successive transformations were to depend on the activists' assessment of the cultural *and* organizational reality of the communities, on the one hand, and of the balance of forces—from the local to the international levels—between the communities, the social movement, and other social sectors, economic groups, and centers of power, on the other.[16] While the principles are continuously refined and debated, their basic orientation and structure have remained the same.

Finally, PCN's main organizational strategy, at a formal level, is simple: (1) a set of regional *palenques*, corresponding broadly to each of the main regions in the southern Pacific (Valle, Cauca, Nariño), plus the Norte del Cauca and Atlantic Coast regions; these palenques gather together the existing ethnoterritorial organizations within the region; (2) a national coordinating committee; (3) technical teams at national and, in some cases, regional levels. Originally designating the autonomous territories of maroon or freed slaves in colonial times, today's palenques are spaces for discussion, decision making, and policy orientation in each of the regions with important black presence. They operate in conjunction with the ANCN and as a group constitute the Consejo Nacional de Palenques. Regional palenques are composed of two representatives from each of the region's organizations. The National Coordinating Committee is in charge of coordinating actions, implementing the decisions of the ANCN, and representing the PCN in national and international fora; it is also intended as a space to discuss the various *tendencias* within the movement and to generate consensus on important matters. The committee also coordinates the technical teams and nominates the palenque representatives to special projects or commissions at the national and international levels. The technical teams contribute technical elements in the making of policy decisions in economic, development, environmental, and ethnoeducational matters.[17]

It would be a mistake, however, to see this structure as a rigid set of norms independent of the day-to-day practices of the activists. There is agreement among social movement theorists about the need to avoid the dichotomy that has prevailed between structure-oriented explanations and agency-focused ones. In the newer models (e.g., Scheller 2001; Peltonen 2006), agency and structure are inseparable and mutually constitutive; even when formalized, as in PCN's case, structures are not ready-made and waiting to be filled in by activists. The structure itself is made of movement and enacted in practice. The structure, if anything, is the result of patterned movement over time. PCN activists' awareness of this fact is visible in the way they view the organizing principles. As one activist put it, "The principles have been the motor of our collective identity. We no longer construct in a vacuum—one is no longer just any black person. The principles confer coherence between discourse and practice, between organization and the everyday. *Everything we do is done out of the principles*; it amounts to creating structures and strategies out of a political project and not out of a prefabricated scheme" (emphasis added).[18] The five principles, then, link structure and agency in everyday practice. PCN activists have generally been successful in developing their approaches and strategies—from the very small to the important ones—from the perspective of these principles.[19]

The centrality of the principles also has to do with the reproduction over time of the collectivity called PCN. PCN is the product of an intense set of practices developed around local contentious struggles (some of which we have seen in some detail, especially in the environmental arena). These practices involve ongoing communication among activists at various levels: horizontally, at the national, regional, and local levels; vertically, across levels. Communication is particularly intense at the national and regional levels. At the national level, there is a high degree of face-to-face and electronic communication among the National Coordinating Committee members and with some of the main members of the technical teams. This group forms a tight collective whose members are in permanent contact, with active, not infrequently heated, debates on ongoing, concrete issues, decisions, and so forth. The consensus on decisions is often arrived at only after the disagreements (*los disensos*) have been discussed substantially. For some outside observers, this intense level of debate and communications hinders effective action. Be that as it may, this practice has enabled a core group of activists to remain steadfast in their resolve and commitment to the collective identity of PCN and what it represents. Communications are weaker at the local level, and so are identities. While national and some regional activists have internalized the principles as a political vision, those at the local level for the most part have not; activists at the national and regional levels are very aware of this fact. In some regions, the lack of internalization has led to the dismantling of local organizations and, coupled with the armed conflict, to the decimation of the regional palenques. At the national level, as activists put it, "we construct on the basis of the collective [*lo colectivo*], not from the individual [*lo individual*]. Our aim is not to be 'me' but to be 'us.'"[20]

There is a tight connection between culture and identity in the understanding of the activists. As one of the best-known Tumaco activists put it, "It is not us who will save culture, it is culture who will save us" (quoted in Alvarez 2002: 13). This does not mean, however, that the activists see culture as a static variable; on the contrary, "culture is a process of construction that is constantly being enriched and fed back into the situation" (OCN 1996: 262). There are, of course, tensions in this conception. The collective identity construction by PCN bears similarities to the construction of Caribbean and Afro-British identities analyzed by Stuart Hall. For Hall (1990), ethnic identity construction entails cultural and political negotiations characterized by a certain doubleness: on the one hand, identity is thought of as being rooted in shared cultural practices, a collective self of sorts; this conception of identity involves an imaginative rediscovery of culture that lends coherence to the experience of fragmentation, dispersal, and oppression. On the other hand, identity is seen in terms of the differences created by history; this aspect of identity construction emphasizes becoming rather than being, positioning rather than essence, and discontinuity as well as continuities at the cultural level. For the activists, the defense of certain cultural practices of the river communities is a strategic question to the extent that they are seen as embodying not only resistance to capitalism and modernity but elements for alternative world constructions. Although often couched in culturalist language, this defense is not essentializing to the extent that it responds to an interpretation of the challenges faced by the communities and the possibilities presented by a cautious opening toward forms of modernity such as biodiversity conservation and alternative development. Identity is thus seen in both ways: as anchored in traditional practices and forms of knowledge and as an ever-changing project of cultural and political construction. In this way, the movement builds upon the submerged networks of cultural practices and meanings of the river communities and their active construction of life-worlds (Melucci 1989), although it sees such practices in their transformative capacity. To the fixed, static, and conventional notion of identity implicit in the Constitution of 1991, the movement thus opposes a more fluid notion of identity as political construction (see Grueso, Rosero, and Escobar 1998 for further elaboration).

The Productivity of Identity Constructions: In, Against, and Beyond the State? ◈ The period 1995–96 saw the appearance of a variety of organized black sectors with different and at times conflicting agendas, seeking to bank on the space created for black people's rights.[21] Over the years, the conflicts and contradictions among all of these groups impinged upon important issues such as the composition and work of the High Commission, the formulation of the National Development Plan for Black Communities, the negotiation of environmental conflicts, electoral representation, and so forth. Notwithstanding, PCN's concrete achievements have by no means been negligible. Some of them have already been mentioned, such as the central role played in the formulation of Ley 70, in the reformulation of PBP, and in other areas of environmental conflict and cultural policy. PCN also made significant contributions over the years to the creation of community organizations in a number of rivers in the southern Pacific, to the configuration of community councils and the titling process, and to the funding of specific projects. In recent years, PCN has been at the forefront, along with the Association of Displaced Afro-Colombians (AFRODES), of the organizing against displacement, the free trade agreement with the United States, and various human rights causes; they have done so nationally and transnationally, becoming highly visible again (say, in their work with the U.S. Congress Black Caucus in Washington, D.C.).

One can assess these engagements from the perspective of their contribution to identity. First, the history and agency approach shows the extent to which activists take the current historical conditions as resources for collective self-authoring. It is certainly the case that by using the tools of modernity the activists also further entangle themselves in the worlds from which they seek liberation. In doing so, however, they attempt to redraw the existing hierarchy of power and privilege and to keep alive the heteroglossic potential of all world-making practice. In the process of struggling with modernist discourses, activists' discourses distance themselves, if in minor ways, from the authority of the dominant norms and in so doing produce differentiated voices—what I have called alternative modernities, and a glimpse of alternatives to modernity. In other words, to take the notion of dialogism seriously one has to bracket the ultimate one-directionality that characterizes state- and capitalocentric perspectives and that makes the idea of genuine difference impossible.

Indeed, from the very beginning PCN activists were keenly aware of the limits of negotiating with the state. The discussion of their participation in PBP, for instance, always broached the dangers of clientelism and co-optation. Most of the activists' engagement with the state was influenced by two factors: the need to buy time (that is, to lessen or slow down the cultural and ecological damage of the Pacific) and organizational strengthening. There are high costs of time and energy in becoming involved with the state. This is why each situation involves a conscious decision. As activists put it, "It is not only through direct confrontation that fissures in the structures of power can be created; we can, and should, work within the structures of the state. Here lie possibilities to heighten contradictions and create trouble, at the risk of being out of a job every six months and of the movement having to take care of the problem of daily sustenance for family and children" (OCN 1996: 264). Influenced by older idioms of the Left (the analysis of the contradictions), this principle nevertheless reveals a political strategy of working "in and against the State" (Mueller 1987), but also despite and beyond the state. The state, as I have repeatedly shown, does not dominate the time and imagination of the activists. If in the late 1990s PCN decided to pull back from relations with the state, after 2004 (during the second Uribe administration) it decided, in light of a counteroffensive by the state on many vital fronts, particularly a series of deleterious legal reforms on forests, water, rural development, and the TLC (Tratado de Libre Comercio) or free trade agreements, that this was a space it could not afford to abandon (see the conclusion).

I have attempted here to use the framework of "history-in-person" to interpret the experience of the collectivity called PCN. This collectivity constituted its identity through dialogical processes of various kinds, some of which involve intrapersonal relationships inside the group, others encounters with a host of actors (from state actors and experts to armed actors) in local contentious practice concerning the control of local territories, the defense of particular cultural practices, the struggle for the right to difference, and so forth. These local conflicts are related to broader struggles concerning the destruction of the humid forest, racism, development, neoliberal capitalism, free trade agreements, and so forth. As a collective identity constructed around a particular figured world—the social movement of black communities—this collectivity constitutes a community of practice that makes possible the production of discourses, performances, activities, and so forth. In so doing, and despite its ups and downs, it has achieved some durability. In the period 1993–2000, this figured world was able to construct a relatively stable audience, including other sectors of the movement, environmentalists, members of NGOs, state programs, and so forth. It can thus be said that the collective identity called PCN is the outcome of intense participation in communities and activities organized by figured worlds—its own and those of others.

III ◈ Activism as History Making: The Personal and the Collective

The emphasis on *lo colectivo* does not mean *lo personal* is neglected. The question of the personal project has began to be raised within PCN only recently, and with the general understanding that even if the personal is important it cannot be construed at the expense of the collective; the starting point is that the personal also has historical and political dimensions—it is, in short, history-in-person all the way down. This section introduces some elements that are useful for thinking about the personal dimension of activism in the context of a strong collective identity such as PCN's. It will also allow me to underscore a certain ethics of activism.[22]

In the personal narratives of activists, the personal dimension of collective action starts with their early experience of difference, discrimination, and the sense of injustice. Many of the PCN activists were born in small river towns of the southern Pacific or in the port cities of Tumaco and Buenaventura, where they spent the formative years of childhood and adolescence, sometimes in their grandparents' homes. At this level, the memories are generally of happy times, pleasant remembrances of life by the river or by the sea, under attentive parental or grandparents' care, enveloped in local culture (food, the drumbeat of music and dance, the carelessness of childhood lived in river and forest, and so forth). Perhaps the most common memory of their first encounter with difference is that of traveling to the Andean cities and there becoming aware of their blackness, with various degrees and forms of discrimination directed at them. In the larger towns of the Pacific, the formative experience of difference often had to do with the difficult relationships with the paisas, although in a few instances it was related to nearby minority groups of *indígenas*. A few were moved to activism by their family history (e.g., parents who were union activists or mothers who were schoolteachers), and a few others by becoming involved in their late teens or early twenties with the leftist parties of the 1970s and 1980s or through the reading of Marxism. In many cases, the progressive church—what is called the Pastoral Afro-Americana, a movement inspired by liberation theology that focused on the rights of black people, chiefly in the Pacific—played an important role. Most PCN activists, however, came to activism through their personal encounter with injustice and difference; the experience was usually, but not always, of racial character. With the emergence of black identity as a social fact in the 1980s in the cities (especially with the Movimiento Nacional Cimarrón) and throughout the Pacific in the 1990s, many blacks turned to ethnic activism as their main form of political engagement.

One activist's recollection of her first awareness of being black is illustrative. Growing up in a river community that was predominantly inhabited by blacks, she says that

> "to be black" was not a determinant fact; it wasn't something that conditioned me negatively from the outside, it did not mean anything different from being a person. I belonged to a community with whom I shared a way of life, beyond racial conditioning. It was when I traveled to Popayán—a "white" Andean city—when I first learned the meaning of blackness for Colombian society. The racial discrimination I experienced there greatly shaped my perception of being black, changing my process of identity formation, which until then was something quite positive and largely unconscious. (Hurtado 1996: 332)

Years later, attending university in Bogotá, she began to make sense of her situation, first by participating in the Movimiento Nacional Cimarrón and, after 1990, in the social movement of black communities. As she explains, "The encounter with ethnicity involves a reencounter with oneself, by countering the dehumanization that arises from society's denial of difference based on skin color. The constitutive elements of the Afro-Colombian person are reinforced by affirming her humanity through the fact of difference. . . . Identity is constructed and mobilized through a series of encounters and misencounters, it's learned and unlearned, since we are not finished beings but in constant change in relation to others" (Hurtado 1996: 332).[23]

Another activist, who was about thirty years old, returned to her native rural town in the Tumaco area after living for many years in Cali. The urge she felt to do something for those around her had been present from an early age. "Some say," she recounts, "that my desire to work on behalf of the community was present since I was a child." In high school, she was known for having "revolutionary ideas." For her, however, these ideas "were just an expression of the need to be equal to others; we live under deplorable conditions, and we have the right to be better off." Her experiences in Cali and Tumaco with progressive priests and nuns were important in shaping her work as a cultural activist and entrepreneur on behalf of her communities. Her conviction, however, was rooted in the community itself. In the early 1990s, she said, "we worked for Ley 70 from a perspective called 'pastoral,' but I always said that we also need to work from the community's own vision." Passionate about injustice ("when I hear mention of a palmicultora I feel as if my blood is boiling in my veins, and I feel as if I were myself in the plantation, and I would not want any of my people to be there"), she found that by the early 1990s culture had become the focus of her political work. One of her first actions was to start a dance group with young girls and boys "to counter the acculturation introduced by the media." She was convinced that "if it is true that it is important to be part of other cultures, if we lose ours everything is gone, we would be nothing; the Pacific is recognized for its cultural manifestations, for its way to work the land, its chants and dances; this is what enables us to keep on living. If it ceases to exist we will be nothing." Indeed, while working with AT-55, she learned that those who had lost their lands realized that "with the [loss of] land they lost their lives." She seeks to combine cultural work with concrete projects to improve people's living conditions. Together, despite tremendous challenges, these two aspects gave her unshakable courage and commitment. "Today," she said in 1993, "I feel as strong as a rock."

In retrospect, one can see that for this particular activist, as for the previous one, the personal dimension was important from the outset. As she put it, "The important thing was to be able to recognize myself, to know who I was as a person; this brought me closer to the people . . . If today they [the local politicians] think I am crazy, this gives me greater strength to work for the people."[24] These meaningful personal experiences can be of many kinds, from happy to the painful. For another woman activist in her early forties who was taken away from the river when she was twelve to make a living in the city and returned to her river almost twenty years later, the memories of and identification with life on the river (fishing collectively, sharing of food, planting and harvesting, etc.) are particularly strong. After working for years as a domestic in Cali and enduring abuse ("You try to serve your bosses as best as you can," she said, "but they always treat you badly, and this makes you feel bad"), she decided to return to the river and work for PCN as a grassroots activist. For her, the fight for rights is important because "it seems that everything is denied to us, including the opportunity to study, because we are black." That is how, she explains, "I became PCN." Today, this woman is one of the most effective local leaders and organizers on one of the rivers in the Buenaventura rural area.

That memories of life on the river can be important in shaping activism is also illustrated by the experience of a prominent PCN activist. She spent her first years as a child on a river in the Buenaventura area. The following account is from one of her more important memories:

> My mother, who was a rural teacher, was always organizing activities in the community, such as projects around the school; any important celebration became a big community affair. I remember with special fondness a great *minga* [community work project] to prepare a plot of land to plant trees and food crops, in which children and adults participated. On that day there was a huge community meal [*olla comunitaria*] and the main dish was *mico tití* [a local monkey]. The hunters had brought this meat from the *monte* especially for the minga. I did not eat that day because I saw part of the monkey in the pot. I can say that I learned in this community the value of collective work, the great satisfaction of solidarity. I have had the inclination to work with people ever since.

For this activist, the river was also the site of cultural practices that she would find out later on were profoundly different from those in the cities (e.g., the *chigualo*, or ritual observed on the death of a child). These differences were to become more and more central with the growth of ethnic consciousness. As a young college student, her influences were her father (a workers' organizer on the docks of Buenaventura) and her work with poor black people in Buenaventura as well as the teachings of Gerardo Valencia Cano, the so-called red bishop of Buenaventura and one of the main advocates of liberation theology in Colombia, who died in an unexplained helicopter crash in the late 1960s. The language of this period spoke of the betterment of conditions for poor black people. One could say that in this language and practice an ethnic consciousness was being born. This consciousness would blossom fully once the language of ethnic identity became available.

This consciousness and these experiences ground powerful visions of the struggle and of the future. PCN activists' conviction is that their struggle goes well beyond the issue of rights for black people. Their ultimate goal is to contribute to the search for more just and viable societal options for the country as a whole, if not the world. The idea was clearly stated early on:

> Anthropologically speaking there are multiple groups among the black people, but politically speaking we are an ethnic group. We struggle [*reivindicamos*] for the right to have a different vision, one that constitutes a

> possibility and an alternative to today's enormous crisis of societal models. This does not mean that there cannot be other groups within the black movement. We cannot conceive of a movement only for the black community, nor do we overlook the fact that the great problems faced by Colombia and Latin America are not only the product of the *mestizo* mind; there are black people who have adapted to this system and contribute to our oppression. Our discourse does not focus only on the ethnic question. There are problems in common with other people, and we know we have to work toward the construction of a project that encompasses many sectors, black and nonblack, who share similar problems. (OCN 1996: 255–256; see also PCN's participation in transnational networks, discussed in chapter 6)

This orientation has been present in the many facets of the movement's work—for instance, in the persistence of the vision of the Pacific as a region-territory of ethnic groups; in its alternative frameworks of development and sustainability, which envision a Pacific that retains much of its cultural and ecological aliveness and diversity; and in the PCN's contribution to the broader project of self-definition for the black groups of Latin America. The issue of self-definition, under development in recent years, is based on the idea of a model of solidarity with people and nature, "as a contribution to society as a whole and toward the recuperation of more dignified ways of existence for all living beings on the planet" (Grueso 1996: 7).

One of the activists stated her vision of the future of the Pacific:

> I imagine a Pacific that preserves its landscape and its people; a Pacific with all its trees, all its rivers, all its animals and birds, all its mangroves. I imagine a Pacific where people live well according to their cultural vision, where money serves to facilitate exchange and does not become life's goal. I imagine a Pacific where music and happiness accompany all activities in individual and collective life. I imagine a Pacific where black and indigenous peoples are able to contribute their cultural values to the construction of societal options based on respect for the other's difference.[25]

The important point is not whether the memories are idealized or whether the visions of the future are romantic. The point is to understand how memories and visions come to be integral elements of a sustained, coherent political practice. Perhaps these activists are dreamers, but they anchor their dreams with great intelligence in a caring and courageous political practice. They sing songs of freedom, of emancipation from the mental slavery that has seemingly become commonplace in the age of the total market—the market as the ultimate arbiter and framework of life. And, as the other well-known song says, they are not the only dreamers. Many others today have come to believe that another world is indeed possible. In this, as some philosophers would have it, they are just retrieving their history-making skills, out of their place-based (not place-bound) commitments. Activism can also be seen in this light, and this is the last theoretical point I want to make.

The notion of the retrieval of history-making skills has been developed by a handful of phenomenology-oriented scholars and has great promise for understanding activism. Phenomenology enables a view of identity as an expression of the profound historicity of one's encounter with the world. According to this view, humans are constituted on the basis of concrete encounters with the world in their everyday coping, and knowledge is built from small domains and tasks that make up microidentities and microworlds; this is a different way of looking at history-in-person, one that brings to the fore the embedded and embodied character of all human action.[26] Varela builds his argument by bringing together new trends in cognitive science, his own phenomenological theory of cognition as embodied enaction, and the traditions of Taoism, Confucianism, and Buddhism. His goal is to articulate a theory of "ethical know-how" (an embedded understanding of action), as opposed to the Cartesian "know-what" (abstract, rational judgment) that has become prevalent in modern worlds.

PCN activists could be seen as having ethical expertise (know-how) of the sort Varela describes. To be sure, this expertise is coupled always with rational judgment (know-what), but what most defines PCN is a continuous engagement with the everyday reality of Afro-Colombian groups, grounded in the last instance (although with layers of mediation, as we have seen) in the experience of the Pacific as a place. Activists are skillful at responding to the needs of their own collectivity and to those of others. This process involves ethical expertise more than rational deliberation. Ethical expertise, as I tried to show in the brief personal vignettes, is cultivated in all aspects of the activists' lives. For those whom PCN activists call traditional authorities (*autoridades tradicionales*, roughly, local wise people who always know what to do because they are profoundly rooted in the community, usually elders) embedded ethical know-how predominates. As the narrative of an elder leader of the Yurumanguí river demonstrates, these "natural leaders" do not see themselves as contributing anything to the process as individuals because they do not see themselves as individuals contributing to something separate (the community) and because they do only what they know to be good (see don

Antonio's life history in Cogollo 2005). For the more cosmopolitan PCN activists, political practice is a combination of both types of ethics. Commenting on Confucian and Buddhist traditions, Varela suggests that although everybody has ethical know-how of a certain type because of the fact of being part of a collectivity, true expertise at virtuous action comes after a long process of cultivation. The process involves pragmatic, progressive learning that goes well beyond the intellectual process and results in nondual action that refuses to separate subject and object. Such nondual action becomes well grounded "in a substrate both at rest and at peace" (Varela 1999: 34). This groundedness explains what many outsiders describe as the steadfastness and inner peace that seem to characterize many activists, even those who face horrendous conditions, including many of those in PCN.[27]

In contrast to the detached view of people and things instilled by modern science, the phenomenological perspective highlights the retrieval of history-making skills, which implies building on a contextualized, embodied, and situated notion of human practice (Spinosa, Flores, and Dreufys 1997). In this argument, humans live at their best when engaged in acts of history making, meaning the ability to engage in the ontological act of disclosing new ways of being, of transforming the ways in which they understand and deal with themselves and the world. This happens, for instance, when activists identify and hold on to a disharmony in ways that transform the cultural background of understanding in which people live (say, about nature, racism, sexism, homophobia). There is also a connection to place in this argument, to the extent that the life of skillful disclosing, which makes the world look genuinely different, is possible only through a life of intense engagement with a place and a collectivity. Skillful disclosing requires immersion in particular problems and places, with the real risk taking that such rootedness entails. Only under these circumstances can the kind of interpretive speaking worthy of attention to a community be exercised. Place-based activists, intellectuals, and common people do not act as detached contributors to public debate (as in the talk show model of the public sphere or as in attempts to explain problems in terms of abstract principles) but are able to articulate the concerns of their constituencies in direct ways. Identities are thus the result of engagement with cultural worlds; they arise not out of detached deliberation but out of "involved experimentation" (24). This is the role of disclosure properly speaking, which requires sensitivity to problematic practices that might have become habitual or to marginal or occluded practices that could be fostered or retrieved. In sum, historical identities are neither rigid or essential nor fully contingent. They are grounded in a familiar style of practices, and it is out of this contextual grounding that they change.[28]

The partially embedded and place-based character of activism is a feature of social movements that often goes unacknowledged (Harcourt and Escobar, eds. 2005). It should be evident that in places like the Pacific the place dimension of movements is of paramount importance. Movements are situated in place and space, and this situatedness is an important component of their practice. In some instances "involved experimentation" in cultural worlds is often as important as explicitly articulated strategies, if not more so. This type of cultural shift happened, for example, with gender in the case we are considering. More than being articulated through a detached discourse of rights, even acknowledging the importance of such a discourse, the gender dimension in PCN has been advanced by women—and, in some ways, by men too—through daily strategies of positioning and of challenging and changing practices. This is the last important aspect of activism to be dealt with in this chapter.

IV ◈ Women, Gender, and Ethnic Identity

Studies of black women in Colombia have been characterized in terms of "eloquent silences and emergent voices" (Camacho 2004). If black groups in general were largely invisible in academic studies until recent decades, black women were even more so. The visibility of Afro-Colombian peoples over the past two decades has been paralleled by an expansion of Afro-Colombian studies, including gender, particularly in anthropology, history, and environmental studies. Among the topics covered are the socioeconomic aspects of black women's contribution to production and reproduction; women's role in the re-creation of culture, including religion and healing practices; and women's protagonism in social organization, particularly the family and extended kinship. Historians have given attention to the place of black women in colonial society—e.g., from sexual representations to matrilineality; anthropologists have described a certain gender complementary of roles in production and in other social domains; and ecologists and ecological anthropologists have shown in detail women's key contributions to food security, biodiversity conservation, and territorial appropriation. Finally, some attention has been give to the participation of women, or lack thereof, in political processes. It is this last aspect I would like to discuss in this section, and the extent to which such participation has led to the explicit articulation of a gender perspective or not.[29]

Many commentators have remarked on the unusually high profile of women in some sectors of the black movement of the Pacific. This is particularly the case with PCN, where women are among the most prominent, articulate, and well-regarded leaders at all levels in contrast with many other ethnic and popular movements. This feature is not lost to PCN male leaders, who are very much aware that some of their women comrades might be among the most important popular women leaders in the country. This prominence, however, has not translated easily into an explicit articulation of a gender dimension of the struggle within the movement; the reasons for this are complex. In order to explain this absence—and taking a cue from the framework developed by Flórez Flórez (2004, 2007)—it is useful to differentiate among three levels: explicit articulation of a gender dimension of the struggle or of a separate gender struggle; second, forms of positioning and strategy by women activists (and, to a lesser extent and in a different way, by men) that constitute de facto challenges to existing gender relations, albeit not articulated as such; and daily practices that unsettle gendered cultural patterns, thus politicizing gender relations. In other words, it is important not to remain at the first level of analysis, as is usually the case, but to probe deeper into the daily strategies and practices of activists vis-à-vis gender relations.

These levels could be detected at the end of the first phase of our research, as the following quotes from our day long interview of January 3, 1994, indicate. One of the women activists offered the following statement on the matter:

> Here [in the movement] the discussion has taken place on the basis of blackness and the black person. Many of us work with young people and with women; but these particularities are set aside when one privileges the ethnic factor [*lo étnico*]. Today, however, we begin to see the need to identify these particularities so we can move forward with the global process of strengthening the black community. . . . In relation to the affirmation of identity we start to think about the role played by women at home, the neighborhood, the place of work, the street, and their central role in the socialization of children. . . . To the extent that they begin to see themselves as black women, they are going to play a fundamental role in the formation of the new generations. (OCN 1996: 256)

At this moment, then, there was a clear privileging of ethnic identity, with a nascent interest in gender. The theorization of gender initially took the form of looking at women's "particularity," including their role in creating a sense of belonging to the territory and shaping the social group, since men (for historical reasons having to do with slavery, and thus governed by a different logic of territorial appropriation) are often away and maintain several households on different rivers. The need to change this sort of practice (e.g., polygyny) was noticed, although there was an early insistence that the mechanisms to change it had to be endogenously generated: it is by women and men becoming conscious of the needs of the black community as a whole (that is, as an ethnic and cultural group) that these practices will change. As the same activist continued:

> In the Pacific, women's groups are gaining ground and becoming consolidated, responding to concrete needs and taking on concrete responsibilities. This does not mean that things should remain as they are [in terms of conventional gender relations]. . . . The idea of gender is important, especially when it becomes conscious; women have demonstrated tremendous organizational capacity and strength; they are the anchor and structure of the process. . . . Before, women did not see this issue in terms of the black community; they only saw the economic and institutional problems, since institutions promoted an economistic view . . . now they are beginning to see the problem in a more encompassing manner, since they have problems with institutions, so they now see the issue of black women and black people as a whole, and they realize that the social and territorial gains of the black communities should solve their problems as black women as well. They also see the need to be part of the movement of black communities because besides being women they are black and they feel it that way. (257)

Again, this does not mean that the activists believe the ethnic struggle will solve women's needs automatically. As another activist concluded, "We live in a sexist society. A couple in which both people have equal possibilities and responsibilities in the [political] process, but where one has to devote more time to taking care of the children, thus having less time for the process, is clearly a loss for the process. We rarely discuss these issues, but we assume them; we don't discuss them openly too much, yet we feel and live them every day" (258). In other words, as activists, women conceive of their struggle as women as a daily political task they need to advance both within the movement and within the community at large. In the initial years, men saw gender as an externally imposed discourse that does not reflect the experience of the black community. For a male leader, "the social movement of black communities has to face the gender problem, but it has to do it from within itself, not from the outside; and it will do so, as with any other political question, when it becomes a pressing issue, or a possibility. . . . The issue of women is not a women's question but one of the entire black community" (260). In a dialectical fashion, the same leader recognized that the very institutions, through their developmentalist attempts focused on women, were making the issue of women an important one for the movement.

For both women and men activists, arguments about gender relations have to be articulated politically, and this discussion has to start with the recognition of the multiple forms of subordination faced by black people; these include a panoply of socioeconomic subordinations caused by racism and exploitation most aptly summarized in the saying, recalled by one of the women activists, that "blacks equalize at the lower end of the scale" (*el ser negro iguala por lo bajo*), that is, there is a common denominator of oppression which equalizes them all at the bottom of the social hierarchy and that has to do, above all, with blackness. There is also the belief that in the first half of the 1990s political expediency dictated that the cultural and ethnic question be given priority. Even in the late 1990s, most women and men activists did not see gender as "the principal problem." Indeed, women of ethnoterritorial organizations tended to resist the articulation of an explicit "gender perspective," as it is usually known in the developmentalist sector (*perspectiva de género*). Asked why this is so, one of the main women leaders replied, "This resistance is obvious, since we are not convinced by the potential political gains of women claiming rights on their own [*por su lado*] when the response to the cultural and socioeconomic demands of the black communities has been so minimal; also, it is not convenient to air one's dirty laundry in public when the outside pressure is so strong."[30] Again, lest this statement sound like a complete dismissal of the need to change gender relations, the same activist went on to emphasize how women actually transform gender relations in their practice, even without an explicit discourse.

As the Venezuelan-Colombian social psychologist Juliana Flórez Flórez (2004, 2007) argues, women are often able to open up spaces of resistance to challenge existing norms while pushing the collective project and the organizational principles in new directions. In time, the issue of women became more and more explicitly discussed. Flórez explains this transformation in terms of the incorporation of *disensos* into the social movement dynamic, that is, of those aspects that might bring about conflicts and divisions. For Flórez, social movement theory has not dealt with this dimension of collective action because it privileges a modern understanding of power that expels disensos and subjectivity from the inquiry. Yet these two dimensions are crucial to movement transformation; a strong identity as a political subject, she argues, demands after a certain point that activists face differences and power relations within the movement; this started to happen, for PCN, after a period of steady work on ethnic identity, particularly 1993–98. The processes of disenso took place around various forms of difference, with gender occupying a prominent role (although other aspects, such as age, religious practice, and rural versus urban origin and orientation, have also been important). For Flórez, dealing with differences such as gender brings to the fore the existential aspects (*aspectos vivenciales*) of political activism, including desires, passions, emotions, friendship, and the ludic (*lo lúdico*).

Flórez's important contribution highlights the "tactics of de-subjectivation" (*tácticas de des-sujeción*) that women engage in at many levels in their day-to-day activities in order to shift gender relations (e.g., in the spheres of production, labor, personal and family relations, the armed conflict, the production of knowledge, and the relation with the racism of the society at large). The end result is that the limits of both black identity and black political identities are shifted along the way, if in a slow but steady manner. As she concludes, "After more than a decade of struggle, we may conclude that although PCN has not explicitly included working on gender in its political strategy, it has dealt with this subject in that some if its activists have enacted creatively, and slowly but cumulatively, various *tácticas de des-sujeción* in order to shift the borders of black identity from the perspective of their experience as women" (2007: 344). By creating more room for internal differences and desires, new spaces of construction of the collective identity are created. What emerges is another possibility for the movement to shift around its own borders. After a degree of consolidation of the agenda of ethnic identity had been achieved and after some years of internal crisis, the acknowledgment of differences is one of the main directions in which PCN seemed to be moving in recent years. In the meantime, women activists continue to engage in a cultural politics that has widened the sphere of the political.

Could these challenges be construed as a gender perspective in any way, or as contributing to such an articulation? Is the everyday displacement of some practices enough to accomplish a more visible and lasting social transformation at the level of gender? Could this be achieved without a discourse that articulates explicitly the existing relations of subordination as relations of oppression, as Laclau and Mouffe (1985) would have it? There are two interrelated issues to be further discussed in this regard. The first is whether modern liberal or radical discourses of gender—as they circulate transnationally among movements and institutions—can provide an appropriate basis for a gender articulation in cases such as the Pacific; the second, and closely related, is how activists explain the adequacy or inadequacy of the available languages. The inadequacy of liberal idioms of women's liberation, particularly women's equality, is explained by activists in terms of the different logic that underlies the cultural practices of the river communities. This is a relatively well known discussion in anthropology (although far from having been dealt with satisfactorily), but it is often taken up in activist and policy circles in a somewhat simplified manner. Simply put, black and indigenous river communities are seen as characterized by gender relations that are different from those of modern societies, and patriarchy is not seen as operating on the same basis, given the very different configuration of family, territory, and kinship. The most common idiom to explain this difference is that of vernacular gender relations, or gender complementarity of tasks. How have PCN activists, relying on anthropological and ecological studies, attempted to theorize this process?

For the river communities, gender complementarity is seen as the basis for the historically specific forms of territorial appropriation based on kinship and traditional production practices; this process entails having intimate knowledge of the environment, including gender specialization of knowledge (e.g., gendered healing practices based on different knowledge of plants or types of illness). Given the spatiality of this process—the spatiality of the life project, in PCN's conceptualization, as discussed in chapter 2—women and men have complementary tasks according to their respective long-standing knowledge and roles. While men engage in those tasks that require greater physical strength and time away from the familial environment, women are in charge of those productive aspects that take place around the river and the domestic space, thus ensuring the socialization and integration of the family, the feeling of belonging to a place, and cultural identity. Use spaces, as we know from the chapter on nature, are also deeply gendered, which is another aspect highlighted by activists to signal the gendered and culturally specific forms of appropriation. Gender complementarity is thus seen as important for the socioeconomic and political project of the region-territory.

In recent years, women activists have introduced a distinction that brings them closer to articulating a gender discourse. As two women activists from PCN write, "Unlike what happens in the socioproductive space, the political and familial spaces are characterized by a lack of gender complementariness" (Grueso and Arroyo 2005: 105). This is an interesting move to highlight inequality in some domains. Again, the response by women activists to this new formulation has been to struggle to widen their political spaces of participation, given that "at the political level, the lack of complementariness is reflected in the difficulties faced by women in having access to public spaces of analysis, planning, and organizing" (63); this is particularly the case for grassroots women. Consequently, it is now acknowledged that "any political project must make explicit the role of women. The political project needs to evaluate traditional values and practices, to be sure, it must open itself up to supracultural rights, and it has to propose ways to change sexist practices, including those that might even be condoned by women"; this women-centered discourse is followed by the assertion that even if "the transformation of these practices needs to originate with women," it should take as a point of departure "the culturally established gender relations and how these are reflected in the political domain" (107). Work by these activists and others within the Black Women's Network of the Pacific has yielded a detailed and well thought out strategy that includes a set of problems and concrete actions and goals to overcome problems specific to women. Each of these problems and actions, however, is carefully developed out of and related back to the five organizational principles of the movement (see table 5). This is a way for women activists to ensure a close articulation with the ethnic struggle. The table shows the women's critical assessment of internal and external obstacles to be overcome in relation to each of the principles (adapted from Grueso and Arroyo 2005: 110–12).

This exercise revealed many of the concerns that could also be arrived at from a liberal perspective (e.g., naturalization of women's subordination, forms of violence, low educational levels and political participation), albeit with some noticeable absences (e.g., goals concerning sexuality) and with emphases often not found in liberal perspectives (e.g., critical reflection on culture, community-oriented participation, collective aspects of gender relations, organizational needs vis-à-vis other black and black women organizations). A crucial issue is the structuring role of the principles. As a woman activist put it, "The *principios* have been the driving force of our collective identity; we no longer operate in a vacuum, but on the basis of the principles; one is not any black person, but our actions are in all cases (even in dealing with the state on human rights violations) guided by the respect for difference. The principles ensure consistency between discourse and everyday practice. Everything that has to do with PCN we construct on the basis of the principles, whether we are considering urban areas such as Aguablanca [in Cali] or a river in the Pacific."[31]

Table 5. Analysis and Strategies of the Black Women's Network for Black Women and the Black Communities of the Pacific

Problems identified in relation to the politico-organizational principles of PCN		Strategies to overcome problems from the perspective of the Black Women's Network	
Principle and meaning	Problem from black women's perspective	Action objective	Line of action
The right to a black identity			
• It means recognition, appreciation, and acceptance within the communities and in the rest of society, as black people with a distinctive culture. • It means self-recognition and self-valuation as black women; it implies the need to raise consciousness of their ethnic-cultural and gender territorial rights as a distinctive ethnic group.	• Racial, social, ethnic, and gender discrimination • Low self-esteem as black women; loss of ethnic identity • Relations of subordination and submission as black women assumed as almost natural • Victims of physical and psychological abuse and domestic violence • Responsibility for the home and the upbringing of children • Lack of prominent black women in past and present history as role models • High rates of maternal morbidity and child mortality, limited access and low quality of health services • Victims of forced displacement	• To raise the level of recognition and appreciation of black communities as a specific population group differentiated from the rest of Colombian society • To promote awareness of the situation of discrimination and undervaluing of black women and girls, as well as of the cultural obstacles that prevent their full development and affect their personal dignity	• Incorporation of ethnic and gender rights perspectives into projects and programs that advance the Black Women's Network (Red de Mujeres Negras) and those arranged with other institutions • Documentation, information, and analyses of human, economic, social, and cultural rights of black communities and their impact on women, girls, youth, and the elderly • Design and implementation of a plan of education about ethnic-cultural and territorial rights of black communities, and about the rights of women and children • Information and training about gender and ethnic biases in the upbringing and development of black girls and women

(Table 5. continued)

Problems identified in relation to the politico-organizational principles of PCN		Strategies to overcome problems from the perspective of the Black Women's Network	
Principle and meaning	Problem from black women's perspective	Action objective	Line of action
The right to the territory (space for being)			
• To develop and strengthen their capacities as women to be a part of the decision-making processes that affect us as a distinctive ethnic group, ensuring our permanent affirmation vis-à-vis the "other"	• Low social and political participation of women in the community's decision-making sphere • Cultural obstacles that prevent or restrict women's participation • Low level of knowledge and abilities for the exercise of politics • Women's participation takes the form mainly of claims for better life conditions for the family. It does not include claims to assert ethnic rights. • Their involvement in processes of organization is of an operational quality. • They are very rarely elected to represent the community or the organization in activities that imply travel outside of the community. • Women can more easily discuss "political" issues among themselves and in domestic spaces.	• To promote women's access to participation in ethnic-territorial decision-making processes, and in the implementation of decisions that affect their lives • To strengthen women's capacity to participate in decision-making and leadership spheres • To promote and strengthen women's organization for effective networking with other community-based groups, as well as with state agencies and NGOS • Raising consciousness and encouraging reflection on cultural limitations that curtail black women's political participation	• Adoption of measures to ensure fair qualitative and quantitative levels of participation of black women in ethnic-territorial organizations and in groups concerned with territorial use, management and defense • Encouragement of black women's organization for dealing with the state and civil society organizations • Dissemination of information to stimulate participation of women • Strengthening of links, sharing of experiences, development of joint projects, reinforcement of solidarity, and appreciation and respect among black women's organizations • Qualifying black women's participation through capacity building and practic experience

(Table 5. continued)

Problems identified in relation to the politico-organizational principles of PCN		Strategies to overcome problems from the perspective of the Black Women's Network	
Principle and meaning	Problem from black women's perspective	Action objective	Line of action
The right to our own vision of the future It means to develop our own life project within a framework of recognition and respect for difference, and the redefinition of the relationship between the black community, the state, and the rest of society.	• Internalization of a concept of development based on "having" and "doing," and in which "being" a community and a black woman is denied • Low levels of ethnic and gender identity that would allow for an identification of needs and interests of the community and black women which are necessary for the design of their own development programs	• To encourage discussion and reflection among black women on the characteristics and implications of the neoliberal development model for the Pacific and for women • To promote and strengthen incorporation of women's interests and needs in the design of the Life Project for the Black Community (*Proyecto de Vida para la Comunidad Negra*) as a political proposal of the PCN	• Identification of and support for women's production projects that are environmentally and culturally sustainable and that favor territoriality and ensure food security • Elaboration of ethnic-sensitive educational proposals for the socialization and upbringing of children in the domestic and school environments, and that favor gender equity and recognition of own ethnic and territorial worth
The right to be a part of, and participate in, the struggles of black people throughout the world This is related to the recognition, currency, and experience of our ethnic and gender rights at the national and international levels.	• Very low knowledge of the struggles of black people throughout the world and of the movements for recognition of ethnic rights and women's rights	• To promote reflection and appropriation of ethnic-cultural and territorial rights and their impact on women's lives	• Capacity building for PCN participation in the Red Continental Afroamericana and the Binational Colombian-Ecuadorian Committee and invigoration of the Latin American Peoples' Global Action

(Table 5. continued)

Problems identified in relation to the politico-organizational principles of PCN		Strategies to overcome problems from the perspective of the Black Women's Network	
Principle and meaning	Problem from black women's perspective	Action objective	Line of action
	• Very low appreciation and recognition of black women's contribution to the construction of Colombian national identity	• To raise consciousness about the organizational processes and the mobilizations against all forms of exclusion of black peoples in Colombia and throughout the world	• Design of strategies for the participation of black women's organizations in campaigns against forced displacement

Taking a cue from Strathern (1988), one could follow these women's theorizing to examine one's own anthropological and feminist theorizing as practices of the Western imagination. Postcolonial feminist critics have long pointed out the ethnocentrism of much liberal and radical Western feminist discourse. My hypothesis, provisionally substantiated with this limited ethnographic analysis of women activist practices, is that women activists have something to teach us about the very styles of theorizing. I am talking not so much about the specific content of the proposals they are making—which might seem insufficiently reflective to our deconstructive eyes, although below I will make a point about the theoretical pertinence of their position on gender difference—but about the anchoring of their inquiries in a shared political project (often the case of feminism but in a rather different way). This anchoring largely shapes their knowledge practices, *rather than the other way around*. Women activists, for instance, borrow strategically and selectively from anthropological and ecological theories about gender complementarity, but they resituate this knowledge in terms of a politics of difference, rather than, say, of ethnographic truth or knowledge for conservation; and they adopt a critical position regarding developmentalist women discourses, while borrowing from them, up to a certain point, ideas about discrimination and rights. From this position, they imagine "their" cultural-political position as a discursive counterpart to "Western culture" (what I have called here logocentric modernity), refusing the project of conversion into it (developmental feminism), and thus opening up a space of difference. If they refuse to accept the separation of gender and ethnicity (as one that is pertinent to them neither theoretically nor politically), they use selectively the modern/different dichotomy to make a cultural-political point.

This space of difference is a border from which to think about politics and culture in another way. Women activists can be seen as speaking from the colonial difference (in terms of both gender and ethnicity) and in doing so they bring to light an unsuspected angle of the modernity/coloniality/decoloniality equation. To fully develop this argument would require a much more thorough treatment than I can attempt here. I would like to give two interrelated indications in this direction to end this section. These involve a reading of the literature on vernacular gender, on the one hand, and certain works on gender difference, on the other. Briefly, *vernacular gender* is an anthropological notion that refers to the difference—oftentimes asymmetrical—that existed between women and men in nonindustrialized societies and that can still be seen at play in many settings. Most controversially proposed by Ivan Illich in the early 1980s, vernacular gender points at "the eminently local and time-bound duality that sets off men and women under circumstances and conditions that prevent them from saying, doing, desiring, and perceiving 'the same thing' " (Illich 1982: 20); Illich differentiates this regime from the modern one of "economic sex," which, for Illich, reigns in societies built around the idea of scarcity, the individual, and, hence, individual rights; economic sex assumes the disembeddness of social life from place and the dominance of the market form; in the long run, it makes women and men interchangeable, thus obliterating their difference, resulting in genderless individuals that are equalized to consume (or, as Trinh adds, to be consumed [1989: 108]). Illich's proposal has been echoed by some feminist theorists and anthropologists, by Trinh, for instance, to buttress arguments against feminist ethnocentrism (although warning of the possible use of the idea of vernacular gender for reactionary purposes), and by Apffel-Marglin and Sánchez (2002) to scrutinize developmental feminism among Andean indigenous communities in Bolivia. For Apffel-Marglin and Sánchez, developmental feminism trains women to relate individually to the market, commodities, and their own bodies, becoming a dominating force to the extent that for these women individuality and autonomy are principles alien to Andean communities. With neoliberal style development, as the Chilean feminist Verónica Schild put it (1998), women and men are socialized into market citizenship.

Strathern (1988) has explored systematically the reasons and consequences of such a discursive problematic. She sees at the basis of it the most fundamental assumptions about society and the individual embedded in Western knowledge practices, which put limits to the extension of these frameworks to societies in which such categories might not be valid descriptions of social life. Melanesian groups, for instance, are not concerned with the relation between the individual and society (a veritable "Western metaphysical obsession" [29]) or with an opposition between men and women. This impossibility should, according to Strathern, serve as a call to question the often conceptually conservative constructs adopted by many of our politically oriented fields of study. "In universalizing questions about women's subordination," she writes, "feminist scholarship shares with classical anthropology the idea that the myriad forms of social organization to be found across the world are comparable to one another. Their comparability is an explicit Western device for the organization of experience and knowledge" (31). Her careful ethnography of the social relations of groups structured around gift rather than commodity economies in Papua New Guinea enables her to show the profoundly relational character of these societies (hence, the absence of individuals in the modern sense, yet the existence of complex notions of personhood), positing a notion of analogic gender, somewhat akin to Illich's, according to which "male and female are analogic versions of each other, each acting in its own distinctive way" (299). In other words, behavior is deeply gendered, and gender ineluctably relational, based on difference.[32] This leads Strathern, in the last instance, to question the imputations of male dominance in standard Western terms of hierarchy and control of social relations and to open up the inquiry for a notion of domination that is more attuned to this society's practices and interpretive frameworks.

The notion of an irreducible gender difference is at the heart of certain feminist theories, particularly Luce Irigaray's. Irigaray unambivalently asserts a radical, indeed incommensurable, difference between women and men. That Western culture has tended to efface this difference is not only a reflection of the existence of the norm of a single (male) subject, but of the entire logocentric tradition that privileges abstract logic over cultural, including sensual, experience. Logocentrism has entailed the construction of the world in a masculine register (hence, phallogocentrism) and the destruction of female subjectivities—a monosexism as a false universal. This is why for her "to put the other on trial can only maintain hierarchical and alienating relations" (2000:79; an assertion echoed by PCN activists). This should not be misunderstood as a disavowal of the aim of, say, equal rights, but as saying that this aim is insufficient to construct a positive identity for women. In Irigaray's view, this identity can be constructed only if women (actually, both sexes) assume their share of the negative, of the different, thus really yielding two autonomous subjects and a regime of veritable intersubjectivity. In other words, it is the assertion of *difference from* (nor merely difference, since this still supposes the yardstick of a male world [see 2000: 86]) that should ground the project of ending the subjugation to the other gender *and* of reconstructing the gender relation. The patriarchal subject, who never sees a subject different from himself, can never be the foundation of this intersubjectivity. Even more, it is necessary to respect the duality between the sexes to achieve this goal, the singularity of a relationship of two beings that are irreducible in their alterity.[33]

Irigaray's position poses challenges not only to phallogocentric modernity but to conceptions of vernacular gender as well. "I belong to a cultural tradition," she writes. "My relation to the world, to others, and to myself, is shaped by it. I had to, I still have to, effect a gesture that is at least double: deconstruct the basic elements of the [patriarchal] culture which alienate me and discover the symbolic norms which can at the same time preserve the singularity of my nature and allow me to elaborate its culture" (2000: 148). The last part of the statement is an injunction for the double task facing societies such as the Pacific, namely, deconstructive and transformative. The deconstructive task needs to operate in two registers: that of the dominant Euro-Andean patriarchal culture, with its growing influence; and that of styles of male dominance in long-standing vernacular relations. Herein lies a double critique by black women activists, who at one and the same time resist both the gender and the race colonialist acts, refusing to enlist the challenge to one at the service of the other. In other words, it is not in the name of the monosexual discourse and male world of Western modernity that they will conduct their struggle against the gender/race colonialist act (as in developmental feminism); nor is it by simply sheltering an alleged gender complementarity within the black world that they will struggle against the race/gender colonialist act. This is to say that the twofold face of race/gender coloniality/decoloniality in the case of black women of the Pacific, because of its very cultural specificity, places activists in a radical position in the politics of alterity. Only difference, says Irigaray, can protect existence (e.g., 2001: 16). Indeed, "if our culture were to receive within itself the mystery of the other as an unavoidable and insurmountable reality, there would open up a new age of thought, with a changed economy of truth and ethics" (110). Activists and some scholars seem to converge in this tenacious journey. It supposes an ethical know-how and a retrieval of history-making skills by activists.[34]

It could be hypothesized that third world and postcolonial feminisms are particularly attuned to the articulation of both of the dimensions just discussed, that is, the question of gender difference and the idiom of gender complementarity. And this seems indeed increasingly the case, as Flórez argues (2007: 362–69), based on a provisional review of the literature on the subject by writers from Africa, Latin America, and the Caribbean. On the basis of these works, Flórez raises pertinent questions such as the following: Is it possible, as PCN women activists seem to do, to speak about the oppression of women without appealing to the category of feminism? Could it be that some proposals for gender difference do not amount to sheltering long-standing forms of oppression but, on the contrary, might constitute genuine discourses for the reaffirmation of identity? Could the distance that many African and Latin American women activists maintain from Western feminisms—seen as constructed on the basis of opposition to men—be made into a rallying point for other ways of dealing with gender? If conceived from a perspective of the relational epistemes that often prevail in many communities, can women's struggles be fully acknowledged without undermining the joint struggles of women and men? What are the limits to thinking about women in contexts such as the Pacific in terms of patriarchal power relations only? In short, what Flórez envisions is a series of "non-Eurocentric feminist approaches," which in turn implies "the creation of more localized conceptual tools" (2007: 368), such as those than can be gleaned from the practices of desubjectivation of PCN women activists.

Conclusion

It may be the case that social movements are forced to reflect systematically on their practice in times of crisis. In the case of PCN, there was the recognition of internal crisis beginning about 1999. Since then, PCN has been going through a process of reflection and critical examination of its practice. The sense of crisis has multiple sources, chiefly, the extension and recrudescence of the war into all corners of the Pacific, a certain exhaustion on the part of activists after an intense period of dealings and *concertación* with the state and NGOs, and a growing awareness of the need for theoretical, political, and generational renewal coupled with a set of conditions that are inimical to this need. One response, as we shall see in the next chapter, was a period of intense international activity and connection with various networks since the mid- to late 1990s. This process, while valuable to the movement, did not assuage the sense of crisis. By 2002, the collective was again focusing on itself; several steps were taken in this direction, including paying attention to the need to attract and train younger activists, dealing with differences and diversity within the movement at a new level of complexity, debating and consolidating the political thought of the movement, and building more autonomous and sustainable organizations. This process included two intensive two-day workshops on evaluation and strategic planning in summer 2003, led by the most important regional palenque, Palenque El Congal from Buenaventura, with the participation of over sixty-five PCN activists (see Palenque El Congal 2003a, 2003b). The workshops were intended to conduct a diagnosis of the movement (a sort of balance sheet of achievements, problems, and bottlenecks), renew its vision and commitment, and draw a strategic plan for the organization for the years to come. Here is the restatement of the organization's mission as it emerged from this exercise:

> We [Palenque El Congal] are a regional organization of the Process of Black Communities, PCN, in the southern Pacific and the inter-Andean valleys; we uphold [*reivindicamos*], promote, and defend the rights to the identity, ancestral territory, autonomous participation, and collective well-being of the black communities; we do so on the basis of our organizational strengthening, social mobilization, and negotiation capable of generating the conditions for the permanence and development of the cultures of the black communities as a people. (Palenque El Congal 2003b: 5)

Affirming the preeminence of the political process, they asserted, "We are part of and embody the Process, we appropriate it and suffer and enjoy the Process, we identity with it." A vast range of problems at many different levels—from the organizational to the individual and from the technical to the political—were clearly identified in the diagnosis, including the stagnation of territorial goals, lack of articulation with other pertinent actors, technical limitations, excessive concentration in a few geographical areas, insufficient capacity to share information in a timely fashion, internal disorganization, lack of political clarity on the part of some members, concentration of tasks in a few leaders and low levels of incorporation of community-based leaders, lack of a program for leadership training and development, and administrative deficiencies. A careful diagnosis of the organization followed this preliminary balancing of problems and strengths, locating the organization in the regional and national context of the moment, but also in relation to its internal working and to other organizations. The armed conflict, of course, figured prominently as a conditioning factor in the organization's development.

What emerged in the end was an agreement on a series of areas of concentration for the immediate future and for strategic planning. This agreement started by distinguishing between the plan itself and the mission, vision, principles and ethical values of the organization underlying the plan (the "strategic dream," as it was called). The double rootedness of the Process in both organization and communities was reasserted, as was the basic definition of PCN as a political organization for the defense of the ethnic rights of the black communities. Established and newer areas of work emphasized included a reassertion of the cultural-territorial basis of the movement, articulated around the notions of ethnic and cultural identity, the region-territory, and the goal of constructing a new societal model; a focus on food autonomy as a pressing need for river communities to resist displacement; the development of culturally appropriate educational models; increasing work with women and youth; new models of participation with local (e.g., mayors of towns), national, and international institutions, including alliances with actors at these levels; maintaining the commitment to the ethical values of the struggle; and developing an autonomous and sustainable organizational and institutional basis, capable of sustaining itself at all levels. Most of the rest of the exercise was devoted to issues that are often the Achilles' heels of social movements: the financial, administrative, and technical elements—that is, those elements that facilitate, and sometimes even make possible, a sustainable movement organization or network.[35]

The *taller* (workshop) did not entail a significant redefinition of the movement or its strategy. All of the topics under discussion fit within the space crafted by the basic organization established in the early 1990s. The resulting plan was an attempt to respond to the conditions of violence and displacement (see the end of chapter 1). Some topics were newly emphasized in this regard, such as food security and alternative economic projects as a strategy to resist in place. Similarly, there was unprecedented recognition of some issues, such as individual needs and aspirations as an important aspect of activism and the need to work with young people and urban groups; but as a whole what took place was a reaffirmation of the basic organization of the network by its most important and active regional subnetwork. Already in ascension by 2003, however, were a series of issues that were becoming more visible and pressing in the aftermath of the Durban conference: racism, persistent discrimination, rights, and poverty. Although the emphasis on these problems over the past few years could be seen as a return to more conventional agendas, the way these issues have been taken up, including a very active debate on reparations and economic, social, and cultural rights, has shown again the capacity of PCN and a growing number of black activists, intellectuals, and academics to construct a sophisticated framework for theory, strategy, and action.

What emerged from this exercise was a more mature organization. A number of features have to be emphasized in this regard, and in connection with the themes of this chapter. A core group of activists remains committed to the Proceso (PCN), to the black cultural and territorial struggle, to environmental goals, to an alternative model for the Pacific and the country. By 2003, the movement had produced a substantial body of knowledge, which is being systematized under the rubric of the "the political, cultural, and ecological thought of PCN."[36] The figure world they have created over the years has been amplified and enriched through the process of encountering difference and disensos. The decade extending from the Asamblea of 1993, where the basic principles were stated, to the workshops in 2003 has signified a long, difficult journey, full of achievements but also of frustrations and difficulties. While the principles continue to be a guiding light, the new context and needs require that they be met through a different combination of practices. So while the political thought of the movement has grown in complexity, the growth of organizational capacity has been much more limited. Despite the renewed commitment to the process on the part of most of the main leaders active since the early 1990s, there has also been a reaccommodation of the collective identity to make room for individual needs. At times this reaccommodation has been motivated by pragmatic concerns (e.g., personal safety, family financial needs), at times by more systemic pressures, such as in the case of gender. Ten years later, however, the collective identity called PCN continued to be a viable presence—or, as the activists would say, *propuesta*, or proposal—among the complicated play of forces in the Pacific and black cultural politics in Colombia and elsewhere. It continues to be so today.

6 networks

> The solidarity among peoples is an *estrategia de lucha* [strategy of struggle] and of exchange of experiences about the most effective ways to oppose monsters that often have no face on the national scene, such as the WTO [World Trade Organization], or Monsanto. International mobilizations and experiences enable us to see the world differently. . . . However, it must be stressed that the global actions need to have local repercussions in terms of changing concrete policies. This remains a problem within global organizing: the lack of articulation between the global and the local. Many groups do their part on the global scene, but the feedback to the local rarely happens. . . . The main costs are still born by the locals. This is why we at PCN always emphasize going from the local to the global and not the other way around—whether it is in Europe, India, Colombia, or what have you.
>
> —PCN activist, 2001, on the participation of PCN in antiglobalization mobilizations in various parts of the world, particularly those coordinated by the People's Global Action Against Free Trade (PGA)

In mid-October 1995 three PCN representatives traveled to Europe to participate in a series of activities geared toward creating international awareness of the plight of the black peoples of the Colombian Pacific and gaining support for their struggles. Originally facilitated by a progressive church-affiliated Swedish NGO, the two-month-long European tour took the PCN members to many cities and put PCN in touch with a variety of groups. One of the activities, held in Bern on October 20 (in which I participated), was a one-day symposium organized by the Swiss NGO SWISSAID under the title, " . . . and if the south refused our conception of the environment?" The main speakers addressed the assembly in the morning, and the afternoon was devoted to simultaneous workshops on three environmental projects, one in India, one in Guinea Bissau, and one in the Colombian Pacific. Run by the three PCN representatives and me, the workshop was held in Spanish and focused on the Proyecto Biopacífico (PBP), which, as we know, was by then undergoing a thorough reexamination. Despite their lack of international experience, the PCN activists delivered their message eloquently and clearly; it concerned two central aspects of the process: the need for NGOs to recognize the central role the black communities should play in any conservation effort, with the consequent need for real participation and control of the resources allocated for conservation by the communities; and the need to strengthen the organizing processes, particularly given the rising levels of aggression and disregard by the state. In a move that was still new in the international solidarity and NGO development field, PCN was adamant about one more thing: NGOs must learn to recognize the political (social movement) organizations in the regions in which they work, to clarify their own political position in relation to them, and to support them wholeheartedly and without the ambivalence shown in the past. The call meant that NGOs such as SWISSAID would have to support the Operational Plan for PBP drafted by the organizations over and against what the government project staff might think of it.[1]

Three years later, a PCN representative arrived in Finland to participate in a "global dialogue" under the rubric "Expanding People's Spaces in the Globalising Economy," held at the Hanasaari Cultural Center near Helsinki on September 5–9, 1998. This gathering was organized and convened by the International Group for Grassroots Initiatives (IGGRI), a network created in 1985 and made up of individuals connected with the alternative development movement and with grassroots struggles in the global south.[2] I participated in the event as an IGGRI member, coordinating with the PCN representative, Libia Grueso. By then, IGGRI had decided to transform itself into a network of grassroots movements, and toward that end convened the meeting with the aim of sharing accounts of popular responses to development and globalization and of discussing innovative experiences with alternative economies by grassroots movements. Over 120 participants from five continents, the large majority members of grassroots organizations and representatives of social movement networks, answered the call; they arrived from, among other places, Chiapas, Mexico, and Russia, from Tanzania, India, and Thailand, from Bolivia, Senegal, Colombia, and Chile, from Sri Lanka, New Zealand, Nigeria, the United States (including a well-known Native North American activist and homeless rights advocate, Cheri Honkala), and, of course, Finland. The range of topics was vast, from Davos to the regeneration of local economies, from commodification and Eurocentrism to the spiritual values and the emergence of countervailing forces, and from the ravages of globalization to a panoply of cases around community struggles, local currencies, people's knowledge, women's projects, cultural resistance, and so forth, all of which could be seen as contributing to crafting people's own autonomous spaces.

After a spiritual-cultural opening act by an African musical group, the Dialogue was opened formally on Saturday, September 5, by one of the Finnish hosts and the chair of the IGGRI group, the Indian activist Smitu Kothari. Then followed the two main speakers of the day, Susan George, one of the main intellectual forces behind the global justice movements, and Orlando Fals Borda, a longtime Colombian activist and intellectual well known for his leading role in the participatory action research (PAR) movement. George's words were harsh, likening globalization to a vampire, with its then-projected treaties such as the infamous Multilateral Agreement on Investment (MAI), a veritable "bills of rights" for transnational corporations to plunder the world. Her recommendation was for organizing beyond the state in both north and south and to build alliances horizontally and vertically. Fals Borda started by emphasizing the inherent diversity of places and went on to note, presaging the spirit of the World Social Forum process that started three years later, how social movements were attempting to move "from the micro to the macro and from protest to proposal." He spoke about the need to deconstruct development and modernity and to build on the PAR tradition in order to catalyze popular counterpowers to globalization and capitalism.[3] Four days later, after many workshops were held centering on various regions and problems, followed by plenaries, the closing ceremony featured a meditation on friendship and solidarity facilitated by an activist from Bangalore, Siddhartha. The weather in Helsinki was still pleasant, and the participants assembled in a circle on the lawn, in the middle of which a mandala was drawn. People were asked to come forward in silence to fill up the mandala with leaves, pebbles, wild berries, and flowers. When the mandala was complete, people shared their responses; a Maori invoked the blessings of the ancestors, and the Maori stone was handed around from participant to participant, so that they could pass on their energy for the future. Then the Indian activist led the delegates in a ceremony of remembrance; a Thai spiritual leader recited a prayer invoking the earth and the sky, a young Mexican couple told of a Huichol poem that summons people to share their knowledge, and there were songs from various countries. Everybody held hands in gratitude, and then we all departed, the facilitator wishing us outer and inner peace and harmony.

The Hanasaari gathering was unusual in many ways, prefiguring many of the qualities of similar gatherings opposed to globalization that were to multiply after the Seattle demonstrations of a year later; but it also evidenced older practices. Attending the meeting was a group of activists from the People's Global Action Against Free Trade (PGA). The PGA had been formally created only a few months earlier and had circulated its first powerful manifesto against neoliberal globalization in March of 1998. The influential role this network was to play as perhaps the main facilitator of "a convergence space" for the antiglobalization movements (Routledge 2003; see also Juris 2004) and of the great demonstrations that followed after Seattle was still ahead.[4] However, the practices favored by the PGA—including its emphasis on decentralization and autonomy, direct action and civil disobedience, and a radical rejection of states and institutions linked to neoliberal globalization and free trade, all within an overall anarchist philosophy—and IGGRI's design for the meeting—workshops and plenaries, with largely older third world men and northern women at the table for the latter—soon clashed. The split was triggered by a report by one of the working groups which suggested that under certain conditions movements should be willing to negotiate with the state and with corporations. While the comment was made in part in the context of the Finnish state and Finnish (e.g., timber and electronic) global corporations, with whom the working group thought it might be possible to establish a critical dialogue, the fact that the entire gathering had been funded by the Finnish government (progressive and critical of the neoliberal model as it was at the time) was an additional sign of fatal compromise, in the eyes of the PGA. By the end of the third day, the PGA made its critique vocal. Grueso had arrived as part of the PGA group, and she spoke first for the group that day:

> We do not share the attempt at establishing a dialogue between representatives from the TNCs [transnational corporations] and popular organizations, nor can dialogue be established with governments that maintain repressive systems. Second, the methodological limitations of time etc. have made it impossible to overcome difficulties. We gain only superficial knowledge of popular struggles in the working groups. We emphasize that our interest is to learn from the experience of those present here, but the methodology does not allow it. We need to create a space for this to happen, that is, for a collective construction that strengthens us mutually.

Then followed one of the European PGA activists, who questioned the legitimacy of the entire gathering and what he saw as an attempt on the part of the IGGRI core group to manipulate the meeting's deliberations and findings. In short, the clash was not only over contents but also over the very form of the event and its politics. By now, it has become clear that many movements worldwide favor a style of politics that emphasizes the principles of autonomy, nonhierarchy, self-organization, and the like. This feature has been variously referred to as "prefigurative politics"—that is, a way of acting that enacts in everyday practice the characteristics of the future world desired (e.g., Graeber 2002; Grubacic 2004)—or discussed in terms of the concern with form, particularly those decentralized forms enabled by new digital technologies (e.g., Juris 2004; Waterman 2004; Escobar 2004b; King 2006). Above all, this form of politics is evoked by the concept of network, the central concern of this chapter. Before discussing networks, however, it is important to mention that if the IGGRI meeting still exhibited older styles, it also brought with it concerns arising from third world experiences that are still to be incorporated fully into the global justice movements. The concern with policy and the state remains central, albeit not defining, for most social movements in the global south; Eurocentrism is often analyzed from southern perspectives and missed altogether in northern discussions. But even what was called "the struggle of wisdom over information" by Kothari at the meeting (valuing place-based knowledge, or know-how besides know-what in Varela's terms [1999]) could serve as a corrective to what can be seen as an exaggerated concern with disembodied information in the northern global justice movements.

Part I of this chapter reinterprets part of the discussion about PCN in light of notions of self-organization and networks; it also examines the experience of internationalization of PCN in the period 1995–2005. Part II starts the theoretical discussion of networks. Network approaches have blossomed over the past decade and a half. In some cases, networks are summoned to reconceptualize globalization and social movements; in others, networks appear as an element of a larger effort at social theory development. Part III applies this discussion to another empirical domain discussed in previous chapters, the discourses on global biodiversity. The aim of this part is to show how power operates in networks; as this case shows, whereas subaltern actors influenced significantly the shape of the biodiversity debate for a time, in the long run conventional agendas asserted themselves as the biodiversity network became more hierarchical.

Part IV, finally, returns to the theoretical register. It starts by telling a history that is rarely told in accounts of networks: the influence of sources that range from 1950s information theory, systems science, and cybernetics to complexity theory since the 1980s. It then goes on to review a current wave of critical theory related to networks and complexity that sees itself as being less concerned with questions of epistemology (the conditions of knowledge, as was still the case with poststructuralism) than with ontology, that is, with basic questions about the nature of the world. As Manuel de Landa, one of the main sources for this part, explains in his approach to "social ontology," today's critical theories are fueled by a fundamental scrutiny of the kinds of entities that theories assume to exist and, concomitantly, geared to the construction of theories based on different ontological commitments (2002, 2006). These trends are resulting in approaches variously called flat ontologies or assemblage theories that, as I will argue, have much to tell us not only about networks but also about prefigurative politics and the dynamics of organizing. This part of the chapter is largely synthetic and could be read almost as a theoretical appendix, although I hope it will be useful in mapping the complicated field of networks research. It is an attempt to point to politically relevant reconstructions of social theory.

I ◈ Assembling Movements: PCN and Transnational Self-Organizing Networks

I have suggested elsewhere (1992) that social movements may be described as autopoietic entities—that is, as self-producing, autonomous entities whose basic internal organization, despite important changes, is preserved in their interaction with their environments through structural coupling. The view of social movements as self-producing is counterintuitive, to the extent that they are often seen as responding to structural forces, strategizing on the basis of their interests, accomplishing particular functions in society, shaped by contentious politics, deeply embedded in networks that take them in all kinds of directions, or what have you. These are also apt descriptions of movements. However, there is a growing sense that many of today's movements can be analyzed through approaches that build on theories derived from looking at the dynamics of natural processes, particularly self-organization. This has been argued in relation to antiglobalization movements in particular (Waterman 2004; Escobar 2004b; Juris 2004; Chester and Welsh 2006). But is it also true of those other, "humbler" mobilizations that account for the vast majority of movements in the world today? Do they create a "network of their own," albeit connected to larger networks worldwide? In this section,

I suggest that these insights also apply to movements such as PCN, up to a certain point. From the perspective of the new theories, this entails giving a sense of PCN's emergence as a network and assemblage of sorts, its maintenance through time, the constitution of its basic organization, its recurrence at various scales, interaction among components resulting in self-organization, changes via structural coupling triggered by the environment, systemic orderings, and open-endedness, and so forth.[5]

As described in chapter 5, PCN grew out of a specific set of conditions in the late 1980s. Its emergence can be seen as a process with important components of self-organization that took place around the time of the National Constituent Assembly (1990–91). *Self-organization* here does not mean an automatic process that happens by itself, regardless of context; on the contrary, it often entails aspects that are both dependent and independent of context and environment, self-organized and other-organized, with both linear causality and nonlinear explanations, in which agents and structures are inseparable, and in which the emergent patterned movement is best explained as the result of interaction between on-the-ground recurrent activity and surrounding conditions. In explaining the rise of environmental movements in Finland, based on activists' life narratives and drawing on theories of self-organization and heterarchy in biology, Peltonen characterizes the early phases in a way that applies to PCN:

> Thus we can hypothesize about a movement striving, in its emergent phase, towards operational closure, that is, towards a phase where it can enter into an evolutionary mode of development through self-organization. When we see language as an essential mechanism in the self-organization of a social movement, it is possible to look for linguistic mechanisms that provide for operational closure, allowing for the co-evolution of a social network and a code of conduct. . . . One could perceive discursive frameworks as the construction of a "discursive membrane." These membranes are used to differentiate the network and the code it carries from its operating environment *and therefore allow for operational autonomy*. Like biological cell membranes, these discursive membranes have a double function, as they both isolate cells and connect them to others. As in the case of any organization, the permeability of the membrane is a delicate matter, and it should keep the organization intact and operational, without endangering its capacity to adapt to changing circumstances. (Peltonen 2006: 163–64; emphasis added)

This explanation is remarkably in tune with the view of autopoiesis. It also takes into consideration the place-based dimension of movements and the role of individual trajectories in emergence. "The dynamics of social movements," Peltonen (169) concludes, "can be studied as comprising trajectories of systems operating as heterarchies at different, intermeshed levels of structures, always building on their previous states. . . . Instead of pre-given entities, individuals and movements may be understood as creations of their own history, transactions, and complex genealogies." As de Landa would have it (see part IV), each social movement needs to be treated as an individual entity on its own, besides being made up of individuals.

An activist's decision to join a movement often results from a combination of intuitive, emotional, and rational considerations, but it happens more by "drift" toward attractive situations or agents than by a fully explicit set of decisions. What takes place can be described as a mutual selection between attractors and a movement toward agreed-upon codes and meanings (e.g., the centrality of cultural difference, which eventually resulted in the five principles, etc.). In other words, there was no ready-made "PCN" structure awaiting to be occupied by activists. To be sure, activists came into the process with their own histories and concerns, some of which I discussed in the previous chapter; however, without obeying a single logic or root cause, they started to converge around what complexity theorists call a possibility space, which, through their action, they in turn help to shape. As they explored this space, their actions gave rise to an emergent phenomenon, resulting in what I called a figure world (PCN). Each activist gravitated over this period around a set of "attractions" that would be impossible to explain as a result of a single driving force (e.g., as an effect of state restructuring, economic conditions, the sudden development of ethnic consciousness, or what have you).

This is just the beginning of the story. In what ways did the "co-evolution of a social network and a code of conduct" take place in PCN? How about change over time with conservation of basic movement organization? This requires an additional brief reinterpretation of the main periods I discussed in previous chapters as well as the introduction of some new elements. In chapters 1 and 3 I discussed the framework produced by PCN's linking of territory, biodiversity, and culture. As I emphasized, this framework was produced in the encounter with a host of actors at all levels, including the state, NGOs, universities and research centers, the global biodiversity network, and other social movements. What I did not emphasize in these chapters was the main practices accounting for the production of knowledge in the period 1993–98. These ranged from work with the river communities to international traveling, and included (a) exercises with river communities in support of the demarcation of territories and application for the collective title (called *monteos*, literally "traversing the bush")—these exercises, which lasted several days, included reconstructing the history of the settlement of the river, the elaboration of local maps, involving young and old people in the community, collecting "traditional knowledge practices," and the like; (b) more specialized *talleres*, or workshops, with local leaders and activists on issues ranging from Ley 70 and the political aspects of identity to the more technical issues of river basin design, territorial appropriation, etc.; (c) participation in negotiation exercises with the government, especially around development and conservation projects (as we saw for the case of the PBP); (d) political and conceptual speeches at these and many other meetings; (e) preparing documents for other movements, the government, and NGOs; (f) writing more elaborate papers for academic presentation, chiefly nationally but also internationally.[6]

It is the sum of these activities that accounts for the construction of sites or nodes in a network. Organizations such as PCN should be seen as creating a network of their own and a powerful discursive field of action; this entails making connections with other networks and, conversely, constitutes a space for other actors to construct their own linkages. In other words, the PCN network does not exist by itself but as inevitably imbricated with other networks, from the rain forest itself to the global biodiversity network (see part III) and other social movement networks. (The articulation with the rain forest is built by paying attention to the cultural-territorial dynamics embedded in peoples' practices, that is, following the "logic of the river," as I explained in chapter 1.) There are aspects of this dense networking, however, that have remained intractable and underdiscussed; it involves much work by activists which is both inward- and outward-oriented and an articulation between both kinds of work; PCN activists indeed speak of work that needs to take place *para adentro* (directed toward the inside, a critical reflexive work), and *hacia afuera* (outward and action-oriented, e.g., in relation to the state; see, e.g., PCN 2007). From the most intimate practices on a day-to-day basis and the explicit attempts to draft a stable identity to the internationally oriented actions, this work of articulation is always going on.

As we saw in chapter 5, from its early days PCN activists agreed on a basic set of principles (being/identity, territory, autonomy, self-defined view of the future, solidarity with other struggles worldwide). These principles continue to this day to guide all actions by activists (as in the case of gender and the approaches to displacement already discussed). Another important emphasis, present since the mid-1990s, is the need to construct autonomous ways of thinking, being, and doing. The idea of a *pensamiento propio* (the group's own thought) has been important throughout, and it is seen in relation to the need to defend the life project of the communities. There is, then, a tight articulation between the intellectual project, the political project, and the life project of the communities, as seen by the movement. For outside critics and observers, the decision by PCN to proceed by consensus and to discuss all matters of strategy broadly often has been a waste of time and has resulted in ineffective or belated action. For PCN, this practice has been essential to their intellectual and political strategy and to maintaining a high level of internal democracy and horizontality. As Libia Grueso put it, this strategy requires "an intense and permanent collective intellectual work that does not necessarily operate on the basis of the written word and that tends to be inwardly oriented, in terms of self-understanding, self-valorization, and self-definition of our role" (2005: 22). PCN sees this orientation as crucial to its efforts to differentiate itself from black struggles against discrimination that are not based on cultural difference.

To be sure, activists are very much aware that this process is also triggered from the outside. We have seen in detail the extent to which they engaged actively with environmental and development situations over the period 1993–2000 and with the problematic of displacement ever since. Sometimes this engagement entailed open confrontation, other times *concertación* with state institutions (negotiation toward minimum consensus). In the case of the Proyecto Biopacífico (PBP), concertación was crucial to the movement strategy. What I did not note about this process in chapter 4 is that there are multiple layers of concertación: first, within PCN, to agree by consensus on the strategy to follow; second, between PCN and other black and indigenous social movement organizations participating in the same projects; third, between all of these and PBP. Always at play in these practices there is (a) the orientation provided by the five principles; (b) the articulation between the movement's intellectual and political project and the communities' life project; (c) the need to maintain internal democracy, through a careful methodology of discussing disagreements with the aim of reaching consensus. It is the basic system of relations among these components—principles, relation to communities, relations among PCN activists and with other movements, internal democracy, and knowledge production—that is maintained in each collective action. This amounts to an autocentered logic driven by the goal of maintaining the movement's internal organization—in other words, an autopoietic strategy.

The same logic has determined the pattern of transnational networking developed by PCN. What follows is a brief account of the main international activities for the 1995–2007 period.[7] As I mentioned above, the first international experiences yielded two contacts that were to prove of particular salience in the ensuing years. The first was with the PGA; the second was with Danish groups, including the International Work Group for Indigenous Affairs (IWGIA), in Copenhagen. The contact with IWGIA provided the entry point for the participation of PCN activists in several meetings of the United Nations Working Group on Indigenous Populations on the status of groups such as the black river communities of the Pacific. From then on, the participation of PCN activists in international networks rapidly multiplied, particularly in the following overlapping arenas: (a) transnational antiglobalization mobilizations, particularly PGA-organized events; (b) meetings of Latin American black movements, particularly in the nascent Red Continental de Organizaciones Afroamericanas (Continental Network of Afro-American Organizations), and the Afro-Latin American Women's Network (e.g., events in Colombia, Uruguay, Peru, Bolivia, Ecuador, and Chiapas), which included preparatory meetings for the UN World Conference on Racism, Racial Discrimination, Xenophobia, and Related Intolerance, held in Durban, South Africa, in September 2001, in which PCN representatives participated; (c) events related to environmental and development issues, especially in the context of the Convention on Biological Diversity (e.g., COP [Conference of the Parties] 3 in Buenos Aires in November 1997): forest policy (including meetings on green certification of forest products, the impact of free trade on deforestation, etc.; meetings in Chile and Costa Rica); genetic resources and intellectual property rights regimes (particularly trade-related intellectual property rights [TRIPS]); the experience of collective territories (e.g., meeting at the Inter-American Development Bank in Washington, D.C., to discuss a report on collective territories in Latin America, 1999); (d) academic events at which PCN activists presented formal papers, even if not necessarily adhering to conventional academic canons (e.g., LASA [Latin American Studies Association] meetings in 1995 and 2001; conferences at universities in the United States and in Brazil, and other parts of Latin America); (e) events dealing with particular situations seen as key to either enforcing globalization or articulating resistance to it (e.g., Chiapas, Plan Colombia, the free-trade agreement with the United States).

Why do movements such as PCN engage in these networking activities despite the obvious trade-offs in resources, time, and energy? The following reflections by a PCN activist who participated in many of the activities tell why:

> Many of the decisions that affect the black communities are made at the international level—to given some examples, those concerning forests, timber, and genetic resources. This is why demands directed solely to the national government fall short. The same happens with decisions about megadevelopment projects, where the interests of multinational corporations are always present. This is why international cooperation and solidarity are important, in terms of creating options for societal alternatives, e.g., to defend the Pacific for the world as a whole. We need international alliances because capitalists and governments are not really interested in these goals. What the PGA offered was a call from the perspective of multiple struggles against something that affects us all in diverse forms, namely, free trade and the WTO. This goes on largely unnoticed at the national level. Only international mobilization can focus attention on these processes and galvanize public opinion and opposition to them. The solidarity among peoples is an *estrategia de lucha* [strategy of struggle]) and of exchange of experiences about the most effective ways to oppose monsters that often have no face on the national scene, such as the WTO and Monsanto. International mobilizations and experiences also enable us to see the world differently. We not only see the devastation caused by development in other parts of the world (e.g., India), the other forms of poverty people have, but are able to identify points of encounter and common interest, how to construct healthier anticapitalist social relations and modes of living. However, it must be stressed that the global actions need to have local repercussions in terms of changing concrete policies. This remains a problem within global organizing: the lack of articulation between the global and the local. Many groups do their part on the global scene, but the feedback to the local rarely happens; there has not been a minimal strategy in this regard. The main cost is still born by the locals. This is why we at PCN always emphasize going from the local to the global and not the other way around—whether it is in Europe, India, Colombia, or what have you.[8]

As PCN activists explain, the aims of joining antiglobalization movements are to contribute to resistance against global transformations and policies to the extent that these affect struggles for the defense of the territory and the cultural project of the communities; and to contribute to strengthening particular struggles or redress particular situations (e.g., struggle against Narmada, Chiapas, Plan Colombia). Additional goals include possibilities of international cooperation and worldwide dialogues about alternative models and about the most effective forms of struggle that would not take place otherwise. As PCN activists who participated in the PGA-organized actions in south India (which included an international group of forty-five people for twenty days) put it, these experiences "made us see the world in an entirely different way, and to realize that some of the struggles [e.g., *adivasi* struggles for territory] are similar to ours." Similarly, learning firsthand about the actions of the British group Reclaim the Streets for the control of public space and about movements against consumerism and in favor of organic foods in Europe was meaningful for PCN in terms of connected issues, models of organizing, and so forth.

International networking is not without its tensions. Latin American groups linked to the PGA, for instance, emphasized a strategy that moved from the place-based to the transnational, in contradistinction to the emphasis of the European activists, and openly discussed issues in which they differed from the Europeans, for example, issues of paternalism, control of resources, power relations, and political vision. The agendas of PGA Latin America meetings include topics absent from European meetings, from militarization and antidrug policies to structural adjustment and racism. Nevertheless, what takes place in these networks are forms of convergence (Routledge 2003) or "broad umbrella spaces" (Juris 2005) in which movements from many parts of the world and diverging place-based experiences strive to craft collective visions and strategies of mobilization while actively negotiating differences.

The imbrication of PCN with transnational networks exposed PCN to topics and demands that it accepted as its own. The productivity of some of these topics was to show only years later. A case in point was the early exposure during the European tour in 1995 to the issue of the definition of *black communities*, particularly in the ambit of the United Nations. A year later, this translated into an interest in "self-definition." Conversations with IGWIA in 1996–97 opened the possibility of describing the black communities as indigenous or native. Despite the fact that this move could have made possible funding from organizations such as IGWIA, whose mandate specifies working only with explicitly defined indigenous constituencies, PCN rejected the idea. A key set of events in the process of self-definition were the "binational *encuentros*," or meetings, among black communities of Ecuador and Colombia, held since the mid-1990s. Besides inventing a new geography encompassing both sides of the border (the Gran Comarca Afropacífico, or Greater Afro-Pacific Region), the encuentros were crucial to the process of self-definition of the black peoples of South and Central America and the Caribbean, as ideas from these binational meetings were taken to meetings of the Afro–Latin American Network and the preparatory meetings for the UN meeting in Durban. As a result of this bi- and transnational process, the positioning of the black communities as a *pueblo* (people) and the use of the term *afrodescendientes* became common. By 2003, a series of issues became more visible in the aftermath of Durban: racism, persistent discrimination, and the *derechos económicos, sociales y culturales* (DESC, economic, social, and cultural rights). Although the emphasis on these problems over the past few years could be seen as a return to more conventional agendas (as we saw at the end of chapter 5 in the context of the 2003 Evaluation and Strategic Planning Workshop by Palenque El Congal), the way it has been conducted has shown again the capacity of PCN and a growing number of black activists, intellectuals, and academics to construct a sophisticated framework for theory, strategy, and action, again in dialogue with transnational agendas and networking.[9]

Activists in PCN and AFRODES (Asociación de Afrocolombíanos Desplazados) have lately, in alliance with a number of human rights, advocacy, church, and solidarity groups based in Washington, D.C., taken the lead in articulating a position that has sought to influence U.S. policy toward Colombia, particularly by lobbying members of the Congressional Black Caucus. This perspective includes conceptualizations and concrete provisions in the areas of territorial and human rights, social and economic development, and strengthening of the institutional participation of black peoples and communities throughout the country. This coalition has led the struggle to transform Plan Colombia, to resist the U.S.-Colombia free trade agreement, and to make visible the cynicism of such maneuvers as the expansion of African palm as a putatively sustainable development strategy for biofuels, as propagandized by the Uribe administration.[10]

Another area in which PCN has been one of the main groups taking the initiative is that of reparations. The reparations challenge was faced head-on at a three-day conference entitled "Afro-Reparations: Memories of Slavery and Contemporary Social Justice," held in Cartagena in October 2005.[11] The conference, attended by well over one hundred activists, intellectuals, and academics, discussed a range of topics, from memories of slavery and colonial histories to the national, regional, and gender dimensions of discrimination, conceptions of reparation, varieties of ethnoracial discourses, the armed conflict, and implications for public policy. Echoing trends elsewhere, the debate highlighted the need to advance simultaneously emphases on difference and equality, identity and racism, territory and poverty, and individual and collective social, cultural, and economic rights. But the Afro-Colombian proposals took a specific form in ways that can only be hinted at here. First, as one activist put it, there was an emphasis on *establecer la ruta del pensamiento afro* (establishing the path for black thought), as found in all kinds of knowledge and *saberes*, past and present, from writers and intellectuals to the agriculturalists' knowledge of biodiversity. As a PNC activist added, "The focus on the *pensamiento* is the first condition for resistance; otherwise we end up reproducing the dominant thought of racism and slavery." Second, there were adamant expressions of the need to resist the conventional models of development and "forced globalization," given that the neoliberal state can only redeploy racism through other forms of exclusion. Third, there was imaginative hybridization of concepts—for instance, the transnational concepts of patrimony and race were recombined with local ethnoterritorial discourses in unique ways. Finally, there was a deep historical sense of the struggle, metaphorically indicated by the survival of the marimba and the *cununo* (traditional drum from the Pacific). These emphases are often absent from mainstream debates on reparations.[12]

On all of these fronts, PCN has shown a tremendous ability for articulatory politics. From the rivers to Washington, D.C., from Buenaventura and Bogotá to European capitals, activists of this network have provided, in a sustained and increasingly complex manner, discourses and practices of articulation with a large variety of groups and constituencies—NGOs, solidarity groups, river communities, international organizations, other black movements, other social movements within Colombia, global justice movements, and so forth—and on a variety of issues, including biodiversity, sustainability, displacement, reparations, rights, violence, neoliberal policies, free trade agreements. They have thus acted as an important node and force in local, regional, and national networks and in the self-organizing meshworks of transnationalized struggles.

To sum up: PCN created and is anchored in a "network of its own," with geographical basis on the rivers and towns of the Pacific region and with a number of regional and national sites that are maintained through embedded and embodied practices. This network became increasingly tied to other networks, both dominant and subaltern, in various ways. The presence of some catalysts has been important to the construction of the PCN as a network and of the PCN network as part of the larger meshwork. This includes some progressive church, environmental, and popular communications NGOs and, of course, some dominant NGOs and government programs; information and communication technologies (ICTs), particularly the Internet; some academic spaces; and groups of activists beyond those belonging to the PCN. What circulates in the meshwork are activists, cultural models, information, frameworks, e-mail messages, communiqués, declarations, agreements, concrete actions and mobilizations, and feelings and expressions of outrage and solidarity.

From the European tour in 1995 to Durban and beyond, PCN became increasingly tied to transnational networks. This process, however, was preceded by the creation of a place-based organization and network that started to emerge in the early 1990s as a result of the convergence of a number of trajectories; by 1993, a basic organization had been laid down bringing together a series of elements—principles, practices, individuals, concepts, strategies—that enabled the movement to embrace transnational concerns from a position of operational autonomy. This networking certainly changed the organization (via structural coupling with its environment, including with other networks), while enabling it to conserve its autopoietic character. New agendas came into the picture, agendas that would not have appeared otherwise; at the same time, ever since the 1995 tour it has been clear that some debates, e.g., pluriethnicity and multiculturality, were more advanced in Colombia and Ecuador than, say, at the Unied Nations, and they were far from being on the radar of European activists. In this way, the entire network changed, a feature of self-organizing meshworks that I will study in more detail theoretically later in the chapter.

One corollary of the transnationalization of Afro–Latin American movements is that considerations of diaspora have become very important in the strategies of many of them. The African-American scholar and activist Joseph Jordan has suggested that the efforts by Afro–Latin Americans are among the most farsighted at present in terms of transformational processes linked to the worldwide dispersion of African descendants. For Jordan, Afro–Latin American struggles and thought today, because of their more radical character, occupy the place that Pan-Africanism and the struggles against apartheid occupied in the 1960s and 1970s. Afro–Latin American struggles, in Jordan's words, "produce the conditions for sustained, critical intervention. . . . For the first time in history, the histories of their struggles have been linked up with those of other African descended communities in the Americas in practical and meaningful ways" (2006: 9). This is a hopeful development not only with respect to the resistance to antiblack racism, but also for making other possible worlds visible and more viable.

Theories of Social Movements and Networks ◈ I have argued elsewhere (2000) that one can differentiate between two kinds of theories of networks. In the first type, the concept of network fits into an existing, taken-for-granted social theory. In the second group, social theory itself is reconstructed out of, or on the basis of, the concept of network. Castells's application of networks to contemporary society is the best-known case among the first set. Central to Castells's theory of the network society is a distinction between the space of flows (the spatial structures relating to flows of information, symbols, capital, etc.) and the space of places (1996, 415–29). The space of flows is composed of nodes and hubs hierarchically organized according to the importance of the functions they perform for the network. For Castells, places have to network or perish. This structural schizophrenia can be avoided only by building bridges between the two spatial logics, that of flows and that of places (428). From his globalocentric perspective, power resides only in or with flows and strategic nodes. In this book I have emphasized that social movements affirm the centrality of the space of places in the constitution of societies.[13]

Actor-network theories (ANT) are the most well known example of the later type; ANT "aims at accounting for the very essence of societies and natures. It does not wish to add social networks to social theory but to rebuild social theory out of networks" (Latour 1997: 1). The theory of actor-networks asserts that the real is an effect of networks. Reality arises in the bringing together of heterogeneous social, technical, and textual materials into patterned networks. No matter how seamless it might look at times, reality is the end-product of actor-networks which have put it together after a lot of work. As in Foucault's notion of the microphysics of power, reality comes into being after a lot of "dissemination, heterogeneity and careful plaiting of weak ties. . . . through netting, lacing, weaving, twisting of ties that are weak by themselves"; concomitantly, analysis must start "from irreducible, incommensurable, unconnected localities, which then, at a great price, sometimes end up in provisionally commensurable connections" (Latour 1997: 2; see also Law 2000 [1992]).[14]

Most theories of social movement networks belong in the first class of network theories—that is, they assume a particular social theory into which networks fit. Examples are theories of social networks of activism, such as Smith, Chatfield, and Pagnucco (1997) and Keck and Sikkink

(1998); broader attempts at theorizing social movement networks (e.g., Diani and McAdam, eds. 2003; Alvarez 1998, forthcoming); and ethnographic studies of particular movement networks, such those associated with Zapatismo (e.g., Leyva Solano 2002, 2003; Olesen 2002) or with antiglobalization movements (e.g., Juris 2004; King 2006; Osterweil 2005b, 2006). To date, few studies have sought to apply notions of complexity to social movements, although interest in the topic is rising rapidly (e.g., Chesters 2003; Chesters and Welsh 2006; Peltonen 2006; Notes from Nowhere 2003; Summer and Halpin 2005; Escobar 2000, 2004b).

The primary aim of theories of activist networks is to explain how the networks emerge, how they operate, and how effective they are. For Keck and Sikkink, networks are "organized to promote causes, principled ideas, and norms, and they often involve individuals advocating policy changes that cannot be easily linked to a rational understanding of their 'interests'" (1998: 8). The environment, human rights, and women's rights movements exemplify advocacy networks, which are often composed of NGOs, foundations, churches, consumer groups, and so on who have a common set of values. They operate through information sharing and "frame alignment"—the construction of shared frames of meaning; they link local, national, and transgovernmental political arenas (Smith, Chatfield, and Pagnucco 1997). These models are nevertheless based "on the belief that individuals can make a difference" (Keck and Sikkink 1998: 2); as such, they are located within liberal traditions and are limited in terms of understanding movements that have a more collective character and a style of action that goes beyond issue campaigns and policy reforms. However, these models underscore important network elements, such as the centrality of NGOs, struggles around particular policies, and the role of resources and shared interests in building alliances. A more general model for research on network-centered social movements has been proposed by Diani, McAdam, and coworkers (Diani and McAdam, eds. 2003); the research focuses on networks linking individuals, organizations, and events; on network configuration (boundaries and ties, nonhierarchical relations, centralization and decentralization, temporality, etc.); and the research requirements for mapping networks.

In revisiting simultaneously prevailing notions of both "social movements" and "political contestation," Sonia Alvarez (forthcoming) underscores the centrality of NGOs for social movements such as Latin American women's movements; yet she aims at a broader conceptualization of processes and actors that incorporates some elements of a flat ontology to be discussed in the last section. Her call is for a reconceptualization of social movements as expansive, heterogeneous, and polycentric discursive fields of action which extend well beyond a distinct set of civil society organizations. This field is constructed, continuously reinvented, and shaped by distinctive political cultures and distributions of power. Movement fields configure alternative publics in which dominant cultural-political meanings are refashioned and contested; these publics are parallel discursive arenas in which subaltern groups reinvent their own discourses, identities, and interests. These fields are potentially contentious in two ways: they create and sustain alternative discourses, identities, and challenges; and they maintain an internal contestation about their agendas in ways that enable them to respond adequately to their own ethico-political principles. It is easy to see how the concept of social movement field and the double contestation that structures it may apply to the social movement of black communities of the Pacific or to antiglobalization movements, to the extent that their networks/meshworks can be seen as apparatuses for the production of alternative discourses and practices, on the one hand, and as bringing together cultural politics that find articulation in dispersed networks, on the other (Alvarez, Dagnino, and Escobar eds. 1998). Alvarez's work also calls attention to the relation between the internal democracy of local social movements and their transnationalization, or the impact of differential access to cultural, political, and material resources on local network nodes.

One final example of social movement networks research concerns the ensemble of networks that has emerged throughout the years around the Zapatista movement in Chiapas, Mexico. What is interesting in the analysis of this case provided by the Mexican anthropologist Xochitl Leyva (2002, 2003) is that she treats neo-Zapatismo as precisely that: an ensemble of articulated networks emerging from broad political contexts, many of them with deep historical roots in the region and the nation. In her network ethnography, Leyva distinguishes among six interrelated but distinct neo-Zapatista networks: those based on historical agrarian and peasant demands; democratic-electoral and citizen-based networks; *Indianista*-autonomist networks, largely focused on indigenous peoples; women's rights networks; alternative revolutionary networks, promoting antistate ideology and radical change; and international solidarity networks. All of these networks are both sociopolitical and cybernetic at the same time; after 1994, they became articulated around the armed Zapatista movement, Ejército Zapatista de Liberación Nacional (EZLN, Zapatista Army of National Liberation), in some fashion but are not by any means restricted to it; they all emerged from local historical and regional and, in the case of the sixth network, global conditions; they share moral grammars (e.g., concerning rights, citizenship, land, autonomy, etc.) and construct cognitive frameworks through which they have impact on power relations, institutional politics, and daily life. The image of Zapatismo that emerges from this conceptualization is very complex—a manifold of actualizations of a field of virtuality, to borrow from de Landa's concept (see below). Each network may be considered as an assemblage in relation with other assemblages; each represents a multilayered entanglement with a host of actors, organizations, the natural environment, political and institutional terrains, and cultural-discursive fields that may be properly seen as a result of assemblage processes.

Networks and Complexity ◈ Self-organization, assemblage theory, and autopoiesis constitute relatively new forms of thinking about the organization of the living, including networks and social movements. They contrast sharply with long-standing models of theory and social life. Applying arguments of complexity to historical social processes, the argument is made that over the past few hundred years economic and social life have tended to be largely organized on a logic of order, centralization, and hierarchy building. So have social theory frameworks. The approaches discussed here aim at making visible a different logic of social organization that resonates clearly in two domains of concern to this book: digital technologies (cyberspace, as the universe of digital networks, interactions, and interfaces), and social movements. To start with cyberspace, while modern media operate on the basis of a top-down, action-reaction model of information, I argue that the model enabled by ICTs is based on an altogether novel framework of interaction—a relational model in which all receivers are also potentially emitters, a novel space of dialogical interaction (as in the best examples of net.art). As a space for intercultural exchange and for the construction of shared artistic and political strategies, cyberspace affords unprecedented opportunities to build shared visions with peoples from all over the world (the World Social Forum process can be seen partly as a result of this dynamic). This view emphasizes the micropolitics of the production of knowledge made possible by the "fluid architecture" of cyberspace, emphasizing the "molecular" (as opposed to molar, or characterized by large, homogeneous conglomerates) nature of cyberspace. This micropolitics consists of practices of mixing, reusing, and recombining knowledge and information.[15]

Such a vision resonates with the principles of complexity and self-organization, which emphasize bottom-up processes in which agents working at one (local) scale give rise to sophistication and complexity at another level. Emergence happens when the actions of multiple agents interacting dynamically and following local rules rather than top-down commands result in some kind of visible macrobehavior or structure. These systems may be adaptive in that they learn over time, responding more effectively to the changing environment. Networks constitute the basic architecture of complexity. Physical and natural scientists are currently busy mapping networks of all kinds and trying to ascertain network structures, topologies, and mechanisms of operation. Social scientists have also been jumping on the bandwagon of complex networks research.[16]

Manuel de Landa (1997, n.d., 2003) has introduced a useful distinction between two general network types: hierarchies and self-organizing meshworks. This distinction underlies alternative philosophies of life. Hierarchies entail a degree of centralized control, ranks, overt planning, tendency toward homogenization, and particular goals and rules of behavior; they operate largely under linear time and treelike structures. The military, capitalist enterprises, and most bureaucratic organizations have largely operated on this basis. Meshworks, on the contrary, are based on decentralized decision making, self-organization, and heterogeneity and diversity. Since they are nonhierarchical, they have no overt single goal. They develop through their encounter with their environments, although conserving their basic organization, as in the case of autopoietic entities. Other metaphors used to describe these phenomena are "strata" (for hierarchies) and "rhizomes" or "self-consistent aggregates" for meshworks (from the philosophers Deleuze and Guattari 1987). The metaphor of rhizomes suggests networks of heterogeneous elements that grow in unplanned directions, following the real-life situations they encounter. These two principles—hierarchies and meshworks—are found together in most real-life examples. They could also give rise to one another (as when social movement meshworks develop hierarchies; or the Internet, which can be said to be a hybrid of meshwork and hierarchy components, with a tendency for the elements of command and control to increase). The reverse could be said of the global economy in that today's corporations are seeking to evolve toward a networked form with flexible command structures.

The model of self-organization constitutes a significantly different form of ascertaining the creation of biological, social, and economic life. Building on the field of biological computing, Terranova adds useful elements to the conceptualization of networks as self-organizing systems which engender emergent behavior. For her, networks can be thought of as "abstract machines of soft control—a diagram of power that takes as its operational field *the productive capacities of the hyperconnected many*" (Terranova 2004: 100; emphasis added). This leads to a view of social phenomena as the outcome of a multitude of molecular, semiordered interactions between large populations of elements. Individual users become part of a vast network culture. There is a particular focus on "the space-time of the swerve" leading to emergence (117). These systems allow only for soft control (as in cellular automata models); it is in this sense that Terranova's definition of network as "the least structured organization that can be said to have any structure at all" (118) makes sense. The open network (such as the Internet or network of networks) "is a global and large realization of the liquid state that pushes to the limits the capacity of control of mechanisms effectively to mould the rules and select the aims" (118). This network culture emphasizes distributed/autonomous forms of organization rather than direct control—soft control that does not imply full knowledge of all parts. In short,

> the biological turn is, as we have seen, not only a new approach to computation, but it also aspires to offer a social technology of control able to explain and replicate not only the collective behavior of distributed networks such as the internet, but also the complex and unpredictable patterns of contemporary informational capitalism. . . . The biological turn thus seems to extend from computing itself towards a more general conceptual approach to understanding the dynamic behavior of the internet, network culture, milieus of innovation and contemporary "deregulated markets"—that is, of all social, technical and economic structures that are characterized by a distributed and dynamic interaction of large numbers of entities with no central controller in charge. (121)

These notions contrast sharply with concepts of control based on Taylorism, classic cybernetics, and even governmentality, although these have by no means become irrelevant. Like de Landa, Terranova sees pros and cons in the new situation; on the down side the multitude/mass cannot be made to unite under any common cause; the space of a network culture is that of a permanent battlefield; yet the benefits are clear in terms of opportunities for self-organization and experimentation based on horizontal and diffuse communication (again, as in the case of antiglobalization social movements and the experience of PCN). In the best of cases, the simultaneous tendencies of networked movements to diverge and separate, on the one hand, and to converge and join, on the other, might lead to "a common passion giving rise to a *distributed movement* able to displace the limits and terms within which the political constitution of the future is played out" (156). The logic of distributed networks thus amounts to a different logic of the political. In his study of the anticorporate globalization movement in Calalunya, Spain, Juris (2004, 2005) puts it in terms of the intersection of network technologies, organizing forms, and political norms that accompany the cultural logic of networking.

To sum up, a number of theories of networks in the past two decades have tried to make sense of what is perceived as a novel logic of the social and the political. The trends based in self-organization and complexity, it seems to me, do it most clearly because they articulate a network concept from the perspective of an ensemble of new logics operating at the levels of ontology, the social, and the political—what I will in the last section call flat alternatives. Flat alternatives make visible a set of design principles based on the interoperability among heterogeneous networks and information systems and open architectures conducive to expansive internetworking enabled by decentralization, resilience, and autonomy. The political implications are manifold. Since, as I have insisted, places and embodiment have by no means ceased to be important—and one must not forget that there are many embodied aspects to activism; activism is not only about sharing information or technology—one needs to think anew about the productive capacity of places and bodies for difference, that is, how difference is mobilized politically; second, struggles over ICTs and the world they help create become crucial at many levels, from bodies to nature and economies (e.g., Harcourt, ed. 1999; Consalvo and Paasonen, eds. 2002; Bell and Kennedy, eds. 2000; Fernandez, Wilding, and Wright, eds. 2002). This contested character of ICTs involves experimentation with appropriations of the fluid architecture of networks, new forms of collaboration, and so forth—in short, network politics linked to emerging cultural and material assemblages. The risks are real, to the extent that information networks and ICTs are also part of the infrastructure of imperial globality.[17]

III ◈ Dominance and Subalternity in Social Movement Networks

It seems sensible to assume that most social movements are a mixture of newer and older forms, of hierarchies and self-organization. Structural and spatial conditions, the weight of expert knowledge, political expediency, inimical environments and conventional media, internal struggles, and so forth foster practices and forms of power that by and large do not operate on the distributed principle. Most network approaches and flat ontologies are not very explicit about these power dimensions, and it is important to think about this issue further. Elsewhere (Escobar 2000), I introduced the notion that one can differentiate between two kinds of networks: subaltern actor-networks (SANs) and dominant actor-networks (DANs). Most theories reviewed so far do not make this distinction, and for good reason, since SANs and DANs overlap and often coproduce each other or maintain relations of contiguity via structural coupling; yet they can be analytically differentiated on political grounds and in terms of contrasting goals, practices, modes of agency, mechanisms of emergence and hierarchy, and time scales. Social movement networks undeniably constitute a wave of confrontational social engagement at many levels, so their oppositional character is hard to deny; it is important, however, to avoid falling back into modernist notions of opposition—that is, into thinking about movements as discrete entities independent of their enactment and self-production (King 2005). In other words, in characterizing networks as dominant or oppositional, it is important to remain within a "flat" terrain. A simple criterion is to say that DANs are networks in which elements of hierarchy predominate over those of self-organization; conversely, SANs are those in which the opposite is true. I have already illustrated the subaltern networks in which self-organization is noticeable by revisiting some aspects of PCN and narrating their experience of internationalization. Here I want to demonstrate the predominance of hierarchy in the area of biodiversity conservation.

Assembling Nature: The Biodiversity Network ◈ The field of biodiversity conservation constitutes a good example of the emergence of a transnational network with central elements of hierarchy.[18] The concept did not exist before 1980. The biodiversity network originated in the late 1980s and early 1990s out of conservation biology, where "the idea of biodiversity" (Takacs 1996) first flourished. It soon articulated a master narrative of biological crisis, launched globally at what has been called the first rite of passage to the "transnation state," the Earth Summit in Rio de Janeiro in 1992 (Ribeiro 1998).

From the perspective of actor-network theory, the biodiversity narrative created obligatory passage points for the construction of particular discourses. This process translated the complexity of the world into simple narratives of threats and possible solutions. This simplified construction was perhaps most effectively summarized by the conservation biologist

Daniel Janzen in his aphorism about biodiversity: "You've got to know it to use it, and you've got to use it to save it" (1992). In a few years, an entire network was established that amounted to what Brush (1998) aptly called an "invasion into the public domain." Yet the biodiversity network did not result in a hegemonic construction, as in some instances of technoscience. Countersimplifications and alternative discourses produced by subaltern actors also circulated actively in the network with important effects. However, the result was the creation of a relatively stable network for the movement of objects, resources, knowledge, and materials.

I have argued that "biodiversity" is not a true object that science progressively uncovers, but a historically produced discourse (Escobar 1998). This discourse is a response to the problematization of survival motivated by the loss of diversity. As one of the main voices of the movement, E. O. Wilson, put it, "Biological diversity is the key to the maintenance of the world as we know it" (1993: 19). This kind of assertion resulted in the irruption of biodiversity on the world stage of science and development in the late 1980s. The textual origins of this emergence can be identified with precision: the publication of *Global Biodiversity Strategy* (WRI/IUCN/UNEP 1992) and the Convention on Biological Diversity (CBD), signed at the Earth Summit. Scientific approaches to biodiversity are largely geared to assessing the significance of biodiversity loss to ecosystem functioning and to ascertaining the relation between biodiversity and the "services" that ecosystems provide.[19] Established definitions of biodiversity thus do not create a new object of study that is outside of the existing definitions in biology and ecology.[20] Rather, "biodiversity" is the response to a concrete situation that is certainly preoccupying but that goes well beyond the scientific domain.

There is a little-known chapter in the history of the emergence of biodiversity and the CBD that reflects the struggle between self-organization and hierarchy; it has been researched recently by King (2006).[21] This history starts with alternative development proposals of the 1970s, particularly the movement surrounding the concept of "another development" launched by the Dag Hammarskjöld Foundation in Sweden. Following in the wake of the declaration of a New International Economic Order by the nonaligned nations, the Conference on the Human Environment held in Stockholm in 1972, dependency theory, and notions of self-reliance, among others, the Dag Hammarskjöld Foundation supported the creation of the International Coalition for Development Action (ICDA) in 1975. This was the heyday of the Green Revolution; at ICDA, a pioneering group which included Pat Mooney (who was to create the Rural Advancement Foundation International, RAFI, a few years later, now Group ETC) and Henk Hobbelink (who would establish the NGO Genetic Resources Action International [GRAIN] in Barcelona also a few years later) started to focus on agricultural biodiversity, linking it to biotechnology and markets. From this pioneering work came the first sounds of alarm about plant genetic erosion and the loss of autonomy of farmers. The focus on seeds was crucial at this stage and resulted in Mooney's influential book *Seeds of the Earth* in 1979.

From then on an intense and rich debate developed over such subjects as plant genetic resources, the establishment of progressive NGOs in many parts of the world devoted to seeds, agroecology, and alternative development, and the initial confrontations with agribusiness corporations, often in the ambit of UN agencies such as the Food and Agricultural Organization (FAO). For King, civil society organizations (CSOs), particularly RAFI and GRAIN, fulfilled a definitive role in shaping the early biodiversity movement and the structure of the CBD itself. Crucial in this regard were the well-known Keystone Dialogues held in Colorado in the years prior to the Earth Summit, which brought together activists, scientists, international organizations, and transnational corporations to discuss plant genetic resources, farmers' rights, and the like. It seemed that for a time there was an open dialogue among this disparate set of actors, the CSOs playing a defining role. This moment of openness, however, was brought dramatically to an end soon after the convention, when the neoliberal agenda became consolidated and when corporate biotechnology adopted a counterstrategy that would bring about the dominance of transgenic agriculture, TRIPS, and market-driven approaches to agriculture and conservation. As King wrote,

> The buildup to the United Nations Conference on Environment and Development (UNCED) in Rio 1992 was important in summoning together civil society groups from the world over in preparation for the meeting. . . . These groups were essential in the planning for the meeting and in formulating an agenda for consideration. In essence, the participation of civil society groups reoriented the direction of the negotiations. . . . Groups like GRAIN and RAFI had an important role in drafting the wording of the documents related to biodiversity. Of all the international agreements derived from the Rio meeting, the CBD tended to attract those concerned about the environment and local development. Thus, the CBD became the political space that offered the possibility for the most participation. . . . Operating from a memory of the Alternative Development Agenda [of the

> 1970s and 1980s], progressive CSOs were advocating wholesale challenges to the economic system in order to enact the Rio mandate. (2006: 236)

This dynamic, however, was not to last for long. In the end, the network, in the face of the decisive influence of biotechnology after the early 1990s and of trade negotiations, particularly after the establishment of the WTO in 1995, turned into a hierarchy. By the mid-1990s, international institutions, mainstream northern NGOs, botanical gardens, universities and research institutes in the first and third worlds, pharmaceutical companies, and the great variety of experts located in each of these sites came to occupy the dominant sites in the network. From a discursive perspective, biodiversity articulates a new interface between nature and society in global contexts of science, cultures, and economies. In chapter 4, I discussed the transformations undergone by the discourse in Colombia over the period 1993–2000, particularly in light of the action of social movements.

Here I want to advance a provisional explanation in terms of assemblages. To this end, I need to give an account of the assemblage's emergence and maintenance through time; the processes of assembly; the recurrence of these processes at various scales; the interaction among components resulting in self-organization or hierarchies; strategies of soft control; the ways in which the assemblage both limits and enables components; and the network topology, including systemic orderings and open-endedness; and so forth. As I mentioned earlier, the textual origins of the assemblage are easily identifiable. Its social origins include a complex set of components and interactions. Chief among these were particular kinds of scientific knowledges and institutions. Among the former, the main ones have been those of biologists (experts in conservation biology, systematics, and landscape ecology but also ethnobiologists and ethnoecologists), although other professionals (e.g., economists) have also been important. In terms of organizations, the mainstream view of biodiversity has been produced by the main northern development and environmental organizations and NGOs (e.g., World Conservation Union or IUCN, World Resources Institute, and World Wildlife Fund, the World Bank, and the United Nations). This dominant view is based on a particular representation of the "threats to biodiversity" that emphasizes loss of habitats, species introduction in alien habitats, and fragmentation due to habitat reduction, rather than underlying causes; it offers a set of rational prescriptions for the conservation and sustainable use of resources; and it suggests appropriate mechanisms for biodiversity management, including scientific research, in situ and ex situ conservation, national biodiversity planning, and the establishment of appropriate mechanisms for compensation and economic use of biodiversity resources, chiefly through intellectual property rights. The dominant view originates in well-established views of science, capital, and management (see, e.g., WRI/IUCN/UNEP 1992; WRI 1994: 149–51).[22]

The CBD underlies, for most purposes, the architecture of the biodiversity network. The CBD operates on the basis of practices that establish the same kind of relations among components at various levels. Among these practices are the creation of particular groups within the CBD structure, for example, the Subsidiary Body for Scientific, Technical, and Technological Advice (SBSTTA); national, regional, and international meetings leading up to COP meetings; and national delegations and reports. The practices also include the steady incorporation and development of new knowledge and policy areas, including forest biodiversity, agricultural biodiversity, marine and sea biodiversity, biosafety; the proliferation of issues, such as genetic resources, benefit sharing, biotechnology, impact assessment, indigenous and traditional knowledge, in situ conservation; and the normative role of scientific knowledge. Finally, a key process in the topology of the network continues to be the participation of NGOs and social movements.

It is through this set of practices that the dominant biodiversity discursive formation is crafted, implemented, and eventually contested. This contestation takes place at many levels. For instance, at COP 4, held in Bratislava in 1998, indigenous representatives reached a consensus on the implementation of article 8j of the CBD, which called for the respect and maintenance of local knowledge practices. The consensus called for the creation of a permanent working group with the full participation of indigenous peoples as the only way in which the defense of their knowledge and resources can be advanced within the CBD. The sustainable development conception at the heart of the convention, however, is rarely problematized, even if critics have long pointed at the impossibility of harmonizing the needs of economy and environment within the existing frameworks and institutions (Redclift 1987; Norgaard 1995; Escobar 1995). Finally, the intellectual property rights discourse dominates the biodiversity debates on benefit sharing and compensation. This is a neoliberal imposition of the industrialized countries that has become central to free trade agreements and the work of the World Trade Organization. Another important practice is that of prospecting and ethnobioprospecting. Under the guise of "gene hunting," bioprospecting played an important and somewhat unfortunate role in the initial years of the discourse (WRI 1993), giving rise to hopes ("gene rush") or fears (biopiracy) that were neither entirely substantiated nor easily assuaged. Later works have analyzed attempts at navigating between these two positions and their varying degrees of success (e.g., Brush andStabinski 1996; Balick, Elisabetsky, and Laird 1996; Hayden 2004). Tied to the patenting of life forms, bioprospecting can indeed result in very troubling developments, including the loss by small farmers and indigenous peoples of rights to their own plants and knowledge (e.g., Shiva 1997; GRAIN 1998).[23]

The network topology is shaped by a set of hubs such as the northern organizations just described and by national biodiversity offices, such as those we saw in the case of Colombia. If the first set of sites produces a dominant view that could be said to be globalocentric—an assemblage from the perspective of science, capital, and rational action—the second creates "third world national perspectives" which often emphasize sovereignty over natural resources. Although there is great variation in the positions adopted by governments in the third world, there are shared positions that, without questioning the globalocentric discourse, are intended to negotiate the terms of biodiversity treaties and strategies. Unresolved issues such as in situ conservation and access to ex situ collections; sovereignty of access to genetic resources; and the transfer of financial and technical resources to the third world are important agenda items in these negotiations, sometimes collectively advanced by regional groups, such as the Andean Pact countries.

A third group of interrelated sites is made up of progressive NGOs advancing a southern perspective. Autonomy and biodemocracy are key notions of this group. Biodemocracy advocates shift the attention from South to North as the source of the diversity crisis. They suggest a radical redefinition of production and productivity away from the logic of uniformity and toward the logic of diversity. Their proposals are articulated around a series of requirements that include local control of natural resources; suspension of megadevelopment projects and of subsidies to diversity-destroying capitalist activities; support for practices based on the logic of diversity; and recognition of the cultural basis of biological diversity.[24] The fourth, and last, perspective is that of social movements which emphasize cultural and political autonomy. As we saw in the case of PCN, they explicitly construct a political strategy for the defense of territory, culture, and identity linked to particular places. In many cases, these movements' concern with biodiversity follows from broader struggles for territorial control. In Latin America, a number of valuable experiences have taken place in this regard in conjunction with the demarcation of collective territories in countries such as Ecuador, Peru, Colombia, Bolivia, and Brazil.

There are basic cultural and epistemological differences among these positions, particularly the extent to which local and modern forms of knowledge entail different ways of apprehending the world and of appropriating the natural. These were discussed earlier as an aspect of the coloniality of nature. A decolonial approach would start by making visible the ensembles of meanings-uses that characterize diverse groups' engagement with the natural world. It would go on to ask the following questions: From a multiplicity of practices of ecological difference, is it possible to launch a defense of local models of nature within the scope of biodiversity conservation debates? In what ways would current concepts of biodiversity have to be transformed to make this reorientation possible? Finally, which social actors could more pertinently advance such a project?

From a network perspective, the biodiversity conservation movement exemplifies well the "small worlds" created by networks. The basic architecture is decentralized, although with important elements of hierarchy—an assemblage with some key hubs that give shape to much of the network activity. The network is not distributed in the sense that it does not present a meshlike architecture without significant hierarchy among nodes.[25] It operates partly as a scale-free network in that a hierarchy of well-connected organizations operates by integrating players at various scales and in many parts of the world, through particular local, national, and international projects. According to this concept, new nodes preferentially attach to the more connected nodes (those with command of resources, scientific expertise, etc.). The preeminence of scientific knowledge is a key mechanism for maintaining this hierarchical yet decentralized network, exercising soft control over it.

This case also exemplifies other aspects of flat approaches, such as how an emergent event leads to the simultaneous production of systemic ordering and open-endedness. As Terranova put it, "The problem of contemporary modes of control is to steer the spontaneous activities of such systems to plateaus that are desirable and preferable. What we seem to have, then, is the definition of a new biopolitical plane that can be organized through the deployment of an *immanent control* which operates directly within the productive power of the multitude and the clinamen [swerve]" (Terranova 2004: 122). In this way, the biodiversity network both enables and limits the range of possibilities, with progressive NGOs and social movements trying to maintain the vector field as wide as possible and the larger nodes attempting to cancel out differences through homogenization or by containing heterogeneity. If one were to look at the subaltern networks in this assemblage, one could conclude that they tend to be based on a design principle of interoperability among heterogeneous organizations (information systems), which allows for interconnection of autonomous components, decentralization, resilience, and autonomy. Like the antiglobalization movements, they resemble the *distributed movement* that allows them to rearticulate politically key questions in a novel manner. The process is no longer the open-ended architecture that prevailed between the late 1970s and the CBD described by King, but rather a ceaseless negotiation between subaltern and dominant actor-networks but with clear power dynamics among sites. These dynamics have been further turned toward hierarchical modes by the progressive dominance of trade considerations in the biodiversity field, a point I cannot discuss here. Rather, I want to take one last look at networks by delving into the trends in social theory that I have referred to as flat alternatives.

IV ◈ The "Ontological Turn" in Social Theory and the Questions of Information, Complexity, and Modernity

There is always a close connection between social reality, the theoretical frameworks one uses to interpret it, and the sense of politics and hope that emerges from such an understanding. One's hopes and politics are largely the result of the particular framework through which we analyze the real. As I pointed out in the case of the different theories of development in relation to their root paradigms, this connection is often overlooked; however, it is brought to the fore in times of heightened struggles. We might be going through one such period at present, in which we are witnessing both unprecedented social processes and an eagerness for novel languages and categories. As is often the case, the better-known theories are not yet at the point where they can say something radically new because the languages at their disposal do not allow it. As we discovered in the case of the theories of capitalism and modernity, many of these languages are indeed disempowering of these intentions.

It would seem as if from the depths of the social an urge to revamp social theory were springing more intensely than is usually the case. There is one crucial difference as compared with similar theoretical moments of the recent past: the cohort of those engaged in the production of new theories has expanded well beyond the usual suspects in the (largely northern) academies. The complex conversations that are taking place among many kinds of knowledge producers worldwide are themselves a hopeful condition of theory at present. A second feature is that this urge addresses not only the need to transform the contents of theory but its very form; in the last instance, what is at stake is the transformation of our understanding of the world in ways that allow us to contribute to the creation of different worlds. More philosophically, this trend means that a main feature of the current wave of critical theory is its concern with questions not only of epistemology, but also of ontology, that is, with basic questions about the nature of the world; in other words, today's critical theories are fueled by a fundamental scrutiny of the kinds of entities that theories assume to exist and, concomitantly, the construction of theories based on different ontological commitments.

Assemblages and Flat Alternatives ◈ The various waves of deconstruction and discursive approaches of the past few decades brought with them a critique of realism as an epistemological stance.[26] Some of the most interesting social theory trends at present entail, implicitly or explicitly, a return to realism; since this is not a return to the naïve realisms of the past (particularly the Cartesian versions or the realism of essences or transcendent entities), these tendencies might be called neorealist. Other viable metaphors for the emerging social theories could be "biological sociologies" (given the influence of developments in biology; this term has been used to refer to Maturana and Varela 1980) or new materialist sociologies (e.g., actor-network theories). The philosopher Gilles Deleuze has inspired some of these developments, and I shall be concerned here mostly with these, in particular, with the reconstruction of Deleuze's ontology by the Mexican theorist Manuel de Landa (2002) and de Landa's own resulting social theory (2006). In de Landa's view, Deleuze, unlike many constructivists, is committed to a view of reality as autonomous (mind-independent); his starting point is that reality is the result of dynamic processes in the organization of matter and energy that lead to the production of life forms (morphogenesis); things come into being through dynamic processes of matter and energy driven by intensive differences. This amounts to "an ontology of processes and an epistemology of problems" (2002: 6). Deleuze's morphogenetic account makes visible form-generating processes which are immanent in the material world.[27]

A central aspect in de Landa's social ontology arises from Deleuze's concept of the virtual. There are three ontological dimensions to the Deleuzian world: the virtual, the intensive, and the actual. The larger field of virtuality is not opposed to the real but to the actual. This is a different way of thinking about the relation between the possible and the real. The possible does not necessarily resemble the real, as in the notion of "realization"; the possible is not thought about in terms of a set of predefined forms that retain their identity throughout any process of change, thus prefiguring the end result (this is one of the most self-serving modernist assumptions). In the actualization of the virtual, the logic of resemblance no longer rules, but that of a genuine creation through differentiation. The actualization of the virtual in space and time entails the transformation of intensive differences into extensive (readily visible) forms through historical processes involving interacting parts and emergent wholes; this leads to what de Landa calls "a *flat ontology*, one made exclusively of unique, singular individuals, different in spatio-temporal scale but not in ontological status" (2002: 47). "The existence of the virtual is manifested . . . in the cases where an assemblage meshes differences as such, without canceling them through homogenization. . . . Conversely, allowing differences in intensity to be cancelled or eliminating differences through uniformization, effectively hides the virtual and makes the disappearance of process under product seem less problematic" (65). This concealment is the result of human action.[28]

In other words, differences have morphogenetic effects; they display the full potential of matter and energy for self-organization and result in heterogeneous assemblages. Intensive individuation processes occur through self-organization governed by singularities (attractors and bifurcations); differences can be amplified through positive feedback, mutually stimulating coupling, and autocatalysis. Differences in intensity drive fluxes of matter and energy; individuals possess an openness and capacity to affect and be affected and to form assemblages with other individuals (organic or not), further differentiating differences through meshing (2002: 161). One consequence is that in a flat ontology "there is no room for totalities, such as 'society' or 'science' in general" (178). What takes place are processes of metricization and striation of space (such as those we saw with capital and development in the Pacific).

Basing himself on a careful reconstruction of Deleuze's concepts, de Landa goes on to propose his own approach to "social ontology" as a way to rethink the main problems of sociology (including notions of structure and process, individuals and organizations, essences and totalities, the nation-state, scale, and markets). His goal is to offer an alternative foundation for social theory (an alternative "ontological classification" for social scientists). The focus of the resulting realist social ontology is on the objective, albeit historical, processes of assembly through which a wide range of social entities, from persons to nation-states, come into being. The main object of study is assemblages, defined as wholes whose properties emerge from the interactions between parts; they can be any entity: interpersonal networks, cities, markets, nation-states, etc. This view conveys a sense of the irreducible social complexity in the world.

Assemblage theory differentiates itself sharply from theories that assume the existence of seamless webs or wholes. In social ontology, what is crucial is not relations of interiority that generate the totality but the components' capacities to interact with other entities. Assemblages are wholes characterized by relations of exteriority; rather than emphasize the creation of wholes out of logically necessary relations among parts, assemblage theory asserts that the relations among parts are "contingently obligatory," for instance, as in the coevolution of species. In this way, ecosystems may be seen as assemblages of thousands of different plant and animal species; what accounts for coevolution is the symbiosis of species and the relations of exteriority obtained among self-sufficient components. Assemblage theory thus does not presuppose essential identities, that is, social entities with an enduring and mind-independent identity. (This position parallels trends in evolutionary theory based on a view of speciation in terms of historical individuation of species and individuals without taxonomic essentialism; in some of these views, evolution is due as much to natural selection as to self-organization [e.g., Kaufman 1995].)

A particularly salient problem for social theory is the causal mechanisms that account for the emergence of wholes from the interaction between parts; this impinges on the question of the micro and the macro. Conventional approaches assume two levels (micro, macro) or a nested series of levels (the proverbial Russian doll). The alternative approach is to show, through bottom-up analysis, how, at each scale, the properties of the whole emerge from the interactions between parts, bearing in mind that the more simple entities are themselves assemblages of sorts. Moreover, through their participation in networks, elements (such as individuals) can become components of various assemblages operating at different levels. This means that most social entities exist in a wide range of scales, making the situation much more complex that in conventional notions of scale:

> Similar complexities arise at larger scales. Interpersonal networks may give rise to larger assemblages like the coalitions of communities that form the backbone of many social justice movements. Institutional organizations, in turn, tend to form larger assemblages such as the hierarchies of government organizations that operate at a national, provincial, and local levels. . . . A social movement, when it has grown and endured for some time, tends to give rise to one or more organizations to stabilize it and perform specialized functions. . . . That is, social movements are a hybrid of interpersonal networks and institutional organizations. . . . All of these larger assemblages exist as part of populations: populations of interpersonal networks, organizations, coalitions, and government hierarchies. (2006: 33)

There is recurrence of the same assembly process at a given spatial scale, and recurrence at successive scales, leading to a different way of conceptualizing the problem of linking the micro and the macro levels of social reality. For de Landa, the question becomes, How can one bridge the level of individual persons and that of the largest social entities (such as territorial states) through an embedding of assemblages in a succession of micro and macro scales? (2006: 33–40). In the case of markets, the issue is to show how differently scaled assemblages operate, with some being component parts of others which, in turn, become part of even larger ones. In his historical work on the development of markets, de Landa (1997) shows how larger entities emerged from the assembly of smaller ones (including town, regional, provincial, national, and world markets, following the Braudelian explanation).

Wholes exercise causal capacity when they interact with one another. Groups structured by networks may interact to form coalitions (or hierarchies). These larger assemblages are emergent wholes as well—the effect of their interactions goes beyond the interaction of the individuals, in a sort of redundant causality. The fact that in order to exercise their causal capacities social assemblages must use people as a medium of interaction "does not compromise their ontological autonomy any more than the fact that people must use some of their bodily parts (their hand or their feet, for example) to interact with the material world compromises their own relative autonomy from their anatomical components" (38). To sum up,

> The ontological status of any assemblage, inorganic, organic or social, is that of a unique, singular, historically contingent, individual. Although the term "individual" has come to refer to individual persons, in its ontological sense it cannot be limited to that scale of reality. . . . Larger social

> assemblages should be given the ontological status of individual entities: individual networks and coalitions; individual organizations and governments; individual cities and nation states. This ontological maneuver allows us to assert that all these individual entities have an objective existence independently of our minds (and of our conceptions of them) without any commitment to essences or reified generalities. . . . Finally, the ontological status of assemblages is two-sided: as actual entities all the differently scaled social assemblages are individual singularities, but the possibilities open to them at any given time are constrained by a distribution of universal singularities, the diagram of the assemblage, which is not actual but virtual. (40)[29]

Flat ontology and assemblage theory are related to important reformulations of concepts of scale in geography. The past two decades in this field have seen intense debate on this concept, intended to move away from the vertical hierarchies associated with the most established theories (e.g., the Russian doll model) and toward conceptions that link vertical with horizontally networked models (e.g., scalar structuration, glocalization). Building on the insights of flat ontology, these latter conceptions have in turn been critiqued for remaining trapped within a foundational hierarchy and verticality, with lingering micro-macro distinctions and global-local binaries. According to these critics, such problems cannot be solved just by appealing to a network model; the challenge is not to replace one "ontological-epistemological nexus (verticality) with another (horizontality)" but to bypass altogether the reliance on "any transcendent pre-determination" (Marston, Jones, and Woodward 2005: 422). That bypassing is achieved via a flat (as opposed to horizontal) ontology that discards "the centering essentialism that infuses not only the up-down vertical imaginary but also the radiating (out from here) spatiality of horizontality" (422). Here flat ontology refers to complex, emergent spatial relations, self-organization, and ontogenesis.

This framework also moves away from the kind of "liberalist trajectories" that fetishize flows, freedom of movement, and "absolute deterritorialization" at larger scales that are also present in a number of theories inspired by Deleuze and actor-network theories. In contradistinction, the geographical application of flat ontology emphasizes the assemblages constructed out of composition/decomposition, differential relations, and emergent events and how these result in both systemic orderings (including hierarchies) and open-endedness. One conclusion is that "overcoming the limits of globalizing ontologies requires sustained attention to the intimate and divergent relations between bodies, objects, orders, and spaces"; for this, they propose to invent "new spatial concepts that linger upon the singularities and materialities of space," avoiding the predetermination of both hierarchies and boundlessness (424). In this flat alternative, "sites" are reconceptualized as contexts for event-relations in terms of people's activities. Sites thus become "an emergent property of its interacting human and non-human inhabitants"; they are manifolds that do not precede the interactive processes that assemble them, calling for "a processual thought aimed at the related effects and affects of its *n*-connections. That is, we can talk about the existence of a given site only insofar as we can follow the interactive practices through their localized connections" (425).

It follows that processes of localization should be seen not as the imprint of the global on the local, but as the actualization of a particular connective process, out of a field of virtuality. Indeed, what exists is always a manifold of interacting sites that emerge within unfolding event-relations that include relations of force from inside and outside the site. This site approach is, needless to say, of relevance to ethnography and anthropology as much as it is to geography. These recent frameworks do provide an alternative to much established state-centric, capitalocentric, and globalocentric thinking, with their emphasis on "larger forces," hierarchies, determination, and unchanging structures. In contrast, the newer visions see entities as made up of always unfolding intermeshed sites. To paraphrase a well-known work (Gibson-Graham 1996), flat approaches spell out "the end of globalization (as we knew it)." To the disempowering of place embedded in globalocentric thinking, these approaches respond with a plethora of political possibilities. As we shall see below, some of these possibilities are being tapped into by social movements.

Information and Systems Theories ◈ Flat alternatives and theories of complexity and self-organization have not emerged in a vacuum, but the history of some of the most important predecessors and antecedents is rarely told. By telling it, I hope to position other elements and traditions of thought that often lie outside the scope of the main social sciences. Some of these have been mentioned in passing, such as Maturana's and Varela's notion of autopoiesis, but telling a larger story here, even if in the broadest strokes, is of some use.

Flat ontologies have been under development since the 1940s. In fact, it would not be far-fetched to assert they were an opportunity lost during the forties, fifties, and sixties. I am not talking here so much about their roots in European philosophy (including phenomenology, and perhaps beyond), but about the intellectual ferment unleashed by information theory, cybernetics, and systems thinking during those decades. A main precursor of the development of network and complexity theories was the systems research movement that took off in the 1940s and 1950s and that sought not only to develop broad interdisciplinary conversations in the academy but also to relate them to the social, ecological, and technological challenges of the day. An early group established around von Bertalanffy's general systems theory was instrumental in the development of systems thinking. This movement constituted a moment of integrative and collaborative work across the physical, life, and social sciences that remains unparalleled to this date. As history would have it, however, this moment of ferment was lost in the context of the Cold War, militarism, nuclearism, and the consolidation of capitalism based on the military-industrial complex. Applications of systems analysis, operations research, and systems science more broadly, including artificial intelligence, were geared largely toward military and industrial goals. That the early work of systems thinkers was intended to illuminate the decision-making process in social systems within an ethical framework was lost on many of its successors. Many of the pioneers of systems thinking were highly critical of the technologies of destruction and the military-industrial complex they saw ominously arising from the late 1940s on. Their position was informed by theory: the critique of the predominant mechanistic and reductionist scientific doctrines of the day and their contrasting emphasis on wholes, the hallmark of systems approaches.[30]

Systems thinking was a key player in an interrelated set of developments which included cybernetics, information theory, operations research, management science, cognitive science, and computer technologies. The term *behavioral science*, coined in 1949, was meant to integrate the biological and social dimensions of behavior, conceiving of humans as active agents in open systems; it was deeply concerned with the management of the large-scale sociotechnical systems of the day. From the beginning, there were two tendencies present in the fields of systems and cybernetics. The first was generally holistic and nonmechanistic, interested in dynamic representations, learning, and self-organization; the second, focused on information and communication, was more functionalist and mechanistic and interested in control and equilibrium, albeit emphasizing mutual causality, feedback, homeostasis, and self-regulation (e.g., Varela 1996 [1988]; Hammond 2003). These various strands have continued to diverge, although there have been important moments of convergence, for instance, around general systems theory in the 1950s and 1960s, strands of cognitive science and artificial intelligence based on emergence and enaction since the 1980s, and the sciences of complexity in recent decades. General systems theory—much like the sciences of complexity today—sought to ascertain and model structural similarities or common dynamics underlying many different kinds of systems.[31]

Varela's concise history of the development of cognitive science since the 1950s is instructive (1996 [1988]; see also Varela, Thompson, and Rosch 1991). For Varela, the main goal of the cybernetics movement was to create a science of the mind, logic and mathematical modeling being its main instruments. This was also the origin of von Neumann's first digital machines. Two products of this early phase were general systems theory and information theory. In a second, "cognitivist" phase, beginning in the mid-1950s, the alleged similarity between mind and the computer led to a view of cognition as the computation of symbolic representations; linguistics, neuroscience, and artificial intelligence (AI) became central (e.g., expert systems, robotics, fifth-generation computers), with computers as the model of thought and intelligence as the ability to manipulate symbols according to logical rules (Winograd and Flores 1986). In the late 1970s, after a twenty-five year hiatus, the ideas of self-organization that were present in the formative years of cybernetics came back on the scene and presented a first alternative to cognitivist orthodoxy. The key idea was that cognition takes place not in terms of a system of rules and symbols but on the basis of vast numbers of interconnected elements which, operating locally, create modes of global cooperation that, depending on the conditions, emerge spontaneously. This transition from simple local rules to global coherence is at the heart of what in the days of cybernetics was called self-organization and that more recently has been talked about in terms of nonlinear dynamics, emergence, and complex adaptive systems. By studying neural networks scientists came to appreciate the dynamic cooperation between local components and global effects. In this view, symbolic regularities emerge out of distributed processes, and this applies even to the notion of a decentered or "virtual" self (Varela 1999).

Varela's fourth phase, enaction, has also been under development since the 1960s. The fundamental idea is that, contrary to the mainstream cognitivist assumption, cognition consists not in the representation of an external, pregiven world by the mind but in the enaction of a relation between mind and world based on the history of their interactions. This is so because of the fundamental fact of embodiment. Unlike the rationalistic assumption of the separation of mind and body, the enactive view of cognition starts with the radically different phenomenological position of the continuity between mind and body, the body and the world. As Maturana and Varela are fond of saying, all doing is knowing and all knowing is doing; indeed, there is "an unbroken coincidence of our being, our doing, and our knowing" (1987: 25). Every act of knowing, in this way, brings forth a world. Cognition is always embodied action in a historical background in which our body, our language, and our sociality are inextricably linked. The world is seen not as something predefined and representable, but as always emerging from one's embodied actions. One can never stand fully detached from the world because one is inevitably thrown into it (from Heidegger). The fact that in this view all knowledge is enactive and brings forth a world means that every knowledge act has an ontological dimension, providing another connection with the idea of ontological turn with which I started the chapter. As I mentioned, Maturana's and Varela's work constitutes a sort of biological sociology (Maturana and Varela 1980: vi) that, early on, aimed at the same kind of materialist neorealism that is flourishing today in flat approaches to networks and complexity.[32]

Information, Networks, and Modernity ◈ Many recent theories of networks assume that networks are about information above all else. This is a restricted (modernist) view that relies on a disembodied view of information; there are in fact many embodied aspects to knowledge and networks. Information, nevertheless, is a central component of networks. Foundational work on information and communications occurred during the early cybernetics period. As Terranova (2004) has argued, the relation was firmly established with the theories of information of the 1940s and 1950s based on thermodynamics and statistical mechanics, particularly in the work of Claude Shannon. Since then, there has been a tendency to reduce information to its technical aspects, overlooking the fact that it always involves practices, bodies and interfaces, particular constructions of the real, and, in general, "a set of relays between the technical and the social" (25), to which one could add the biological (body, nature). For Terranova, an entire cultural politics of information is associated with distributed networks; this requires a critical examination of information and communication technologies (ICTs) that focuses on how they involve "questioning the relationship between the probable, the possible, and the real. It entails the opening up of the virtuality of the world by positing not simply different, but radically other codes and channels for expressing and giving expression to an undetermined potential for change" (26). Akin to de Landa and Marston and collaborators, Terranova envisions a cultural politics of the virtual, understood as the opening up of the reality to the action of forces that may actualize the virtual in different ways.

The relation between networks and information has become central to most network theories, although not in the same way in all of them. Building on the work of Marilyn Strathern, Riles (2001) has made an eloquent case for the anthropological study of networks as self-producing entities that operate on the basis of information. This opens up a serious epistemological problematic that is, indeed, at the center of much work in the anthropology of modern science and technology, already mentioned in the introduction (e.g., Marcus, ed. 1999; Fischer 2003; Osterweil 2005b): how does one study and describe situations in which the objects or subjects are thoroughly constituted by the same knowledge practices of which the ethnographer herself is also a product? The anthropological solution so far has been, first, to recognize that there is no radical outside from which to conduct a completely detached observation and, second, to posit that all we are left with is the possibility of a deepened reflexivity. These conditions have led to various proposals, from "enacting the network" in the ethnographic description (Riles), to focusing on emergent forms of life (Fischer), to heightened reflexivity coupled with novel techniques to render the modern artifacts ethnographic (actor-network approaches, distributed notions of fieldwork [e.g., Fortun 2003]). These are all interesting steps, yet they leave a number of insufficiently addressed questions concerning epistemological circularity, boundaries, the reconceptualization of sites, and so forth.

The problem of the "lack of an outside" is posed by flat ontologies in ways that provide a different, less anthropocentric and representational set of answers; in fact, this particular problematic was present from the very beginning, at least in the work of Maturana and Varela. In his main essay of the late 1960s, Maturana put it thus:

> The knowledge that an observer claims of the unities that he distinguishes consists in his handling of them in a metadomain of descriptions with respect to the domain in which he characterizes them. Or, in other words, an observer characterizes a unity by stating the conditions in which it exists as a distinguishable entity, but he cognizes it only to the extent that he defines a metadomain in which he can operate with the entity that he charac-

> terized. . . . We become *observers* through recursively generating representations of our interactions, and by interacting with several representations simultaneously we generate relations with the representations of which we can then interact and repeat this process recursively, thus remaining in a domain of interactions always larger than that of representations. . . . we become *self-conscious* through self-observation; by making descriptions of ourselves (representations), and by interacting with our descriptions we can describe ourselves describing ourselves, in an endless recursive process (reprinted in Maturana and Varela 1980: xxiii, 14; emphasis in the original).[33]

The general problematic to which they provide an extended answer throughout their work is this: "Thus we confront the problem of understanding how our experience—the praxis of our living—is coupled to a surrounding world which appears filled with regularities that are at every instant the result of our biological and social histories." This is a flat statement, one that, to recall Varela, keeps us "close to biology" (1996 [1988]: 113). More fundamentally, "a theory of knowledge ought to show how knowing generates the explanation of knowing. This situation is very different from what we usually find, where the phenomenon of explaining and the phenomenon explained belong to different domains" (Maturana and Varela 1987: 239). Indeed, "the whole mechanism of generating ourselves as describers and observers tells us that our world, as the world which we bring forth in our coexistence with others, will always have precisely the mixture of regularity and mutability. . . . the fact remains that we are continuously immersed in this network of interactions, the results of which depend on history. Effective action leads to effective action: it is the cognitive circle that characterizes our becoming, as an expression of our being autonomous living systems" (241). In this way, the authors aim at a postrepresentational epistemology through their unique view of the hermeneutic circle in an action-oriented conceptualization.

An important question I will not broach here is whether there are instances of flat, relational, self-referential, autopoietic, or network thinking—or, rather, ways of thinking-being-doing—among social groups "in the real world" beyond the academy and social movements and, if so, what this might mean for social theory. One highly pertinent case in contemporary times (there might be many historical cases) is that of many indigenous peoples, whose epistemologies and ontologies can frequently be described as operating within the overall relational and postrepresentational dynamic described here. Some recent ethnographies of these indigenous worlds, carried out with network awareness in mind, suggest that this is the case. These worlds are veritable worlds and knowledges otherwise, although they always constitute border areas in relation to dominant national and global societies. The implications of this kind of thinking for theories of networks and activism are being explored at present, in part by seeing indigenous "worlds and knowledges otherwise" as instances of another thought that could contribute to envisioning quite different constellations of critical thought and political practice (see, e.g., Blaser forthcoming; Walsh 2007).

Conclusion

The interest in flat alternatives is a sign of the times. "We are tired of trees," famously denounced Deleuze and Guattari, two of the prophets of this movement in modern social theory. "We should stop believing in trees, roots and radicles. They've made us suffer too much. All of arborescent culture is founded on them, from biology to linguistics" (1987: 15). What they mean is that researchers need to move away from ways of thinking based on binarisms, totalities, generative structures, preassumed unities, rigid laws, logocentric rationalities, conscious production, ideology, genetic determination, macropolitics, and embrace instead multiplicities, lines of flight, indetermination, tracings, movements of deterritorialization, and processes of reterritorialization, becoming, in-betweenness, morphogenesis, chaosmosis, rhizomes, micropolitics, and intensive differences and assemblages. If the dominant institutions of modernity have tended to operate on the basis of the first set of concepts, it would make sense now to build a politics of world making based on the second set, being mindful that both sets of processes coexist in contradictory manners (Gibson-Graham 2006). From biology to informatics, from geography to social movements, from some critical theorists to many indigenous and place-based groups and activists this is a message that can at least be plausibly heard.

Not that this message solves all the problems of theory or political action, although, getting back to my reflection on the relation between social reality, social theory, and the articulation of political purpose, it perhaps renews one's sense of hope. The question of scale, for instance, remains an open one. While some geographers would favor the abandonment of this concept altogether because of its inevitable allegiance to hierarchical and modernist ontologies (Marston, Jones, and Woodward 2005), other authors are attempting to reconceptualize it, avoiding the traps of deterministic verticality; in de Landa's view, assemblages at one scale, using the existing population of assemblages as components, may generate larger-scale assemblages, as in the case of markets. Second, the idea that material and biological processes could inspire understandings of social life at more than analogical or metaphorical levels is bound to remain controversial and resisted by many. As Dianne Rocheleau put it (personal communication), many social theorists panic at the idea of extending "biological metaphors" to "the social," but in fact "if this issue is all about assemblages constructed from a continuum of experience and matter, that is, both self-organized and other-organized, then there is no separate biological and social. The lessons would simply come from one kind of theory and study to the other and not from some pregiven biological realm per se." Third, while some, perhaps many, of today's movements seem intuitively or explicitly aiming at a practice informed by flat conceptions (e.g., self-organizing networks), it remains to be seen how they will fare in terms of the effectiveness of their action. Most observers would say that the experience of groups like PCN is ambiguous in terms of this criterion. There is a need for more empirical and activist-oriented research on particular experiences, such as the kind of time series used in some fields to ascertain longer-term dynamics in the globally oriented movements.

Flat alternatives, finally, also contribute to putting issues of power and difference on the table in a unique way. If actual economic, ecological, and cultural differences can be seen as instances of intensive differences, and if, moreover, these can be seen as enactments of a much larger field of virtuality, then the spectrum of strategies, visions, dreams, and actions is much larger than conventional views of the world might suggest. The challenge is to translate these insights into political strategies that incorporate multiple modes of knowing while avoiding the modern dream of organizing (the people) in logocentric, reductionist ways. I do not mean to say, however, that the implication of flat alternatives is that anything goes. On the contrary, as I discussed in the case of PCN, making self-organization work paradoxically takes a lot of work, but of a very different kind. But again, this is just a beginning. I have not said anything about two other fundamental questions: First, should modern social theory be reconstructed on the basis of flat approaches? If so, would these approaches fit with ease within the table of modern social theory (see table 4 in chapter 4), as a fourth column next to liberal, Marxist, and poststructuralist theories? or should the idea of fitting these alternatives within this table be rejected altogether? Second, how does one think about the implications of flat alternatives for social change? Could concepts of articulation, translation, autonomy, and counterhegemony, for instance, be rethought from this perspective? Finally, how does one articulate flat approaches with the actual practices of concrete groups in place and in networks, particularly in cases where these practices evince nonmodern scripts? These will remain open questions for some time, although I will have some additional brief remarks to make about them in the conclusion.

I will end with a recent discussion by the biologist Brian Godwin, a complexity theorist and wise elder of an alternative West, about these trends. "We need to hold a vision of what is just dawning," he said, by which he meant that emergent, self-referential networks are indicative of a certain dynamics, signaling an unprecedented epoch and culture for which a new vocabulary is needed. By a new culture he means something far deeper than any rationalistic understanding of networks might suggest. For one thing, this dynamics is not something one invents but something one experiences. If the shadow of modernity is death—its greatest fear—the message of biological worlds (from neurons to rivers, from atoms to lightning, from species to ecosystems and evolution) is that of self-organization and self-similarity. If language and meaning, as some of these biologists suggest, are properties of all living beings and not only of humans—that is, if the world is one of pansentience—can activists and others learn to become "readers of the book of life" and avail themselves of this reading to illuminate their reveries and strategies? How do humans learn to live with and in both places and networks creatively? In a manner that echoes the critical opalescence that happens at phase transitions, activists may envisage something new coming into being and learn to nourish it. In the words of Maturana and Varela, the lesson of this deeply relational biology is that "*we have only the world that we bring forth with others, and only love helps us bring it forth*" (1987: 248). In bringing forth this world, it is perfectly fine to use our rational minds, but it certainly means as well embracing ways of knowing other than the rational and the analytical. These other ways of knowing are not naïve romanticism and would have to be included, in Goodwin's view, in any new foundation of realism and responsibility.[34]

conclusion

What happens in the Pacific is not accidental; it is a planned process. Colombia has the largest humanitarian crisis in the West, and it affects Afro-descendants disproportionately; this is not random. As the pressure over the territories mounts, and more and more *compañeros* are threatened, the conditions for resistance become more difficult. The fight will now be gauged in terms of our ability to influence public policies from the perspective of the real situation of the communities. To move along this path is a test for us. But in this context, the right for the cultural objection against development and in favor of the development vision of the communities becomes even more important.
—PCN activist, Cali, August 2007

Yurumanguí

On July 2, 2002, a PCN activist from Buenaventura who was working with an alternative development and organizational strengthening project in the Yurumanguì river wrote as follows in an e-mail:

> I just got back from Yurumanguì; it was great to be able to see again and *dar un abrazo a todos* there [hug all the members of the PCN-affiliated local organization, APONURY], who have been *secuestrados* in [constrained to] their own *vereda* [hamlet], since as you know they can barely move along the river without putting their lives in danger. We had an *asemblea general* [general community meeting] to inaugurate the *trapiche* [low-tech sugarcane mill], along with the trained people to operate it and everything else, according to the terms of the Solsticio project. César (pseudonym, one of APONURY's leaders], along with the young people involved, explained how the rice project is going, and it was really exciting to see the rice field about to be harvested, for the first time in over 25 years. The paramilitaries and the army continue to block the entrance of food to the river; this is why rice cultivation and the trapiche have been so enthusiastically received by the people. . . . despite feeling overwhelmed by the stories of the impact of the armed conflict in the river we danced and sang and joined in the

agricultural activities. We were accompanied by [*hubo un acompañamiento por*] representatives from the Swedish Embassy, UNHCR, the Red de Solidaridad, and the Defensoría del Pueblo; this responded to the efforts we [PCN] have been carrying out since 2000 to achieve international *acompañamiento* and support for the *Territorios de Vida, Alegría y Libertad* [Territories of Life, Happiness, and Freedom, a strategy of autonomous territories in the midst of the armed conflict] that we announced in 2000 in Yurumanguí, in response to the massacres in the Naya and Cacarica rivers.

The project in question was called "Fortalecimiento de las dinámicas organizativas del Proceso de Comunidades Negras del Pacífico Sur Colombiano, en torno al ejercicio de los derechos étnicos, culturales y territoriales" (Strengthening of the organizational dynamics of Proceso de Comunidades Negras in the southern Colombian Pacific in regard to the exercise of ethnic, cultural, and territorial rights). The project began to be implemented in January 2000 after receiving funding from the Solstice Foundation in Copenhagen, and it was carried out in two locations over 2000–2005, the Yurumanguí River and Pílamo, a black community in the Norte del Cauca Andean region, south of Cali. Pílamo had grown up around a recovered hacienda and, like the Yurumanguí, was to receive its collective title in 2000 through Ley 70 and with the support of the project.[1]

From its inception the project had twin goals: to support the life project of the black communities, as seen from an ethnoterritorial perspective, and to strengthen the organizing process vis-á-vis the appropriation of the territory. The key elements activated to carry out these goals were a series of projects geared toward food autonomy (*autonomía alimentaria*) and self-governance, based on long-standing productive practices (the "traditional production systems" outlined by Proyecto Biopacífico; the projects could thus be seen as a radicalization of the work of the PBP). It was out of the intensive work carried out in Yurumanguí and Pílamo that the conceptualization of the "Basis for Culturally and Environmentally Sustainable Development" presented in chapter 3 (figure 6) was developed by PCN and project staff.[2]

The political and technical training of local leaders was done through workshops on various topics, from national and international conjunctures and the armed conflict to Ley 70, participatory planning, social cartography, and project design and implantation. The *Plan de Acción* (Action Plan, or *Plan de Uso y Manejo*) occupied a central place in the project. This was articulated with the application for the collective title (including *mon-*

teos, mapping, oral histories, inventories of use spaces, and conflict over territory, etc.). The collective title for the Yurumanguí was finally granted at the end of 2000 and covered over fifty-two thousand hectares, from the river's source in the Andean foothills to its *desembocadura*, or outlet, into the Pacific Ocean.[3]

The strategy for autonomía alimentaria, as a way to resist in place, relied on two projects identified with river people: the first brought in a sugarcane trapiche to enable local processing of the cane into syrup, a basic foodstuff used for sweetening, alcoholic drinks, and other purposes; the second project was to reintroduce rice cultivation and processing, including a *trilladora*, or thresher, for the milling of rice, which had been absent from the rivers of the Pacific for over thirty years, ever since the Green Revolution dry rice from the Andean interior displaced all other forms of rice cultivation. The projects were seen in terms of the recovery of traditional crops and the lessening of food dependency. While the first project involved many men and women in the community, the second was targeted at enlisting the participation of young people of both sexes. The projects also sought to refine notions that had been introduced progressively into the ethnoterritorial discourse, including region-territory, Territorios de Vida, Alegría y Libertad, *vivir bien* (good living), and *bienestar colectivo* (collective well-being).[4]

Armed actors, including guerrillas, paramilitaries, and the navy, had made inroads into the Yurumanguí on and off since 2000. In response to the initial process of intimidation and threats to leaders and the community PCN designed an international strategy to designate river communities as communities in need of special protection; this idea was to result in the notion of Territorios de Vida, Alegría y Libertad. Because several relatives of a main leader of APONURY had been assassinated there, the Yurumanguí was chosen as the first community to be declared a Territorio; it was hoped that the massacres already perpetrated in rivers such as Riosucio and Naya would thus be avoided. International alarms and solidarity actions continued to take place in the period 2000–2006, particularly in relation to virtual invasions of the river hamlets by the navy (for instance, in 2003 and 2005) and by paramilitary actions often linked to the navy's.[5] Guerrillas of the Fuerzas Armadas Revolucionarias de Colombia (FARC) carried out actions in the foothills of the upper reaches of the river, and local organizations had to deal with them. This situation affected the projects but also made clear the need for self-provisioning in food to resist the strategies of displacement and emplacement pursued by all of the armed actors.

Several aspects of these projects can be highlighted from the perspective of the social movement's knowledge production and political strategy. PCN bet on the Yurumaguí and Pílamo projects to both develop and demonstrate the validity of their ethnoterritorial approach in the face of conflict and displacement. The group sought to implement an economic strategy for the collective territories that articulated ecological, political, and socioproductive projects. A lot was learned in terms of traditional and new knowledge of species and agriculture, viable household community economies, the organizational and technical aspects of a project, and people's openness to an approach that referred to somewhat abstract organizational principles.

Despite undergoing change and assuming different shapes, these projects worked up to a certain point. While armed conflict, massacres, and coca cultivation increased rapidly in surrounding rivers after 2002, for example, in the Naya, Cajambre, Anchicayá, and Raposo, the Yurumanguí at least for a time has been able to resist coca, maintain a fair degree of control over the territory, and keep the organizations going despite the not-infrequent intimidation and selective assassinations that continue to take place. The economic criteria used by the projects resonate with those of the alternative economies discussed earlier. The trapiche and trilladora were constituted as *empresas comunitarias* (communal enterprises) to be collectively run under the supervision of the community council. One of the reports states that "the trapiche should function in the most equitable way possible; every person utilizing it should be treated in the same way" (PCN 2003a). Moreover, the trapiche project instituted a *fondo* (fund) for activities deemed important by the community council "as a component of the political construction of the project of the *comunidad negra*" (43). The project even considered giving subsidies in particular cases, such as to people who had to travel long distances to the trapiche and to those who wanted to grind cane for special *fiestas* like birthdays, godparents' celebrations, and other ritual festivities since "these are cultural festivities that enrich daily life in our territory" (44). To the question What happens to the surplus? then, a partial answer is that it is reinvested in activities that are seen as strengthening people's cultural practices, sustainability, and ability to resist in place. More generally, PCN has made great strides in the past few years in the design and implementation of sophisticated *Planes de uso y manejo* and *Planes de contingencia* for several areas that, building on long-standing practices and forms of resistance, confront the current problems of displacement, armed conflict, and human rights violations. These plans are seen as strategies of self-affirmation; they constitute instances of knowledge production that can be exemplified here only briefly through the case of the Yurumanguí river.

Modernity, Life, and the Politics of Theory

Globality versus Modernity? ◈ The situation in the Pacific reflects global processes that affect many world regions, albeit in particular ways. In some regions, violence takes on a central role in the regulation of peoples and economies for the control of territories and resources. For the most farsighted social movements, whether the situation is read in an ecological or a cultural register or a combination of both, the basic idea is the same: overcoming the model of modern liberal capitalist society has become a must for survival, and perhaps a real possibility. Despite the contradictory and diverse forms it has taken in the present decade, the so-called turn to the Left in Latin America suggests that this urge is felt at the level of some governments (Venezuela, Bolivia, Ecuador, Argentina). Why this is happening in Latin America more markedly than in any other world region at present is a question I cannot broach here, other than to say it is related to the fact that Latin America was the region that most earnestly embraced neoliberal reforms starting in the late 1970s and that applied the model most thoroughly; in addition, Latin America has shown the most ambiguous results, in the best of cases.[6] Whether these countries are entering a postneoliberal moment remains a matter of debate.

Modernity's ability to provide solutions to modern problems has reached a limit, making discussion of a transition beyond modernity feasible, perhaps for the first time (here, of course, the recent public debates on global climate change are an important referent). The discussion of a transition beyond the current order brings together those who call for new anticapitalist imaginaries, such as the long-term critics Aníbal Quijano (e.g., 2002) and Samir Amin (e.g. 2003), and those who emphasize non-Eurocentric perspectives on globality, such as Boaventura de Sousa Santos, the advocates of decoloniality, and Boff's (2002) call for a new paradigm of relinking with nature and each other. What these two groups have in common is an acute sense of modernity's inability to tackle today's problems, which include massive displacement, ecological destruction, poverty, and inequality. Boaventura de Sousa Santos has perhaps most pointedly captured this predicament:

The conditions that brought about the crisis of modernity have not yet become the conditions to overcome the crisis beyond modernity. Hence the complexity of our transitional period portrayed by oppositional postmodern theory: we are facing modern problems for which there are no modern solutions. The search for a postmodern solution is what I call oppositional postmodernism. . . . What is necessary is to start from the disjunction between the modernity of the problems and the postmodernity of the possible solutions, and to turn such disjunction into the urge to ground theories and practices capable of reinventing social emancipation out of the wrecked emancipatory promises of modernity. (2002: 13)

As a way of restating and reinterpreting, one last time, some of the main propositions of this book, I want to pause to reflect on the relation between globalization and modernity. In attempting a state of the art discussion on modernity from the perspective of cultural studies, Grossberg (2007) has suggested that the key problematic of the present is not globalization, as is most often assumed, but the status of modernity itself. Why? Because discourses of globalization are themselves subsidiary to visions of modernity; modernity thus becomes the most important political and cultural question. For Blaser (forthcoming), the present can be described as a generalized struggle to define and shape an emergent order, globality. Will this order be "modernity writ large" (as in Giddens's thesis)? Or can globality be imagined as an alternative, rather than a continuation of, modernity? If there are competing visions of globality, it must be because there are competing visions of modernity.

Schematically, following Grossberg (2007) and Restrepo (2007) (and to recall the discussion at the end of chapter 4), there are four main positions regarding modernity: (a) modernity as universal process of European origin (intra-Euro/American discourses); (b) alternative modernities (locally specific variations of the same universal modernity); (c) multiple modernities, that is, modernity as multiplicity without a single origin or cultural home. In this latter view, modernity emerged from multiple intersecting processes, did not have a single origin, and has followed multiple trajectories. The modern is thus an ongoing struggle to define the real in terms of articulations of time and space, presence and change, lasting structures and the experience of the everyday. Various modernities might have emerged in some parts of the world (e.g., East Asia) before the fifteenth century and intersected with Western modernity; with the colonization of Africa, multiply originated modernities initiated their painful process of adjusting and articulating across power and difference. In other words, not every modernity is Euro-modernity, and multiple modernities can thus be reclaimed as an ontological and political project. By deessentializing modernity more radically than most other works, Grossberg opens up new possibilities for rethinking the modern. It also becomes possible to think about modernity without coloniality. This is another way of delinking modernity from the tight embrace of the West and locating possibilities for remaking it everywhere.[7]

From an MCD perspective, as we know, there is a fourth possibility, one that in my opinion cannot be fully reduced to Grossberg's typology: (d) modernity/coloniality, or rather modernities/colonialities/decolonialities. From this perspective, the Colombian Pacific as much as Europe or any other place in the world is modern/colonial and, of course, potentially the site of decolonial projects (see also Yehia 2006). The "origin" of modernity is always the origin of modernity/coloniality/decoloniality; after all, one of the main postulates of the framework is that all humankind lives in a single, albeit structurally heterogeneous, modern/colonial world system. From this reformulation emerge some important qualifications of the framework, as I explained in chapter 4. First, not all power relations can be described in terms of coloniality; second, subalternization does not exhaust the subject position of oppressed groups like the black groups of the Pacific. On the contrary, in emphasizing their own life projects, they do more than just speak from the position of the subaltern or colonial difference, they affirm an ontological project. What Blaser, Feit, and McRae (2004) say of indigenous peoples in relation to development projects could be said of them, namely, that they stand "in the way of development": this means, first of all, that they affirm their life projects, which have to do with maintaining their networks of reciprocity and relationality, their ontological commitments, their ecological, economic, and cultural difference; and, second, that they relate to development from this perspective, whether to resist it, tolerate it, or go along with it when it supports this or that aspect of their life project.

The meaning, for me, of alternatives to modernity or transmodernity is a discursive space in which the idea of a single modernity has been suspended at an ontological level; in which Europe has been provincialized, that is, displaced from the center of the historical and epistemic imagination; and in which the examination of concrete modernities, symmetrical projects, and decolonial processes can be started in earnest from a deessentialized perspective. This leaves one with a view of multiple modernities or multiple MCDs as coexisting theoretical possibilities to be maintained in tension. It may take some years to advance in the investigation of these two perspectives (Restrepo 2007b). For now, I hope that the notion of practices of difference that underlies this book offers a way of understanding different modernities and perhaps contributes to the advancing of questions about alternatives to modernity, nonmodernity, and the meaning of decoloniality in terms of "the histories, subjectivities, forms of knowledge, and logics of thought and life" that challenge the hegemony of the modern/colonial world system (Walsh 2007: 104).

Questions of Knowledge and Theory ◈ Besides the positions it takes on modernity and development, this book stands in favor of seeing social movements as important spaces of knowledge production about the world and of recognizing the value of activist knowledge to theory. The book was constructed as a conversation between activist and academic knowledges of somewhat corresponding issues. More than this, the book took its main cues from the intellectual and political analyses of the activists. Whether trying to understand the concept of territory, following the guiding role of the five organizational principles, discussing ecological practices, or attending to new issues such as reparations, I elaborated my subsequent arguments by taking the activist articulation as a point of departure. Of importance also were the kinds of academic theories summoned to the conversation. Following a world anthropologies approach, the book privileged—again, at least to a greater extent than usual—little-known theories, noncanonical authors, and Latin American intellectual productions.[8] I also wagered on the value of collective framework building to intellectual work (MCD, WAN, women and the politics of place).

Multiplying the landscape of knowledge production is a way of unsettling the megastructure of the academy as the knowledge space par excellence; the knowledge landscape itself appears flatter, in the sense of being populated by many more sites of knowledge production than in the past and the many networks in between those sites. From MCD and WAN discussions, a new series of concepts appears: *pensamiento propio* (one's own thought), other knowledges and knowledge otherwise, *academias otras* (other academies), activist knowledges, activist research, border thinking, and so forth. One may refer to these forms of knowledge production in terms of what Mato (2005) calls "intellectual practices in culture and power." Questions of what, with whom, how, and from what locations one thinks become of paramount importance to elaborating effective strategies of collaboration. From the activists' perspective, the point of departure for such collaboration can be nothing short of joint consensus about the character of the political project. More than the validation of theories, the goal of collaborative projects comes to be seen as contributing to the goals of particular social and political movements. In the case of PCN, this process involves taking the organizational principles as a point of departure; always taking into account PCN's thought and practices from the perspective of its own reading of such practices; and thinking about the project in terms of the valuation and analysis of the movement's thought (pensamiento propio), as it is collectively constructed (PCN 2007).[9] In a more philosophical vein, one could say that the call for other theories and theory otherwise is itself the result of ever more significant encounters with difference—social, theoretical, epistemological, ontological.[10]

Questions of Nature, Life, and Networks ◈ As some complexity theorists argue, to continue with the discussion at the end of the last chapter, creativity and adaptability are aspects inherent in all forms of life (e.g., Goodwin 2007). For those designing social worlds there are lessons to be learned from how creativity works in the natural world. Relational networks in particular are ubiquitous in biological life—from the brain to the ecosystems; what underlies many self-organizing networks and self-similar formations is the coexistence of the coherence of the whole with maximum freedom for the parts, with minimum energy used to arrive at the formation. From this relational biology, Goodwin draw lessons for collective social forms and relations with the natural world—concerning, for instance, the healing power of relationships, sustainable agriculture, place-based economies, etc. All of this clearly goes against the grain of individualizing modernity, the separation of nature and culture, and the current economic paradigm. Goodwin entertains the idea of "a cultural transformation that will either carry us into a new age on earth or will result in our disappearance from the planet. The choice is in our hands. I am optimistic that we can go through the transition as an expression of the continually creative emergence of organic form that is the essence of the living process in which we participate" (177).

Visions of a transition are not rare among ecologists, but besides gaining renewed urgency they are showing greater awareness of the cultural and economic processes that are bringing about the profound damage to the biological structure and functioning of the planet and, conversely, of the social transformations needed for change. For the environmentalist Thomas Berry of North Carolina, the disturbance of the planet "is leading to the terminal phase of the Cenozoic era" (1999: 4), opening up the possibility of an Ecozoic era in which humans can build a mutually enhancing relation with the planet. To move in this direction would require reversing some of the most cherished tenets of modern society, particularly the radical discontinuity between humans and other modes of being and the utilitarian attitude toward the earth; more generally, it would require revamping the institutions of politics, economics, religion, and culture, which in Western culture have been so committed to the destructive pattern that is bringing about the disintegration of life systems everywhere. Moving from a human-centered to an earth-centered vision entails healing both the earth and the human. Such healing requires thinking in terms of the integrity of places and regional geographies but in the context of the whole planet. Many sets of interests militate against this goal, particularly the tension between the logic of development and that of ecology; this tension has become the primary contradiction in most regions of the world, as we saw it exemplified in the Pacific. One can say that activists in regions such as the Pacific are at the forefront of the struggle for the kind of transition envisaged by ecological visionaries and cultural critics. PCN's Territorios de Vida is a statement of human and ecological viability from this perspective.

Social scientists and humanists have traditionally been skeptical of big claims such as these, for good reasons. However, the current deeply anthropocentric and secular modes of inquiry seem to be falling short in the task of thinking about the kinds of transformations that are needed in the face of the current social and ecological catastrophes. Some biologists seem attuned to this type of inquiry, a development that in itself is a historical development. But there are some parallel ideas in the social sciences calling for a transition, as in Santos's notion of oppositional postmodernism, mentioned above. What I am suggesting is that there is also an oppositional postmodernism in the sciences (it has been there for quite a while, at least since the landmark study by Prigonine and Stengers, 1984) and that it has important and unique elements to contribute to bringing about the transition socially and imaginatively. Critical intellectuals in many fields will find in these works elements of interest: for instance, alternative genealogies of modern science—indeed, the existence of an alternative, less Eurocentric and colonialist West; mutually enriching conversations between alternative Western traditions and nonmodern knowledges and traditions (e.g., those of indigenous peoples, peasants, and Afro-descendants in many parts of the world, many of whom for centuries have knowingly lived relationally and thus ecologically); collective action needed to contribute to bringing about change in the direction of ecologically sensitive worlds, including the question of whether social movements can emulate the dynamical processes and structures of self-organizing networks; and ideas about design, process, and form in many domains, from the economy and agriculture to cities and communities.

To restate a concept from the chapter on networks, one may think of social and biological life in terms of assemblages from a continuum of experience and matter that is both self-organized and other-organized; in this thinking, there are no separate biological and social worlds, nature and culture. One can then read the insights of complexity as lessons passed on from one kind of theory to another, not emanating from some pregiven biological realm per se. At the very least, complexity and flat approaches appear as viable proposals to work through two of the most damaging features of modern theory: pervasive binarisms and the reduction of complexity, which are part and parcel of the coloniality of nature, gender, knowledge, and power; to a greater degree than most other proposals complexity theory and flat approaches enable the reintroduction of complexity into our intellectual accounts of the real.

Regions and Places in the Age of Globality

Until the 1950s, black communities on the rivers of the southern Pacific found ways to develop their life projects in contexts that ranged from slavery to freedom, with many shades of marginalization and resistance in between; they were able to do so through place-based economic, ecological, and cultural practices. Rather than backward, these practices were effective, viable options in terms of life-worlds and identities. When these kinds of adaptation became increasingly untenable in the 1950s, completely different approaches had to be called on, particularly after the late 1990s, when the armed conflict spilled over into the region. Today, as we have seen from my (limited) analysis of PCN's knowledge and political strategies, activists and to some extent communities, for example, in the Yurumanguí, are attempting to reinvent themselves through a new relation to the state, themselves, the environment, and global forces. Such reinvention takes place in the context of the emergence of the Afro–Latin American as a potent social, cultural, and political fact in many parts of the continent. The Afro–Latin American is only partly related to what in my discussion of biodiversity I called the emergence of the biological as a global social fact, but the articulation between the two processes has been important in the case of the Pacific. Discussions on diasporization, conservation, and rights thus find in the Pacific a potent space of articulation, one shaped more by PCN than by any other group.

Historical ecological perspectives suggest that the Pacific region has been more significantly transformed over the past three to four decades than in all the previous centuries of anthropic activity, coca and African oil palm being two of the main vectors of reterritorialization at present. Capital and the state—the state through development discourses, post–Ley 70 legal instruments concerning water and forests, violence, the forced promotion of African palm, and neoliberal policies such as the proposed free trade agreement with the United States—are the main agents of transformation. A crude form of capitalist modernity seems to have become entrenched once again in the shadow of which to speak about alternative production rationalities, diverse economies, respect for place-based models of nature, self-organization, or even about the more accepted goal of conservation might seem naïve or utopian. Perhaps in a few places, such as the Yurumanguí, social movements may have a chance to be real contenders for ecologically and culturally grounded alternative modernities—although even these cases are fragile—but a significant defense of the region as a whole seems out of the question. Even activists recognize that what gives them strength are "those fights we haven't lost . . . even when we play them to a tie we are happy [since] most of the time they win" (see the epigraph to chapter 2). And yet the same activist went on to say, "Today, more than ever, I am convinced that there are alternatives; we will continue waging the battle, whatever the costs."

Are social movement activists alone in thinking this way? This book has sought to demonstrate the validity of PCN's fundamental position, that is, that the construction of the region-territory is a key strategy in conserving the cultural and biological diversity of this strategically important region and ecosystem, in maintaining its viability as place and region (see also PCN 2007). Besides their material and political goals, movements such as PCN contribute to widen the field of the possible. Confronted with the democratic, social, and ecological catastrophes which many regions in the world are witnessing at present—even though they are readily visible in some parts of the world only—a politics of the virtual at this level takes on extreme value (Osterweil 2006). Activists have led the way in this regard (e.g., the entire "other worlds are possible" movement meshwork), and some critical academic proposals resonate with this politics, for example, theorizing modernity as virtual multiplicity; dissolving some of the strong structures of Euro-modernity at the level of theory by favoring flat alternatives; positing the fact that epistemic differences can be—indeed, are—grounds for the construction of alternative worlds; calling on scholars and activists to read for difference rather than just for domination; or imagining that aiming for worlds and knowledges otherwise is an eminently viable cultural–political project. If the struggle of some groups can be seen as decolonial, in the sense that in the very negation of their difference lies their epistemic potential, does this not mean that the "collective disidentification with capitalism" (and with development and the strong versions of Euro-modernity)—which Gibson-Graham summons us to entertain as theorists of possibility—is an act of decoloniality in which all humans can participate? What strategies of de- and resubjectivation would be required on the part of academics, activists, intellectuals, and so forth to activate disidentifications of this sort? (See also Flórez 2007.)

The question of which insurgent movements might actually point at decolonial processes remains a pertinent one, particularly for those movements that arise from below. To what extent do they affirm their life projects–that is, *lo propio*, as many movements phrase it, or, in my parlance, place-based differences? Even more, how might one envision the kinds of decolonial societies one wishes to construct—those capable of admitting greater epistemic and ontological symmetry across multiplicities, that is, across diverse worlds and knowledges? These questions are indeed pertinent; after all, the democratic, social, and ecological crisis of the world at present is not so much a problem of science, but of existence; the crisis calls not for more science, but for different forms of existence. Many social movements today, such as PCN, might indeed be constructing conditions for reexistence.[11]

Perhaps the Pacific has had *la desgracia de la buena suerte* (the curse of good luck), as an activist of the Yurumanguí put it. What he meant is that while its peoples have been fortunate to inhabit forgotten territories these are now coveted by many; because of their riches, these *tierras de negros* (black lands) have fallen into the mire of national and international interests (Rosero 2002: 548) so that Afro-descendants have progressively lost control of their lives and territories. "If war is the continuation of the economy by other means"[12] it is clear that in Colombia "independently of who wields the weapons, they are used to enforce societal and developmental logics that are completely at odds with those of the ethnic groups" (550). I want to end with the following statement by this PCN activist about the dire situation. It restates established principles of the movement; emphasizes a perspective that links past and future, finding inspiration in the dreams of liberty of those who came before them; aims at new collective subjectivities; outlines a politics of articulation; and, while insisting on the specificity of Afro-Colombians and Afro-descendants, opens up to the rest of society in an attitude of caring, hope, and freedom.

> To be able to contribute to a collective construction, and as a possible and necessary aspiration, dignity requires that we remain on the path of our *intereses propios* [our own interests], in the sense that these represent aspirations and alternatives that go beyond the Afro-descendants in the cultural, ecological, and societal domains. The current war in Colombia does not constitute a step ahead on the path of freedom delineated for us by our elders and which did not achieve closure with the legal abolition of slavery 150 years ago. It is our duty to stay within the *mandatos ancestrales* that have guided our resistance even in the midst of the most difficult and adverse times. Today, this mandate has in the defense of Territory, Identity and Autonomy a raison d'être and an opportunity; this opportunity is not new in that it represents a constant search of the Afro-descendants, even if it is expressed in different ways in various times and places. . . . Although our principal strength might have to come up from within the very struggle with ourselves, we also need to assume that given the critical situation of the country "one cannot save oneself alone," as the poem goes, so that the future will also depend on the capacity of the Afro-descendants to join our struggles and desires with those of other subordinated and excluded groups. . . . Failing to assume today our responsibility to our past and to our future will only make more difficult and painful the path ahead for our communities and *renacientes*. With their legacy of Life and Happiness, Hope and Freedom, our elders also left *una senda* [a path] for us to tread; in this sense, what we need to do today does not constitute an entirely new path.[13] (558, 559)

notes

Introduction

1 See the proceedings from the meeting (Fundación Habla/Scribe 1995).

2 Here I also have in mind the works of authors such as Dipesh Chakrabarty (2000) and Ranajit Guha (e.g., 1988). For a fuller treatment of the concepts of imperial globality and global coloniality, see Escobar 2004, and the chapter on Development.

3 The main exceptions are the discussions on epistemology in the Nature chapter, modernity/coloniality in the chapter on Development, and flat ontologies in the chapter on Networks.

4 It may be said that anthropologists and ecologists have always lived with the realization about the colonial difference. However, taken as a whole, anthropology and ecology—at least in their mainstream forms—have tended to domesticate difference rather than to release its epistemic and political potential for alternative socionatural designs (Restrepo and Escobar 2005). There are few ethnographies of coloniality as yet; three recently completed anthropology doctoral dissertations have a partial focus on coloniality: Carmen Medeiros (CUNY, 2005) provides an excellent interpretation of peasant responses to development in the Bolivian Andes from the perspective of coloniality; Mónica Espinosa (University of Massachusetts, 2004) treats the Colombian indigenous leader Manuel Quintín Lame as a border thinker; and José Martínez (University of Massachusetts, 2004) analyzes indigenous ecological knowledge in Yucatán by appealing to the notion of "coloniality of nature." The work of Freya Schiwy (2002, 2003, 2005a, 2005b) with indigenous video makers in Bolivia also has an ethnographic basis.

5 This is what Guattari calls a generalized ecology or ecosophy (1990).

6 The "women and the politics of place" framework also includes the body (and, hence, diverse embodiments) as a central element. I will not develop this dimension in this book. For the full framework, see Harcourt and Escobar, eds. 2005.

7 Leonardo Boff's writings are vast. A good place to start is his recent books linking the critique of capitalism with ecology and an ecumenical notion of spirituality (2000, 2002, 2004). There are English versions of the first and last ones. In his 2002 book, Boff develops a theory of care as a basic ontological structure and as the basis for a new paradigm for relinking with humans across differences and with nature and the spiritual world. It is noteworthy that writers who think deeply about difference sometimes conclude by outlining an ethics of love. See, besides Boff, Maturana and Varela (e.g., 1987); Panikkar (1993); Anzaldúa and Keatin, eds. (2002). This conclusion is more commonly found among those concerned with interreligious dialogue, but not only in these cases. Ecologists emphasize the principle of harmony.

8 The first quote in Spanish is by Carlos Rosero from PCN. Statements about difference of the sort mentioned here are often found in the writings of black and indigenous intellectuals, from Fanon and Césaire on. For recent statements from the perspective of black radical intellectual thought, see, for instance, Casimir (2004); Bogues (2003).

9 For treatments of the current situation in Colombia, see Garay, ed. (2002); Ahumada, et al (2000); Uribe (2004); Robledo (2000); Leal, ed. (1999); and the special issue of *Revista Foro* entitled "Colombia's New Right," no. 46 (January 2003).

10 Plan Colombia is a U.S.-based multibillion-dollar strategy intended to control both drug production and trafficking and guerrilla activity. Spearheaded by the Colombian and U.S. governments, Plan Colombia, according to many critics, also constitutes a strategy of militarization of the Andean-Amazon region. Its first installment of $1.3 billion (2000–2002) was largely spent on military aid. Among the aspects of Plan Colombia most criticized by Colombian and international organizations are the indiscriminate program of fumigation, the increased militarization it has fostered, and the shift of the armed conflict to other regions. It continues to be funded largely by the United States as a centerpiece of both Uribe administrations (2002–06; 2006–10).

11 For well-known statements on political ecology, see the collections by Biersack and Greenberg, eds. (2006); Haenn and Wilk, eds. (2005); Paulson and Gezon eds. (2005). See also Brosius (1999); Bryant and Bailey (1997); Rocheleau et al., eds. (1996); Peet and Watts, eds. (1996); Schmink and Wood (1987).

12 I will not be able to provide even a cursory commentary on very interesting trends in science and technology studies. Of these trends, I have found particularly relevant to my project the proposal for "reconstructive agendas" in STS and Hess's notion of postcontructivist ethnography (2001). What does it mean to develop "near-native competence" in the field of social movement activism? On what basis can the ethnographer claim "better knowledge" and how can she or he bring this knowledge to bear on particular situations? On world anthropologies, see WAN Collective (2003); Restrepo and Escobar (2005); Ribeiro and Escobar, eds. (2006); and the project's web site.

13 In our current project of restructuring cultural studies at Chapel Hill, we aim to develop creative pedagogical and research practices that broach the interrelations among cultures of science and technology, cultures of economies, and cultures of politics in thoroughly interdisciplinary fashion.

14 The implications of decoloniality for cultural studies have been developed primarily by Walsh (2007). See the special double issue of *Cultural Studies* edited by Walter Mignolo, vol. 21, no. 2–3 (2007).

15 I am reporting here chiefly on the Social Movements Working Group at UNC Chapel Hill. The group is made up largely of anthropologists, but also includes participants from sociology and geography. About ten anthropology doctoral dissertations are currently in process focusing on social movements.

The notion of "network ethnography" in connection to social movements has been under development by anthropology Ph.D. students since the late 1990s at places like University of Massachusetts, Amherst and Chapel Hill (Michal Osterweil, Maribel Casas, Dana Powell, Vinci Daro), and in recent anthropology dissertations by Xochitl Leyva Solano (2002) and Thomas Olesen (2005) for neo-Zapatista Networks, Chaia Heller (2004) on movements against genetically modified organisms, Jeff Juris (2004) for antiglobalization social movements in Barcelona, and Mary King (2000, 2006) on biodiversity and antiglobalization movements. For a recent approach to the sociology of social movement networks, see Diani and McAdam, eds. (2003). A current focus is the collective experiences and expertise built across movements, particularly in the context of the global justice movement. All of this amounts to what one of the participants has referred to as a *knowledge turn* in both social movements and social movements research (Casas Cortés 2006a; see also Osterweil 2005b; Casas Cortés 2006b; Yehia 2006; Conway 2006).

16 These were some of the principles discussed within the collaborative project sponsored by the Latin American Studies Association, LASA, *Otros Saberes* (2006–07), which brought together seven teams of activists and academics working collaboratively. Our team included ten PCN activists and four academics (Luis Carlos Castillo, Juliana Flórez, Ulrich Oslender, and me) for a series of discussions and research and writing activities. See the project's final report (PCN 2007); for related approaches, see Rappaport 2005 and hale 2006.

17 One additional word on literatures: Throughout the book, I have privileged works that do not circulate widely in the metropolitan English-based academy; certain works currently in vogue are thus absent. Second, I have been unable to update completely the vast literatures on some subjects, such as biodiversity conservation. I can only apologize to the authors, including some friends, for my failure to include some recent works that should have been here.

18 A brief note on what the book is not about. First, this book is not a study of the black cultures of the Pacific, although one will learn a great deal about them along the way. The book will not make a comprehensive review of the literature on what has been called *Colombias negras* (black Colombias), which has grown steadily since the late 1980s. The book falls within the limitations identified by Restrepo (2005) for studies of the Pacific—"pacificalization" (excessive emphasis on the Pacific) and "ruralization" and "rivercentrism" (privilege given to the rural groups of the rivers of the Pacific). More important, the book is not about race or racism, fields in which I am not an expert. Let me, however, state clearly that the situation of the Pacific evinces the long-lasting and widespread antiblack racism that continues to exist in so many regions and countries of the world. Antiblack racism is one of the most damaging structuring features of modernity and a central aspect of coloniality. I hope the book's cultural-political analysis helps to unveil practices and mechanisms through which such racism is effected and maintained. Third, the book says little about indigenous

peoples of the Pacific. Peter Wade (1997: 35–39) is right in saying that the study of black and indigenous groups in Colombia needs to be undertaken in unison since the dynamics of race and ethnicity addressing both sets of actors are deeply intertwined.

The book, finally, is not about policy and does not explore policy implications. I should mention, however, that activist and policy-oriented aspects pervade this work, although I will not highlight them unless they are particularly pertinent. Among the activist-oriented activities in which I have been involved are the preparation and running of workshops (including a seven-day workshop on ecological river basin design, held in the coastal city of Buenaventura in August 1998 with about twenty-five river community leaders and activists, which I co-designed with PCN, out of which came more refined notions of territory and region-territory); grant writing and fundraising for projects in the southern Pacific (largely through Danish NGOs and a few smaller funding sources in Colombia and the United States); participation in workshops with NGOs implementing projects in the Pacific; helping organize international trips by PCN activists to academic and activist events; collaborating on writing papers and collective works with activist and environmentalists; participating with activists on government and policy-oriented meetings and in human rights, solidarity, and urgent appeal campaigns (e.g., to stop gold mining, denounce paramilitary atrocities, or issue warnings on displacement situations); dissemination of information; and so forth. Face to face and electronically, in Colombia, the United States, and elsewhere these multiple activities over the more than fourteen years since I first went to the Pacific give a particular character to the book, one which surely would have been very different—and I am certain poorer scholarly and politically—without this decided activist and policy-oriented dimension. I would like to think that these activities can be properly seen as integral to one's professional practice, at least within a world anthropologies perspective.

19 The point about life was brought home clearly to me by Larry Grossberg in a conversation in Chapel Hill. *Redes* was suggested persuasively by Marisol de la Cadena. My thanks to both.

1. Place

1 The rethinking of place in the 1980s can be traced back to the critiques of culture as being bounded and discrete. Theorists in anthropology, geography, communications, and cultural studies began to emphasize the deterritorialization of culture. Deterritorialization, diaspora, traveling, border crossing, nomadology, networks and flows, and the like became the metaphors of the day. These important innovations moved the production of culture, identity, and economy away from place; they effected an erasure of place (Dirlik 2001). In the past few years, there has been a countermove in geography and anthropology. The production of place in geography has advanced in political economy and feminist

perspectives (e.g., Swyngedouw 1997, 1998; Massey 1994, 1997). Very important in this regard has been the concept of the "politics of scale" by capital, social movements, and technoscience (Swyngedeow 1998; Peck 2000; Gibson-Graham 1996; Escobar 2001). In archaeology and cultural and ecological anthropology, phenomenological approaches have yielded rich characterizations of place (Bender 1998; Tilley 1994; Ingold 2000a; Jackson, ed. 1996; Feld and Basso, eds. 1996). These approaches call for greater sensitivity in capturing the intersubjective process of shared experience by focusing on the domain of everyday, immediate activity and on the embodied life-world of practical and social life. For these anthropologists, "place is an irreducible part of human experience, a person is 'in place' as much as she or he is 'in culture'" (Tilley 1994: 18). For those studying "senses of place," culture sits in places, and no degree of globalization can ever reduce place to the logic of capital, technology, or transnational media. While these phenomenological orientations tend to overlook larger social forces, some works are already blending phenomenological and political economy perspectives to ascertain the impact of global forces on senses of place and the production of place. Interestingly, these works focus mostly on cases of environmental destruction. They show not only how long-term habitation is unsettled by larger political economies, but how local groups develop "strategic countermeasures to the deterritorialized space" represented by those forces (Kuletz 1998: 239; see also Kirsch 2001; Campbell, ed. forthcoming). This point is important for understanding the localization strategies of subaltern groups in places such as the Colombian Pacific. For a more thorough argument and review of the literature, see Escobar (2001).

2 What de Landa has in mind when he talks about meshworks and hierarchies goes beyond metaphor. He wants to describe common processes behind the formation of structures that cannot be fully represented linguistically. In this vein, he invokes Deleuze's and Guattari's notion of abstract machines or engineering diagrams that would be at the basis of structure-generating processes that yield specific meshworks and hierarchies. His question then becomes, "Is it possible to go beyond metaphor to show that the genesis of both geological and social strata involves the same engineering diagram?" (1997: 59). The answers are quite suggestive, even if they still need to be worked out in more detailed fashion and in other domains besides cities, markets, and linguistic structures, which are the primary examples explored in social science domains. In chapter 6, I will apply this theory to social movements.

3 This sketchy presentation of the Pacific landscape and ecosystems is inadequate. Its aim is to give a sense of the region as a geological and ecological space. Nothing has been said of the fauna, for instance, to some extent because this is not the forte of the region's diversity. But this does not mean it is not important in biological, human, and cultural terms. The interested reader would do well to consult the excellent two-volume set edited by Leyva (1993), which covers aspects from the geological to the cultural. Those interested in plant biodiversity should

consult the work of Enrique Forero and Alwyn Gentry (e.g., Forero and Gentry 1989) as well as the results of the Proyecto Biopacífico, to be discussed in chapter 3. See also CEREC, ed. 1993.

4 One final note about the epistemology of boundaries. The foreword of a book on ecology from the perspective of complex systems starts with the still somewhat disorienting assertion that "complexity in ecology is not so much a matter of what occurs in nature as it is a consequence of how we choose to describe ecological situations. . . . The study of complex systems requires a more even treatment that dissects observer decisions as much as it addresses the world beyond the observer. We appear to have to recast our system descriptions to deal with complexity" (Allen and Hoekstra 1992: xiii). This form of scientific constructivism echoes the earlier affirmation by the biologists Maturana and Varela that "everything said is said by an observer" (1980: xxii; 1987: 27). To be sure, scientists have been working on the relation between frame of observation and observed fact since the theory of relativity and Heisenberg's uncertainty principle. Yet the positivist injunction about the independence of observer and observed world still seems to carry the day in most scientific endeavors, ecology and complexity theory included, despite the various versions of epistemological constructivism that have swept over some of the social sciences and the humanities. In the 1980s and early 1990s, the debate was often couched in terms of modern versus postmodern forms of science (see, e.g., Soulé and Lease, eds. 1995 for the case of ecology). Some have attempted to mediate these varieties by positing a useful distinction between weak and strong constructivisms (Milton 1996). In chapter 6 I discuss a return to realism in the social sciences fostered by "flat alternatives" and theories of complexity. This return is seen by some as influenced by approaches emerging from the sciences (e.g., Santos 1992).

5 This account of the historiography of the region is intended only to highlight some basic issues. Fuller treatments are very recent, particularly the studies by Múnera (2005) and Almario (2005). Almario's two-volume work includes a very useful review of the literature. Restrepo (2006) further discusses the complexity of talking about categories such as race and ethnicity across historical periods and advocates a genealogical approach that "eventalizes" blackness in terms of regimes of blackness, modalities of governmentality, and technologies of alterization. See also the special issue of the *Journal of Latin American Anthropology* devoted to 1990s trends in the study of black cultures and issues in Colombia, edited by Peter Wade in 2002 (Wade 2002). For other regional treatments, see, besides the works of Aprile-Gniset (1993) and Romero (1995, 1997), West (1957), Whitten (1986 [1974]); Olinto Rueda (1993); Leal (2004). More localized or regional histories have been provided by Leesberg and Valencia (1987) for the middle Atrato region of the Chocó; Almario and Castillo (1996) for the Bocas de Satinga region of the Pacífico Nariñense; and Mosquera (1999) for the village of Hunina, Bahía Solano (Chocó). Black intellectuals such as Yacup (1934), Escalante (1971), and Velásquez (see the 2000 edition of his texts from the 1950s)

provided influential accounts of the region and its character in earlier decades of the twentieth century. Historical-political accounts by black intellectuals have become more common over the past two decades (see. e.g., González 2002, plus articles in many of the volumes mentioned in this book). The missionary account of Merizalde (1921) has been useful in terms of both ethnocentric representations of black peoples and a number of historical aspects for the southern Pacific.

6 This account of indigenous settlements, habitats, and peoples is inadequate. Until very recently, anthropological studies focused almost exclusively on the indigenous communities of the Chocó, while black cultures remained invisible. For an introduction to the anthropology of indigenous communities of the Pacific, see Friedemann and Arocha (1984); Leyva, ed. (1993); Ulloa, Rubio, and Campos, (1996).

7 Wade (1993, 2002) has been one of the few researchers focusing partly on the Pacific who has paid significant attention to race. More recently, Leal (2004) has emphasized the centrality of race in understanding landscape and economy in the Pacific. Very little attention has been given to gender in historical researches; this gap remains to be filled in the literature.

8 This model of symbolic territorial affiliation has been developed by Odile Hoffmann for the Río Mejicano in the Tumaco area (1999).

9 It is important nevertheless to caution against a view of a model—whether historical or contemporary—for the entire Tumaco region, let alone the Pacific. As Mosquera warns, not even the "classical" settlement pattern she outlines has been followed in any strict sense of the term. On the contrary, "it would be a mistake to think about a total identity rooted in a common historical trajectory and in ties stemming from an omnipresent ethnic solidarity. . . . It is diversity and not uniformity that characterizes the various sub-regional habitats and societies" (1999: 53).

10 The banana plantations did not last for more than a few years, particularly owing to the difficult commercialization of the product. Interview with Federico Tomiya, the only Japanese-Colombian remaining in Tumaco, October 1993. Most of the Japanese families moved to the Cauca Valley, to the town of Palmira, near Cali where they became important in agricultural development.

11 Interview conducted by Arturo Escobar and Eduardo Restrepo, Tumaco, August 5, 1998. A number of these founding stories, as told by local elders, are reported in Llano (1998).

12 The main international factor was the revised International Labor Organization's Convention on Indigenous and Tribal Peoples of 1989 (ILO 169), which put pressure on governments to recognize traditional indigenous lands and grant a measure of autonomy to them. This convention was instrumental in what has been called the territorial turn in Latin America, that is, a wave of territorial titling of collective lands to indigenous and black communities in a number of countries, including Colombia, Brazil, Ecuador, and Central America. See Offen (2003) for a review of this trend, with special attention to Colombia.

13 See, for instance, the work of Camacho and Tapia (1997) in Tribugá and Nuquí along the Chocó coast, and Vargas (1999) for the San Juan river, involving indigenous and black communities. Camacho's and Tapia's work is exemplary in many ways, including the description and analysis of its participatory methodology; its discussion of the difficulties of using technical cartography with local communities; the substantial and detailed ethnographic information on local conceptions of the natural world, production systems, medicinal plants, etc.; its attention to gender; and its collaboration with a well-known social movement organization, Organización de Barrios Populares del Chocó (OBAPO).

Although technocratic mapping and inventorying were common throughout the 1980s and 1990s, not infrequently the participatory exercises were done with local coresearchers and in collaboration with the territorial-political organizations of the communities in question. These collaborations were an indication of the strength of the social movements during the early and mid-1990s. During this period, activities were largely geared toward supporting the process of collective titling of territories. Ley 70 specified a series of steps to be taken; the first was the creation of a community council and the elaboration of the titling application, which called for detailed census, description and demarcation of the territory, ethnohistory of settlement and habitation, description of forms of social organization, account of "traditional production practices," and identification of conflicts over territory and natural resources. These requirements called for a great deal of activity on the part of communities, facilitated either by ethnoterritorial organizations, NGOs, state experts, or all of the above. The titling process could thus rightly be described as an intense cultural-political process that involved internal and external negotiations, at times manipulation (say, by local elites on some rivers who influenced community councils to their own benefit), exchange of knowledges, historical reconstructions, etc., all of which took place in a heightened political climate, rarely with adequate resources, and not infrequently with obstacles caused by the very same government institutes in charge of facilitating the process (e.g., by delaying necessary funds for activities). The second phase was for the pertinent government institute (INCORA, or Agrarian Reform Institute) to review the application and grant the title. The third entailed the preparation of territorial use and management plans according to local practices. The guide to the collective titling application process for communities is found in INCORA (1998). A detailed example of the application, including all of the information collected by communities and experts, is found in Bid/Plan Pacífico/Acaba n.d., for the Baudó river in the Chocó department.

14 The principles for interethnic relations agreed upon included (a) the acknowledgment of the Pacific as an ancestral territory of ethnic groups; (b) the need for a joint and coordinated strategy for the defense of the ancestral territory from a position of tolerance and mutual respect; (c) the recognition and respect of traditional knowledge as their cultural patrimony, foundation for life, and relation to nature.

15 The importance of knowledge production for the organizing strategies—including the question of traditional knowledge—is evident in the dedication that opens the text that came out of the meeting: "To all those that have contributed their energies to this process of knowledge production; to all those who, like us, dream of a country and a world where multiethnic and pluricultural coexistence become a real possibility; to all those women and men who believe in the respect of difference; to all those who, like us, bet on life" (Fundación Habla/Scribe 1995: 5). The phrase "to bet on life" (*apostarle a la vida*) became a common saying during the mid-1990s. It constituted a cultural-ecological statement without the trappings of formal anthropological or ecological language. The process of territorial politics from the perspective of black communities and the state is also discussed at length in Villa, ed. (1996) and in Ng'weno (2007) for the case of two communities in the Norte del Cauca region.

16 The charts of the loss of territory, and most of the conceptualization included in this section, come from an intensive seven-day workshop with about twenty-five river community leaders and activists held in Buenaventura in August 1998, which I co-designed and ran with Libia Grueso (Proceso de Comunidades Negras) and Camila Moreno, a planner from Bogotá close to the social movement, with the collaboration of Jaime Rivas (Fundación Habla/Scribe). The reason for the workshop was the need faced by river organizations to come up with their own plans of *ordenamiento territorial* (territorial ordering and management) in response to a new government regulation. The first five days of the workshop were devoted to ecological river basin design principles and planning methodologies; the final day was devoted to discussing the territorial ordering plan for Buenaventura itself, with presentations by various representatives of local government, including the plan's head, Viviana Obando. This workshop was very important to me in terms of working out ideas with a group of activists and community leaders. The most important concepts (e.g., region-territory) had been in circulation for quite some time, although we were able to refine some of them and relate them to concrete design questions. I used principles of what I call autonomous design, which, unlike conventional planning, takes people's own definitions and perceptions, rather than those of the experts, as a point of departure for the inquiring system.

One of the daylong exercises we did was mapping the "system that generates loss of territory" in small groups of four or five people. The conceptualization of the system became progressively complex as it went through several rounds and as we reconvened to discuss the various models. Given the purpose of the workshop, which was to enable leaders to prepare their own planes de ordenamiento territorial, the exercise was largely couched in relatively technical terms. There were occasional instances, however, in which the discussion was carried on in more local terms, for instance, when one of the elders stated that "if today we had powerful *brujos* (sorcerers) as in the past, do you think the government would dare to bother us so much now?" Or when somebody else raised the hypothetical question of whether the *tunda* (a well-known supernatural forest entity) could also be found in urban areas.

17 The gender aspects of nature, identity, and political mobilization will be further discussed in subsequent chapters.

18 It is impossible to provide a more adequate treatment of the important process of collective titling in these pages. Two important sources are the evaluation report by Sánchez and Roldán (2001) for the World Bank, focused on the titling program coordinated by PMNR, and the evaluation of collective titling experiences in Colombia, Ecuador, Panamá, and Perú by Plant and Hvalkof (2001) for the Inter-American Development Bank (IDB). Despite generally positive evaluations of the programs, Plant and Hvalkof point at a general contradiction in IDB's policies between indigenous and black notions of territoriality and the bank's emphasis on market-based policies (69). One answer Plant and Hvalkof propose is "development with identity," that is, international and state policies that genuinely support ethnic visions of the future. Sánchez and Roldán emphasized the need for decisive official support for communities' own *planes de vida* (as the resulting plans are referred to by the communities), which in today's context calls for strategies of peace and nonviolence. See also the report of a 1998 request by PCN (in which I participated) to a Danish NGO for financial support for collective titling carried out by the Peruvian lawyer Pedro García Hierro in 1998 (García Hierro 1998). Finally, some analysts have raised the question of why the World Bank would support these projects. Answers include the need to stabilize property regimes, protect biodiversity from strong market forces, and the idea that once land titles are secured communities would engage in sustainable development led by the private sector. Indeed, this happened in some instances in which community councils contracted out the exploitation of resources to third parties (see Offen 2003). Collective titling is neither a panacea nor free of tensions; in some cases, the titling process exacerbated tensions between black and indigenous organizations and made invisible overlapping areas of interethnic habitation. Finally, the title, in some ways, is only the beginning: Will communities find the necessary support needed to effectively control and manage the territories? Would they adhere to the long-standing "logic of the river" in their use of resources (Oslender 2002)? Or would they adopt more conventional development strategies? These were all open questions in the late 1990s and the early part of the present decade.

To mention two cases that exemplify some of the tensions and possibilities: In April 1996, a meeting held to create the Regional Committees (mixed bodies of government agencies and local organizations) for the PMNR titling for the Nariño department identified the black community organizations to be included for each of the rivers. In most cases, these were well-known organizations affiliated with the PCN-coordinated regional *palenque* (network of regional organizations), such as Coagropacífico (for the rivers of the Tumaco area, an organization that I discuss in chapter 2), Acapa (Patía river), and Fundación

Chigualo (Telembí river). These proposals by the *palenque* leaders were an effort on the part of the social movement to neutralize the community councils promoted by capitalists and *narcos* involved in timber, gold mining, African palm, and shrimp cultivation (e.g., in Iscuandé, El Charco, Mosquera, Patía Viejo). At stake were two contrasting interpretations of Ley 70: for entrepreneurs and many government officials, the titling should be done as quickly as possible and be restricted to its legal function (in which case the process would be easier to manipulate, making sure they would have access to the resources down the line); for the organizations, more than the title itself what was important was the cultural-political process, which required the involvement of entire communities and the strengthening of the ethnoterritorial organizations. By 1996, ten community councils had been created linked to ethnoterritorial organizations. By June 2000, 8 collective titles had been approved, totaling 300,070 hectares, and there were 14 more pending amounting to about 350,000 hectares more (including a 45,000-hectare title for the Unión Rosario in the Rosario river), and the obstacles to be surmounted seemed to increase every year. The second example comes from the Yurumanguí river in the Valle del Cauca department. The title to 54,776 hectares, benefiting 13 communities and 529 families, was officially issued in May 2000. However, the unofficial community act to celebrate the title could not take place until a year or more later. The reason was the high level of conflict in the river, which saw repeated incursions by army, guerrillas, and paramilitaries, the assassination of five family members of the leader of the river's political organization, and a politics of fear and intimidation of the population as a whole (more on the Yurumanguí in the conclusion). Source: interviews in Tumaco, summer 1998 and 2000; visits and interviews with local leaders in Yurumanguí and Cali, summer 1998, 2002.

19 "Ay, Dios mío, qué espantajo / me han contado de por allá / donde tumban, queman, matan / sin siquiera preguntar; Ay, Dios mío, qué espantajo / me ha contao don Severino / que la tierra ya no sirve / ni pa'coco, ni colino; Ay, Dios mío, qué será / de la vida de mis hijos/ si la tierra que es su herencia / la volvieron remolino." The poem is by Jaime Rivas, a Tumaco cultural activist, artist, and popular communicator. From the post card collection Mágico Pacífico, produced by Fundación Habla/Scribe with support from Proyecto Biopacífico, mid-1990s. A similar statement by Paul Virilio (1990: 93) in the context of modernity reads as follows: "Deportation has become our daily bread, since, from the weekend to the back-and-forth of work, we delocalize. And as soon as we delocalize, something or someone is there to arrange our mobility for us, to harness the movement of our active lives which—insofar as they necessarily take place in the zone of totalitarian mediation—never escape external control, very simply. . . . because there is no place to stop, to park. . . . All mass must be permanently subject to the dictatorship of movement."

20 For the exhibit catalogue, see Salgado (2000).

21 Complete figures, references, and discussion are found in Escobar (2003a); see also Aparicio (2007) for a treatment of the emergence of the international ap-

paratus to deal with internal displacement and its application in Colombia. The information compiled from the first National Meeting of Displaced Afro-Colombians, convened by AFRODES and PCN on October 13–15, 2000; the Office for Displacement and Human Rights, CODHES; the Red de Solidaridad Social, RSS, the government agency in charge of programs for the displaced; the Latin American Institute for Alternative Legal Services, ILSA, based in Bogotá; and the UN Thematic Group on Displacement, GTD, created in 1999 for the Colombian case in coordination with UNHCR and made up of representatives of nine UN agencies. The GTD reports that about 57–63 percent of recent displacement is caused by AUC/paramilitaries; 12–13 percent by guerrillas; and the rest mostly by unknown groups, plus the state. In April 2001 the worst massacre on one of the rivers of the Pacific, the río Naya, left over one hundred people brutally killed by paramilitaries and many hundreds displaced. It is estimated that 38 percent of all the displaced are ethnic minorities, with an increase of 80 percent in the first quarter of 2001 with respect to 2000, and a further increase in 2002. The gender dimension of displacement remains understudied, despite the fact that women and girls represent nearly half of the displaced, evidence not only of the fact that they are often targeted for physical violence, including rape, but also of their specific reproductive and health needs. Among the few studies of gender and displacement is that of Meertens (2000) with regard to refugees in Bogotá, although UNHCR and other agencies have conducted a few studies.

22 This refers to the Pacific region of the Nariño department, including the municipalities of Tumaco, Francisco Pizarro, Mosquera, Olaya Herrera, La Tola, El Charco, and Santa Bárbara in the coastal zone, and Barbacoas, Roberto Payán, and Maguí in the central and foothills region to the east. It includes the five rivers flowing into the Tumaco bay. The area of this subregion is about 18,000 square kilometers, or about 52 percent of the entire Nariño department. Black organizing was very strong in this region from the early 1990s till the arrival of paramilitaries en masse after 2000.

23 Nearly all of the PCN activists with whom I worked in Tumaco in 1993 and 1998 have been leaving the region steadily since 2000, although a modest level of organizing had restarted by 2006. The same is true of many of the cultural and social activists who had developed their programs throughout the 1990s (e.g., dance and theater groups, radio stations, literacy groups, agricultural producers' and women's cooperatives, and so forth).

24 According to one activist, the coca business has accomplished in two or three years what development programs failed to do in several decades, namely, introduce a culture of commodities and accumulation. The reality of the situation on some of the rivers of the southern Pacific was strongly apparent at a three-hour meeting I held with about fifteen activists and staff of the National Parks Office in Cali in August 2003, at their invitation. At this meeting, I presented the framework of imperial globality/global coloniality and politics of place that in-

forms this book. If anything, the framework seemed to conform only too well to the situation of the Pacific.

25 In the diagnosis conducted by AFRODES (Association of Displaced Afro-Colombians) and PCN in 2000–2001, the main factors causing displacement were: (1) megadevelopment projects such as construction of roads, ports, dams, a planned interoceanic canal, and a dramatic expansion of the African oil palm frontier in the Tumaco region; (2) the spread of illicit crops; (3) the armed conflict; and (4) the existence of natural resources, from gold and timber to tourism. The Proceso de Comunidades Negras made the following additional observations: (1) displacement became accentuated after the titling of collective territories; the displacement experience of the Pacific can be located in the context of a counterattack on the cultural and territorial gains of ethnic communities in the continent, from the Zapatista to the Mapuche, if not globally; (2) displacement is selective and planned; the largest displacement has occurred in zones earmarked for macrodevelopment projects; (3) the aim of terror is to break down the resistance of the communities and negate their difference; in this way, "war is the continuation of economics by other means" (the Salvadoran poet Roque Daltón, cited by Carlos Rosero in personal conversation); (4) displacement has altered the patterns of in- an outmigration that have characterized the Pacific since the 1950s and 1960s, making impossible a return to the home river communities; this ends up modifying the use of land, the traditional production systems, the spatial distribution of population and resources, etc.; (5) the armed actors, particularly the paramilitary groups, have fostered a selective and directed resettlement of river territories, displacing some groups and bringing in others—mostly whites from the interior—who obey the new rules of cultural, economic, and ecological behavior. Conversations with Carlos Rosero and Libia Grueso of PCN, Bogotá, October 16–18, 2001; documentation prepared by AFRODES. See Escobar (2003a) for some of the policy prescriptions of the movements.

2. Capital

1 Some of this information comes from my interviews with Eduardo Peña, an agronomist at the Colombian Institute of Agriculture's (ICA) Research Station in Tangarial, near Tumaco (July 1993). According to Peña, ICA's archives in Tangarial are practically nonexistent: "What the rats have not eaten has been destroyed by humidity." On the circulation of biological material in the colonial world, see Crosby (1986); Miller and Reill, eds. (1996). The role of botanical gardens in this traffic, focusing on the Kew Gardens in Britain, has been analyzed by Brockway (1979). European and North American botanical gardens continue to be central to the circulation of biological materials, as in the case of biodiversity prospecting.

2 A variety of cultural Marxisms, Marxist feminisms, ecological Marxisms, and poststructuralist Marxisms has emerged in the past two to three decades.

It would be impossible to represent adequately here the richness of activity and overlap in these areas. The landmarks are well known: cultural Marxism took off chiefly from Gramsci via Raymond Williams to culminate in the work of diverse authors who also incorporated poststrucurturalist thought (particularly Foucault), from Stuart Hall to Laclau and Mouffe and the Subaltern Studies Group. Marxist feminism has seen extremely creative developments, also in alliance with poststructuralism, from Donna Haraway and Gayatry Spivak to Gibson-Graham and including important works on the sexual division of labor, from the household to the global levels (e.g., Lourdes Benería and Nancy Folbre). The Rethinking Marxism group in Amherst has accomplished an influential reelaboration of key categories of Marxism based on a redefinition of class and class processes (see the journal *Rethinking Marxism*). Ecological Marxism has been one of the most productive sites of reconceptualization of historical materialism, with the most sustained efforts being those of James O'Connor and Enrique Leff (see below). In anthropology, the attempts at developing a poststructuralist political economy, meaning going beyond the dualism of the material and the discursive, the real and the ideal, production and signification, have yielded sophisticated ethnographies (e.g., Comaroff and Comaroff 1991; Taussig 1980, 1987; Ong 1987).

3 Applying these insights to the economic transformations in China today, the anthropologist Mayfair Yang sees not the steady march of one economic model, capitalism, at the expense of another, communism, but a multiplication and radical hybridization of models. Yang makes the claim that the Chinese economy is composed of an array of both capitalist and noncapitalist forms. She challenges her readers to entertain the idea that indigenous economies may survive and sometimes even experience renewal with the entrance of capitalism (Yang 2000).

4 By the time shrimp farming arrived in the 1980s, most of the mangroves of the Pacífico nariñense had already seen a relatively high degree of intervention dating to the tannin years. Generally speaking, the earlier industries (rubber and *tagua*, or vegetable ivory) created conditions for the development of the later ones (timber, oil palm). However, the social and economic relations established by these activities were not necessarily the same. Whereas the tanin and timber industries fostered local industries and a process of proletarianization, other activities included a combination of social relations, as we shall see in the case of the *Palma africana*. The paper industry's raw material is obtained mostly from other parts of the Pacific. This activity, developed initially by multinational capital but now with the significant participation of national capital, is another important contributor to deforestation in the Pacific. Total deforestation in Colombia has been estimated at 600,000 hectares per year, with over 150,000 in the Pacific. Of the close to 9,806,000 hectares of forest with little or no intervention estimated in 1959 in the Pacific, only 4,248,000 remained by the mid-1990s (Sánchez 1996: 190–91; Leal, ed. 1995). The development and transformation of timber extraction in the Pacific is thoughtfully analyzed in Leal and Restrepo (2003). On "the pulping of the South," see Carrere and Lohmann (1996).

5 See the long narrative of don Primitivo Caicedo, from which this account is taken, in Escobar and Pedrosa (1996, eds.). The narrative is constructed as a life history, based on interviews with the author and the project's research assistants in July and October 1993. There are competing stories about the arrival of the first seeds of *Palma africana*. Angulo (1996) reports on an interview with a chemist from Bogotá, Luis Rojas Cruz, who in the early 1960s might have played a crucial role in the spread of seeds among the first capitalist farmers, and with a local ICA functionary, Essiover Mena, who might have distributed gratis the first seeds to a group of local farmers, including don Primitivo Caicedo.

6 Interview with Rubén Caicedo, Tumaco, October 1993.

7 Unless otherwise stated, the information about *Palma africana* and industrial shrimp production comes mostly from my own research in the area in 1993 and 1994, some of which has been updated by short field trips in the late 1990s by Escobar and Restrepo (summer 1998 and summer 2000); from interviews with local government officials; visits to *palmicultoras* and *camaroneras*; conversations with small farmers on the left bank of the Rosario river (1993); the archives of FEDEPALMA and ACUANAL in Bogotá (the African Palm Growers Association and the National Aquaculture Association, respectively); and from specific studies of the palm and shrimp industries in Tumaco, including those by Nianza del Carmen Angulo (1996), Sandra Castaño (1996), Hernando Bravo (1998), and CEGA (1999). Angulo's and Castaño's work focuses on the socioeconomic and environmental impact of the two industries; Castaño's contains a wealth of information on the actual situation of both industries. Bravo's is a detailed study of the shrimp industry; CEGA's includes detailed information and maps of the main palmicultoras, and a feasibility study for the creation of 4,000 hectares of palm by small farmers. Official information on palmicultoras and camaroneras was obtained from the registry of the Tumaco Chamber of Commerce, although there are inconsistencies between this registry and information available from other sources, including local government officials.

8 The most important ones are Palmas de Tumaco, Astorga, Palmeiras, Central Manigua, Palmar Santa Helena, Araki, Santa María, and the Haciendas of J. Echeverri and Alvaro José Lloreda. Astorga, from the Varela group in Cali, controlled more than 10,000 hectares, even though only about half of it was planted in palma by the mid-1990s. Palmas de Tumaco had close to 6,000 hectares under cultivation by 1996 (Angulo 1996). All of the larger plantations also have processing plants. A small plantation, La Remigia, is an exception in that it has its own processing plant. For full information on landholding and hectares under cultivation by size of farm, see the table in FEDEPALMA (1999: 26).

9 The trend toward the production and use of biofuels—bioethanol and biodiesel—as substitutes for regular gasoline in particular is presented as a win-win situation. However, as activists in Latin America and other parts of the South are beginning to assert emphatically, this is far from being the case. This industry implies reinvigorated forms of colonialism for the control of land and labor,

and the social and environmental costs (in terms of forest loss and biodiversity, among others) are staggering. Land is being diverted for the cultivation of such crops as soy, sugarcane, and African palm for production for a market that is estimated to grow at 15 percent per year. See, for instance, Bravo and Ho (2006).

10 This program is currently carried out with funds from Plan Colombia. Ironically, those funds allocated by Plan Colombia to nonmilitary aid have been captured by large-scale capitalists and used in the name of the local people. It was said in Tumaco in 2000 that Cordeagropaz's aim was to increase the current oil palm cultivation area by 30,000 hectares in the next few years.

11 E-mail message on the "African Palm Malaysian Model" sent by Carlos Rosero, from the PCN, March 20, 2001. See also web pages of Forest People's Program and Survival International in the United Kingdom for additional information on Malaysia.

12 On the biology of shrimp aquaculture, see Fast and Lester, eds (1992); Browdy and Hopkins, eds (1995). The environmental impact of shrimp farming on tropical areas is discussed in Tirado's report for Greenpeace (1998). An excellent study of the mangroves of the Pacific is Prahl, Cantera, and Contreras (1990).

13 Data for Colombia come from Castaño (1996), Angulo (1996), Otoya (1993), and interviews and visits I conducted in Tumaco in 1993 at some of the camaroneras, at INPA (in Tumaco), and at ACUANAL in Bogotá. Information was updated by Escobar and Restrepo in Tumaco in the summers of 1998 and 2000. See Escobar and Pedrosa, eds. (1996), chap. 6, for more information on sources. Global production information comes from Tirado (1998) and Rosenberry (1997).

14 Other figures for 1993 are as follows: Maja de Colombia, from another Cali group, had 700 hectares in concession, of which 85 were under shrimp production. The figures for Aquamar were 450 and 185, respectively. Balboa had 250 hectares under cultivation. Other industries included Perla del Pacífico, Exportadora Cali, Produmar, Agropac, El Carmen, Grinuleros, and the community-based camaroneras ASOCARLET and Corpomar (see below). The information is from the registry of the Tumaco Chamber of Commerce and INPA. I thank Cristián de Nogales, director of the INPA office in Tumaco in the early to mid-1990s, for access to this information. We also collected the following interesting information for the late 1990s for two companies: Balboa owned 1,553 hectares of mangrove forest, with 29 ponds totaling 238 hectares; it employed 66 people, including 12 watchmen. The Camaronera San Luis owned 200 hectares of which 92.5 consisted of 16 ponds of 5.8 hectares average size, and employed 26 people. Data are from the companies' own environmental management plans presented to Corponariño, the regional government agency in charge of supervising these plans.

15 This was Idelpacífico, located in the city of Tumaco, created in the later 1980s by the larger camaroneras Balboa, Maragrícola, and Aquamar. By the mid-1990s, there were three other processing plants, El Delfín Blanco, Bahía Cupica, and Maragrícola's own plant. By the late 1990s, all processing plants required

women workers to wear aprons, rubber gloves and boots, and masks, but it was the workers' obligation to acquire them.

16 Stemming the loss of mangroves, social and environmental assessment, better management plans, more careful use of chemical inputs when needed, and regularization of property rights for the communities have also been suggested as important steps in maximizing benefits for communities, ecosystems, and national economies while minimizing the negative impacts (de Walt, Vergne, and Hardin 1996).

17 In other words, these expressions should not be taken as a sign of an ahistorical "primitive environmental wisdom," to use Milton's apt term (1993). Much of the popular discourse, no doubt, is informed by expert discourses of the environment; the appeal to "ecosystems," their "disappearance," "natural resources," and the like are not local categories. Yet to reduce these hybrid popular languages to being a mere reflection of capital or a creation of the state seems to me a gross mistake. The genealogy of popular forms of knowledge certainly needs to be better understood but without overlooking their embedded and place-based character. This evasive balance continues to be a pressing issue in anthropological political ecology (see, e.g., Brosius 1999; Gupta 1998). As we will see in the Development chapter, some of these narratives can be usefully seen as forms of counterwork performed by local groups on development and modernity.

18 The debate on "local knowledge" is far from being resolved in anthropology and ecology. For a summary of these debates, see Escobar (1999a). Ingold (2000a) has provided the most articulate view of local knowledge from the phenomenological perspective of the continuity between being, knowing, and doing. This debate is also at the heart of Deleuze's and Guatari's work, so that a further technical note on the notions of smooth and striated space and nomad and royal knowledge is called for. Deleuze's and Guattari's models for the smooth space are the sea, the steppe, the dessert, and the fractal; one could also speak of a rain forest as a smooth space of this sort. Striated spaces are the spaces of capital and modernity, such as the urban grid and the plantation. What interest these authors most, however, are the passages and combinations established between the two. The two spaces exist in mixtures, constantly being translated into each other. Borrowing from Pierre Boulez, they say that "in a smooth space-time one occupies without counting, whereas in a striated space-time one counts in order to occupy" (477). Striation is the function of axioms, rules, and laws, whereas smooth space is the space of affects, intensities, and local operations (479); "all progress is made by and in striated space, but all becoming occurs in smooth space" (486). In deterritorializing striated space, the smooth can—in fact must—construct organization; the smooth is always being captured, "metricized" by the striated, yet these efforts often result in "retroactive smoothing" by local agents (e.g., place-based agriculturalists "retroactively smoothen" *Palma africana* by subjecting it to different practices of planting and harvesting, more proper of a forest smooth regime; Cordeagropaz's strategy, on the contrary, is a further attempt

at striation). While many local groups try to "voyage [and think] smoothly" (482), capitalist modernity tries to force them to do otherwise: to settle down, specialize, regularize their plots, individualize, engage only with the market, and so forth. What counts most in this provocative analysis is "how the forces at work within space continually striate it, and how in the course of its striation it develops other forces and emits new smooth spaces" (500). Of course, as they warn, smooth spaces are not in themselves liberatory even if it is in them that new possibilities are created. Such is the case of the Tumaco region, a veritable laboratory for this struggle at the heart of modernity and other possible futures. See Deleuze and Guatari (1987), especially chapters 12, 13, and 14. In the chapter on Networks, I will relate this view to the distinction between hierarchies and meshworks.

19 For a multicriteria evaluation exercise in the case of shrimp, see Martínez Alier (2002) and for a similar exercise applied to the case of Tumaco, Castaño (1996: 71–86).

20 This framework has been developed by James O'Connor (see especially his 1998 book). There are many other components to O'Connor's argument, which I have summarized for brevity's sake. Readers interested in the main parts of the argument may consult especially chapters 1, 7, 8, and 15. See also Escobar (1996). I am privileging here the work of James O'Connor and Enrique Leff (see below) because theirs, I believe, is the most sustained effort at articulating Marxism and ecology. See also the debates in the journal *Capitalism, Nature, Socialism: A Journal of Socialist Ecology*, which started publication in 1988 under O'Connor's editorship.

21 I will not discuss here proposals for a more autonomous articulation of a regional economy. Some proponents visualized a network of small and medium cooperative aquaculture enterprises (say, 1,000 to 2,000 square meters) in the rivers, based on the small sector Asian model; local savings and credit systems; and other local cooperative enterprises of small producers in the forest and agricultural sectors. This vision constituted an attempt at regional autocentered development, as opposed to the externally driven accumulation model that came to prevail. There were at least two attempts of this sort in the 1990s, one by a cocoa and coconut commercialization cooperative linked to the social movements (Coagropacífico, to be discussed in the next chapter), another by a well-known and longtime white Tumaco resident, Victor Manuel Mejía; Mejía had support among the local political (black and white) elites. In the early 1990s, an elite-driven movement known as La Primera Fuerza (The First Force) favored a strategy based on the promotion of local innovative and management capacity. Mejia's proposal for an autocentered, regional economy is analyzed in Escobar and Pedrosa, eds. (1996, chap. 6).

22 Interview with Marilyn Wolhman, director of Plan Padrinos Tumaco office, October 1993.

23 With the passage of time, the reforestation of mangroves became an important goal. By the late 1990s, the aim was not only to consolidate the group, but to serve as a potential model for the type of reconstructive (or restoration)

ecology and entrepreneurial activity they see as necessary, one that is anchored in place and that fosters "humane and sustainable development." The reforestation effort is seen as a way to "conserve what is a factor of life" and "to return to nature what has been taken away from her." This project, drawn up in 1999, sought to replant 60 hectares of mangrove and to devise a plan for sustainable use of 200 hectares more. The methodology was somewhat at odds with what technicians favor—direct planting of seeds instead of indirect planting from nurseries, with locals using their knowledge for the selection of seeds, time of planting in the lunar cycle, etc. See ASOCARLET (1999a). The expansion project called for five new 1.5-hectare ponds, that is, to double their capacity. The plan included a projected accounting based on 2.5 cycles per year (each cycle takes about 100 days from seeding to cultivation). Current assets were totaled at about $100,000, and the projected yearly net earnings were estimated at $30,000 (for a project then encompassing 80 families). See ASOCARLET (1999b). There are a few other *camaroneras comunitarias* and *camaroneras artesanales* in the Tumaco region. In 1995, the women's groups Mujeres en Acción obtained COL $ 22 million (about US$10,000) from Plan Pacífico to start their own project. After discarding from consideration the conventional "women in development" programs offered to them (sowing, handicrafts, bakeries), they decided to follow ASOCARLET's example and got some support from them. By 1998 when I last visited them, they already had two 1-hectare ponds in operation, each producing two tons of shrimp three times a year. They expected to build a total of five ponds, which is considered the minimum to generate a modest profit. As in ASOCARLET's case, this group of 15 women took great pride in their accomplishments.

24 See the narrative of doña Ester Caicedo and the interview with Harold Moreno in Escobar and Pedrosa, eds. (1996), chap. 7. All of the information in this section comes primarily from our research team's work in 1993. Additional visits to ASOCARLET by Escobar and Restrepo took place in summer 1998, 2000. Additional information about the camaroneras comunitarias can be found in Castaño (1996) and Bravo (1998).

25 To elaborate a bit, entrepreneurs are people who make historical change by producing both a product that entices people to change the style of their everyday activities and a company that instantiates the new way of life. Genuine entrepreneurs are sensitive to how the problems they see have roots in a pervasive way of living and are able to imagine a different way of dealing with things that is historical and not merely pragmatic, in the sense that they change the style of practices in some domain. Activists in particular hold on to an anomaly leading to innovation which, in turn, leads to the development of a strategy (or enterprise). It is the speech acts of these activists and entrepreneurs that open their peers to new possibilities and persuade them to live according to them. To accomplish this goal, they must make the project at hand seem both strange and sensible in order to give it a new identity. They do so through acts of reconfiguration (e.g., creating new horizons and ways of seeing themselves, the mangroves, and the

outside world but retaining the centrality of their practice as engaged with the natural environment), articulation (bringing into sharper focus the disappearance of their lifestyle as *carboneros* and *leñeteros*), and cross-appropriating (e.g., taking over from capitalism and science the possibility of constructing a camaronera). The result is a new enterprise built around a new way of thinking. In so doing, *they do not disembed their economic practice from community and place*, as in conventional capitalist activity, nor do they abstract general essences from local situations, and yet they modify the style of their subworld and potentially of society at large.

26 Gibson-Graham's conceptualization of the diverse economy shows a convergence with Braudel's triptych (a pyramidal view of the economy with a sea of material life at the bottom, then a broad band of noncapitalist market exchanges, followed by capitalism proper at the top); see Braudel (e.g., 1977), and Gibson-Grahams's "iceberg" of the economy (2006: 70). The "alternative market" and "nonmarket" sectors could be linked to the self-organizing dynamic underscored by complexity views of the economy such as de Landa's (1997). For a view of the development of capitalism out of nonlinear dynamical processes, which progressively led to hierarchical modes of organization of markets and the economy, see de Landa (1997). Following Braudel, Deleuze and Guattari, and some recent works in complexity theory, de Landa develops a convincing view of the self-organizing character of local markets and the progressive encroachment of capitalist hierarchies on the world of everyday life and market exchange. While the early development of the economy benefited from certain nonlinear dynamics and autocatalytic loops (e.g., among money, energy, technology, and markets), it was increasingly captured by linear processes and hierarchical structures, resulting over the centuries in what today is commonly known as capitalism.

27 To fully substantiate Gibson-Graham's argument in the case of Tumaco (and to decide whether these practices qualify more appropriately as alternative capitalist or noncapitalist, actually) would require one to follow up the project in detail for a number of years, which I have not done. A follow-up would document the economics of the surplus—how it is generated, appropriated, distributed, reinvested, etc. I have suggested indirectly that at least in principle Gibson-Graham's argument holds for the camaroneras comunitarias (by referring, for instance, to the focus on livelihood, the reproduction of the base, the nonhierarchical and egalitarian ethics of the groups, and the way in which the surplus is distributed). In an interesting comparison with the Tumaco case, one of Gibson's collaborators has argued that the new economic activities for women made possible in the oil palm sector in Papua New Guinea have enabled these women to strengthen social networks, communities, and smallholder diverse agricultural practices, while making it possible for them to reassert and redefine their traditional power and identity within household and community. The key here is that women redistribute a good part of their income (chiefly as fruit collectors, a very deskilled and badly paid activity in Tumaco, as we saw) to cultur-

ally valued activities and to kin, thus also contributing to reduce household and community disputes. Lest anybody be tempted to interpret this case in terms of the wonders that individual income and market opportunities can do for women, the author makes it clear that women's newly found autonomy is expressed not "in the liberal feminist sense of the free autonomous individual, but rather in terms of how economic independence enabled them to strengthen social connections with their families and communities" (Koczberski 2002: 89). The framework of the diverse economy is fully developed and illustrated with case studies in Gibson-Graham (2006).

28 From an ecological Marxist perspective, the aim would be to bring into place highly social forms of reconstruction of social and material life, including production conditions. This rationalist approach has its limitations. Its potential, however, is to redefine and reconcile "green antiproductivism" (localist conservation politics) with old-fashioned "socialist productivism" (usually in the form of egalitarian developmentalist politics), articulating them into a "red-green politics" that truly redefines productivism. In other words, while ecology and localism may be logical allies, they must transcend localism, as can be seen in the case of most social movements. This is so because environmental problems are trans-scale (local and translocal in scope), which means that while site specificity is important, the "unit of analysis" and political strategy must be much larger. This is an aspect of what we have called "place-based yet transnationalized (networked) politics" (Harcourt and Escobar, eds. 2005). Given the interdependence of sites, ecological problems cannot be solved at the local level alone. In other words, acting locally is unlikely to reach global capitalism by itself and might generate unintended effects, such as green imperialism (O'Connor 1998: 274 ff.). For some, this vision calls for new international alliances, a Fifth International of sorts (O'Connor 1998; Martínez Alier 2002). These proposals imply a *cultural* redefinition of production, a point underscored by Leff but largely left outside of the equation by most Marxists and ecological economists. I will complicate the scalar thinking embedded in these proposals in the chapter on Networks.

29 The full paragraph on this issue reads as follows: "The international capitalist axiomatic effectively assures the isomorphy of the diverse formations only where the domestic market is developing and expanding, in other words, in 'the center.' But it tolerates, in fact it requires, a certain peripheral polymorphy, to the extent that it is not saturated, to the extent that it repels its own limits; this explains the existence, at the periphery, of heteromorphic social formations, *which certainly do not constitute vestiges or transitional forms* since they realize an ultramodern capitalist production (oil, mines, plantations, industrial equipment, steel, chemistry), but which are nonetheless precapitalist, or extracapitalist, owing to other aspects of their production and to the forced inadequacy of the domestic market in relation to the world market. When international organization becomes the capitalist axiomatic [e.g., with neoliberal globalization],

it continues to imply a heterogeneity of social formations, it gives rise to and organizes its 'Third World'" (Deleuze and Guattari 1987: 436; emphasis added). Quijano's conception of structural heterogeneity bears a resemblance to this view (2000). Is the notion of capitalism as a worldwide axiomatic a capitalocentric one? It would seem that Deleuze and Guattari would want to have their cake and eat it, too, as the saying goes. On the one hand, with Gibson-Graham, they admit of a radically diverse economy; on the other, they still posit an all-powerful (although not necessarily totalizing) capitalist machine. For de Landa (2003), Deleuze and Guattari fall into this trap because of their incomplete reading of Braudel. For Gibson-Graham, their lingering capitalocentrism cannot be denied (2003, and personal e-mail communication over spring 2003).

30 "*The mutilation is prior, pre-established*," Deleuze and Guattari state (447). For them, the primitive accumulation described by Marx continues whenever an apparatus of capture is established.

31 See the excellent work by the Colombia desk of WOLA (Washington Office on Latin America) and a binational coalition of solidarity with Afro-Colombian struggles that includes, among others, AFRODES, PCN, American Friends Service Committee, WOLA, Trans-Africa Forum, and other organizations in Washington, D.C. Material on the situation of African palm is found at the WOLA and PCN web pages.

32 Metricizing could be met by retroactive smoothing, as Deleuze and Guattari would say (1987: 481, 486; see also n. 18 to this chapter). Of course, one has to bear in mind that one of the defining features of world capitalism today is its ability to operate more independently of the traditional mechanisms of ordering and striation, from labor to inventories to financing (492). In this way, capital also tends to create a smooth space and to operate on the basis of the network form. As we shall see in my discussion of networks, there are limits to capital's network logic.

3. Nature

1 This section presents, in a very succinct manner, some aspects of the cultural universe of the black groups of the Pacific. The literature on this topic is already significant. I emphasize here particularly those aspects most directly related to conceptions of nature, drawing on the pioneering works of Losonczy (1993), Restrepo (1996a), and Velásquez (1957), and also on Camacho (1998), Ulloa, Rubio, and Campos (1996, for indigenous models in the Chocó), Quiroga (1994), Whitten (1986 [1974]), M. Escobar (1990), and others mentioned in the text. I have not considered important cognitive and practical domains such as the body, sexuality, birth and death, illness and healing.

2 The thermal categorization is pervasive, and it is difficult to say that it corresponds in any way to a Western classification. For instance, honey, oil, alco-

hol, and blood are hot, but so are ice, salt, turtle shells, and plantain; wild plants are generally cold, and so are snakes, soap, and vinegar. The human body is hot above the waist and cold below it. Hot plants are generally beneficial and often have proper names (e.g., *doña Juana, San Juanito*). The hot principle constitutes a paradigmatic chain linking healing plants, fats and oils, alcohol, and the masculine. The cold principle unites night, earth, moon, old age, and the feminine. The dynamic character of these orders is notable: "*El principio 'caliente'* [the 'hot' principle] establishes a correspondence among elements from the animal and plant kingdoms, the magical and the human, that is, those instances that are agents of culture"; the masculine predominates at this level (Losonczy 1993: 51). "*El principio 'frío'* [the 'cold' principle], on the other hand, groups together those elements that must be transmitted and cultured, those that are necessary but ineffective or dangerous without the mediation of the masculine. Masculinity is, in a certain way, a synonym of the human and of culture in its ideal expression. It is, of course, grounded in the natural, and depends on it, but it is threatened by the latter, unless the humanizing function is actualized." In short, femininity is paradigmatic of what needs to be communicated, whereas masculinity is a paradigm of communication. For Losonczy, this reflects "the ambiguous and underlying face of femininity which, at the same time, assigns a relatively egalitarian and free place to women as far as manifest social relations are concerned" (1993: 51).

3 So when ecologists invoke the concepts of use spaces (*espacios de uso*), they reveal not only a logic of territorial organization (articulations between house, village, river, forest, mangrove, beach, and sea, as shown in chapter 1) but also the cultural logics just described.

4 This part of the text is based chiefly on Restrepo (1996a, 1996b, 1996c) and Camacho (1998). Following the usage of these anthropologists, I will place in italics those concepts that exist in the local language, and that are seen as having particular semantic connotations that do not necessarily correspond to the usual Spanish meaning.

5 This presentation of the various worlds and visions is extremely schematic. Don Porfirio Angulo (don Po), whom I introduced in chapter 1, explained his theory of visions, like that of the *tunda*, to us (Eduardo Restrepo and me) in our conversations in August 1998. For him these visions clearly exist as part of a natural world that actively interacts with the human world.

6 Some of the best-known stories have been collected by Vanín (1986). See also Vanín's own poems, of which there are several collections, the most recent being *Islario* (1998). Oral tradition poems and stories are often printed nowadays in ephemeral form as leaflets, single sheets, and in the popular communications materials that have become familiar in the littoral since the early 1990s (see chapter 4).

7 For instance, the distinction between tame and wild is important to conservation, since for locals what is tame is appropriated through domestication,

whereas the wild can be destroyed. And while the category of *renacientes* might be seen as embodying a notion of sustainability, it can also easily lead in local practice to overexploitation of certain resources (e.g., timber) when people are pushed by capitalist interest to do so, as Restrepo (1996b) documents in the case of the *corteros* (small-scale lumberjacks) of the Satinga river.

8 Oyama provides the following definition from biology: "By 'essentialist,' I mean an assumption that human beings have an underlying universal nature that is more fundamental than any variations that may exist among us, and that is in some sense always present—perhaps as a 'propensity'—even when it is not actually discernible" (2000: 131).

9 The complexity of the relation between humans and nature informed much work in ecological anthropology, from Steward to Rappaport. While pervaded by a certain technological determinism, early cultural ecologists like Steward came to acknowledge that "culture increasingly creates its own environment" (1968: 342), thus opening the way for more interactive perspectives. Vayda and Rappaport (1968) attempted a unified approach between ecology and anthropology by introducing a set of categories (individual, communities, populations, ecosystems; niche, species, communities; regulation and homeostasis) that enabled the study of certain cultural traits as, and in connection with, biological phenomena. Theories of adaptation and evolution were reconceptualized with the importation of ecosystems ecology into biological and ecological anthropology (see, e.g., Moran 1991 for a useful review). Rappaport (1991), however, cautioned against the reification of the concept of ecosystem, emphasizing that it must be taken as an analytical category, not as a biological unit. For this influential anthropologist, a key remaining problem was how to represent the systemic character of nature, which he saw as a question of objective natural law, and its articulation with subjective human meaning (e.g., Rappaport 1991). Toward the end of his life, Rappaport came to believe that concepts are virtual interventions in the world, thus approaching a constructivist perspective. Today, one could think of Rappaport as advocating the synthesis of realism and constructivism being worked out by some of his former students (e.g., Biersack and Greenberg, eds. 2006). Most of the work in systems-oriented ecological anthropology, however, continues to uphold the division between nature and culture, knower and the known. Hence Ingold's critique that in these works "we tend to envisage the environment as a vast container filled with objects" and that the very concept of niche, for instance, "entails that niches exist in the environment *prior* to the organisms that fill them." This analogy "ignores the most fundamental property of all animals: unlike vases, they both perceive and act in their environments" (1992: 41). More recently, under the influence of dialectical views of the human-environment relation and political ecology, biological anthropologists have developed a new paradigm of bioculturalism that is critical of positivism and goes a long way toward encountering constructivist political ecologies (see Goodman and Leathermann, eds. 1998; Leathermann and Thomas 2001).

10 This profoundly ecologized view of the social, although largely influential, is not without its critics. For Leff (1998b), Bookchin's unification of nature and society into a general ecological framework—Bookchin's "organismic ontology"—has serious problems: it dissolves the boundaries of what are actually distinct ontological domains (the biophysical and the social), thus reducing the social order to an ecological system; it collapses the concept (e.g., mutualism) with the real, thus bracketing critical reflection, since it overlooks the differentiation of the material and the conceptual; and it falls into an ontological monism based on the generalization of ecological principles. For Leff, in a critique that he also applies to Edgar Morin's version of complex systems ecology, these uses of the dialectical method are not warranted. In my view, while Bookchin's rationalism retains some measure of essentialist realism ("Our first presupposition is that we have the right to *attribute* properties to nature based on the best of our knowledge, the right to assume that certain attributes as well as contexts are *self-evident* in nature" [1990: 56]), these features are constantly problematized. Heller's social ecology, influenced by poststructuralism and feminism, is conscious of these problems. For her, nature is not only constructed—not a thing we can separate ourselves from and know completely (2000: 37)—but knowing nature is itself political. Moreover, Heller acknowledges a certain differentiation between natural and social orders at the level of knowledge. Even if she maintains the notion that the social is developmentally derived from the natural, thus remaining firmly within social ecology, she seems more willing to question the homology between the natural and the social (126) and to emphasize the need for different categories for the social and the natural (e.g., 131). One final note: social ecology's idea of the potentiality of matter to yield ever increasing complexity (e.g., Bookchin 1990: 57) has always seemed to me close to Theilhard de Chardin's idea (1963) of the increasing complexification of life—from inert matter to human consciousness and the noosphere. See, for instance, de Chardin (1963).

11 Oyama's position is influenced by Haraway. Her principle that "to acknowledge our part in constructions of nature is to accept interaction as the generator of ourselves and of our interrelations, of knowledge, and of the world we know" (150) is reminiscent of Maturana and Varela (1980, 1987; see chapter 6). Finally, her critique of dualisms fares well with Ingold's (2000a) sustained attempt at dissolving the distinction between biology, mind, and culture that he sees at the basis of evolutionary biology, cognitive science, and anthropology. For Ingold, these separations create the idea that human beings are the sum of these three complementary parts when in fact—as trends in developmental biology, ecological psychology, and the anthropological theory of practice suggest—the distinctions are spurious divides that must be dissolved through some sort of relational approach (such as phenomenology or Oyama's view of evolution and development as interactive emergence over time). What I find exciting about Oyama's work is her ability to draw from various tendencies, creatively crafting

an original framework of her own. As her work shows, there is a lot of room for synthetic approaches that bring together dialectical, poststructuralist, and phenomenological accounts in biology and the social and the human sciences, thus truly moving us away from logocentric modes of analysis.

12 Ingold's conceptualization can be related to a number of philosophical notions, such as Heidegger's idea of humans' inevitable thrownness and dwelling in the world and their engagement with the ready-at-hand within a background of understanding that is not fully objectifiable (as the natural sciences purport); or Maturana's and Varela's notion of the ineluctable interconnection between the domains of being, knowing, and doing. His appeal to phenomenology is based on this school's nondualist claims. As a philosophy of experience, phenomenology purports to study being-in-the-world, that is, phenomena before they are captured by thought, without reducing them to any transcendental truth or essence. Rather than attempting to come up with an objective view of reality, phenomenology starts with the assumption that the world is first had before it is cognized, and that it is the life-world (as a domain of everyday social existence and practical activity) rather than worldview that matters in ascertaining how the world appears to us. Phenomenology thus prioritizes lived experience over theoretical knowledge, considers various types of knowledge on an equal footing, and acknowledges that the subject is always implicated in the world and in knowledge itself. Finally, and very important, phenomenologists insist that the real can never be completely reduced to any underlying cause, be it discourse, the economy, politics, or what have you. Ingold's phenomenological ecology has given rise to what some of his students call "the dwelling perspective," which I believe constitutes a promising research program (see Campbell, ed. forthcoming).

13 See, for instance, Brosius (1999), Biersack (1999, 2006), Escobar (1999a), and Peet and Watts (1996) for reviews of the trends in poststructuralist antiessentialism in nature studies in anthropology and geography. A very interesting debate on antiessentialism, particularly in relation to the fairness of the critiques of 1970s and 1980s cultural ecofeminism as essentialist, has been taking place in what is variously called ecofeminism, ecological feminism, and feminist environmentalism. See, for instance, Sturgeon (1997), Cuomo (1998). I leave out the development of antiessentialist approaches in other fields with which I am less familiar (e.g., environmental history and historical ecology).

14 De Landa explains at length Deleuzian nonessentialist realism in his recent book (2002). In chapter 6 I will discuss de Landa's theory of assemblages, also derived from Deleuze and Guattari. For now, it is sufficient to place nonessentialist realism (or neorealism, as I prefer to call it, which is also postconstructivist) on the map of the realism–constructivism spectrum. Tim Ingold (2000b), in his critique of the dominant genealogical model of descent, comes close to a Deleuzian model, actually appealing to Deleuze's and Guattari's notion of the rhizome. I will not discuss here another type of neorealist view, actor-network theory; it will make a brief appearance in chapter 6.

15 I am adapting the term "weakening" from Vattimo's idea of the progressive weakening of Being with modernity (Vattimo 1991).

16 An excellent synthesis of the work on traditional production systems (TPSs) in the Pacific is the fourth volume of PBP's final report, written by Enrique Sánchez, the coordinator of the project's socioeconomic area. The report includes a full characterization of the systems in terms of concrete activities, land tenure and uses, production practices, and underlying social relations and their impact on biodiversity. This conceptualization did not emerge in a scholarly vacuum. First, PBP staff were influenced by earlier works on the Pacific (studies of particular production systems such as plants, mangroves, mining, shell collecting, etc.). The single most influential volume in this regard was that of Leesberg and Valencia (1987) on traditional production practices on the middle Atrato river, which included a lengthy, detailed study of local productive systems. From the mid-1980s on there were studies of particular aspects, such as gender, indigenous conceptions, and the impact of development projects on communities, all of which included important information on TPSs that PBP staff built upon; there were also studies of particular river basins (e.g., the substantial research project on the Naya river south of Buenaventura by Universidad del Valle biologists, anthropologists, public health and social medicine faculty and students, which, while largely centered on health problems, including malaria, sponsored studies on agriculture and silviculture). Conceptions in cultural ecology also influenced PBP staff, particularly the concept of adaptation, which they reformulated in a dynamic fashion. Studies of particular systems were influenced by concepts and methodologies of landscape ecology. Finally, PBP staff also had at its disposal the burgeoning literature on agroecology that posited a strong relationship between TPSs and ecological sustainability, particularly Altieri's influential work (e.g., Altieri 1995 [1987]).

The characterization of particular TPSs was often done by subcontractors who focused on particular regions or situations. In many instances, these projects were collective endeavors between researchers (e.g., biologists, communicators, anthropologists, sociologists), social movement organizations, and local people and coresearchers. The projects were often associated with collective titling exercises. See, for instance, the comprehensive study of TPSs among the black communities of Nuquí and Tribugá (Chocó) by Camacho and Tapia (1997); besides a description of TPSs, this report includes a useful analysis of the methodologies employed (collective mapping, identification of landscapes and use spaces, historical narratives of changing environments, etc.). See also Riveros (1996) for a detailed study of TPSs in the Cajambre river, which also follows a landscape ecology methodology (this study includes detailed information on cycles, products and species used, use spaces, economic activities, and local valuation); Tapia, Polanco, and Leal (1997) for a TPS in the black community of the Valle river in Chocó; Ruiz-Palma (1998) for a study on the use of forest fauna in the Cajambre, Naya, and Yurumanguí rivers (which discusses several dozen species of birds and mammals traditionally used, as well as serpents and turtles); and Ruiz-Palma et al. (1997) for a study of traditional forest hunting by Wounaan indigenous communities in the San Juan river, Chocó. Finally, see the lengthy and detailed reports by PCN (2000, 2002, 2003a, 2003b), which contain a description, analysis, and recommendations for the TPSs of the Yurumanguí river.

17 The methodology used by PBP staff for the evaluation of TPSs was simple but creative and can be considered a form of multicriteria evaluation proposed by ecological economists (e.g., see Martínez Alier 2002). They assessed the effort going into different productive activities (e.g., maize, plantain, hunting, fishing) in terms of variables such as money, human effort, knowledge needed, time involved, risks, and impact on nature. In a different chart they classified the "benefits" in terms of food, income, permanence or continuity, exchange/barter value, duration of benefit, and satisfaction. To give an example, maize was seen as requiring little effort (2 on a scale of 10), involving knowledge (4), and producing low impact on nature (1). On the benefit side, maize had a relatively high value in food terms (5 for humans, 8 for animal feed), high exchange value (7), and long duration of the benefit (8). Mining and hunting were listed as having significant impact on nature (9) but important in generating money income (especially mining). Plantain and, surprisingly perhaps, timber extraction were seen as having the highest degrees of permanence (9). And so forth. See Sánchez (1998: 81–86). These assessments are consonant with expert findings in other parts of the world. As Shiva (1993) has shown, the productivity of modern systems of monoculture (e.g., a eucalyptus forest) can be seen as higher when compared to a traditional diverse forest thanks only to an epistemological operation that assesses productivity in terms of a single market commodity (pulp) and that renders invisible the many other values and services produced by the forest, from diversity to use-values for local people, the conservation of watersheds, carbon sinks, and cultural values. Shiva uses this exercise to redefine productivity in a more holistic manner. Similarly, for Toledo (2000b) the strategy of nonspecialized production is conducive to a diverse logic of use spaces that generate many products out of ecological exchanges more than economic exchanges and that tend to maintain (and theoretically even increase) biodiversity (2005). Barkin (1998), finally, argues for the importance not only of redefining the criteria of efficiency in the case of peasant cultivators but of macroeconomic and technological strategies that might make a degree of economic autonomy possible for those who choose to remain in the countryside.

18 I have in mind the sustained critical work on conservation of biodiversity maintained by the Barcelona-based group GRAIN (Genetic Resources Action International) and ETC (formerly RAFI, Rural Advancement Foundation International), with offices in Canada, the United States, and Mexico. See these organizations' web pages for more information, and chapter 6. Since the late 1970s, these groups have developed a progressive view of and strategy for genetic resources and biodiversity that are supported by many social movements through-

out the world. In 2000–2002 GRAIN carried out an important project on experiences with in situ biodiversity in a number of countries in Latin America, Asia, and Africa. For a project description, see GRAIN (2000).

19 A note on methodology: I have been following debates on biodiversity since 1993 in various ways. Between 1993 and 2000, I compiled seven or eight bound volumes on related issues from Internet sources, first at Smith College, with the assistance of Krista Bessinger (from sources such as EcoNet and the Earth Negotiations Bulleting), then at the University of Massachusetts, Amherst, with the assistance of Mary King, focusing on the Convention on Biological Diversity and special topics such as intellectual property rights, the Global Environment Facility (GEF), and genetic resources. I also attended a number of conferences on the subject over the years and collected a voluminous secondary scientific and policy literature. The great majority of these sources have been left out of direct reference or commentary—indeed, these debates could be the subject of an entire chapter, if not a book. I have also left out some research on bioprospecting (including visits to the New York Botanical Garden in the Bronx and to Shaman Pharmaceuticals in south San Francisco), although this topic will make a brief appearance in the chapter on Networks. Second, I have followed actively debates within Colombia (and indirectly the Andean Pact countries) through my research with PBP. My initial contact with the PBP was in the spring of 1994, when I prepared for them a lengthy report on trends in biodiversity thinking with special focus on intellectual property rights. PBP never published the report (it was published as a short booklet in Mexico, see Escobar 1997). I followed PBP closely until 1998, the year of its de facto dismantling. Finally, as it should be clear by now, I have reconstructed the conceptualization of biodiversity developed by the social movement of black communities of the Pacific (particularly PCN) through various ethnographic situations, including workshops, interviews, and numerous activities with movement activists in Colombia and abroad. A first article on the subject appeared in English in 1998, from an original Spanish version presented at a panel in Mérida, Mexico, in October 1997, as part of the Second International Congress of Ethnobotany.

20 According to Wilson (1995: 358–59), the term *biodiversity* was first used at a forum on the subject held in 1986 under the sponsorship of the National Academy of Sciences and the Smithsonian Institution. Wilson confesses to having been a late arrival in articulating this concern in his first piece on the subject in 1980. The concern that conservation biology—as a new field intended to bring together ecology and evolutionary biology in the interest of conservation—articulated in the 1980s had precedents in earlier ideas of, among others, wilderness, nature, and ecological variation. For a thorough history of the concept, see Takacs (1996). Important chapters of the history of biodiversity, from the perspective of scientists, are chronicled in Baskin (1997).

21 Genecentrism is evident in most definitions of biodiversity, including its treatment in the CBD. See, for instance, articles 1 and 2 of the CBD, "Objectives"

and "Use of Terms," where reference to genetic resources and biotechnology is prominent. But examples of genecentrism abound at all levels of the biological literature. Wilson himself includes two cases in his chapter on biodiversity in his memoirs, *Naturalist* (1995). The first is his approving mention of research into the genetic predisposition humans have for an "ideal" habitat (a home perched "atop a prominence, placed closed to a lake, ocean or other body of water, and surrounded by a parklike terrain" [360]). This predisposition is supposedly linked to humankind's original habitat in the African savanna. Ask a Tuareg nomad or a black person in Tumaco, about his or her ideal habitat, and you would get a very different response. The "ideal" diagnosed by the biologist is, I suspect, more closely linked to social and historical factors, such as the colonial penchant for choosing living quarters on a higher ground to gain some sense of control (e.g., Pratt 1992), the development of a view of nature as vista, the nostalgia for paradise lost intensified by modern urban-industrial culture, and a life in the suburbs alienated from nature and people. The second example is the alleged evolutionary genetic basis for "the highly directed reaction against snakes." Again, ask many indigenous peoples of the Pacific who have no trouble recognizing dozens of snakes, many of which they hunt for food. And while there are at times some associations of snakes with danger and fear among black groups, this happens in particular contexts, such as the aversion of a man hunting alone at night.

22 On golden rice, see the cover article of *Time* for July 31, 2000. Debates on bioprospecting have oscillated between heated (e.g., around the Costa Rica–Merck agreement for bioprospecting of 1991 and the work of Shaman Pharmaceuticals) and sedate but continue to be important. Again, the literature on these topics is too vast to include here. Some of it is reviewed in Escobar (1997, 1998, 1999a) and, more recently, in the book-length study of a Mexican experience by Hayden (2004). See also the study of "the new merchants of life" by the Colombian NGO CENSAT Agua Viva (CENSAT 2005).

23 These issues would require an entire separate treatment. I have developed the argument on the impact of genecentrism on biodiversity in an article with Chaia Heller (Heller and Escobar 2003). The debate on intellectual property rights is crucial in today's discussions, as it entails clashing perceptions of nature, property, and the economy. However, the literature on the subject is too vast to cite, and I have been unable to update it. For Latin America, see Escobar (1997), and Lander (2002), who links intellectual property rights to coloniality. Researchers at Cambridge University have been working on a multifaceted project, led by Marilyn Strathern, on types of cultural claims to property, including the very notion of intellectual property. See, for instance, Strathern (1996, 1998), Kirsch (2001). For radical critiques of genecentrism within biology, see Oyama (2000), Ingold (1990, 2000a, 2000b), Keller (1995). A critique of TRIPS in the context of the CBD is elaborated in GRAIN/GAIA (1998a, 1998b); see also the ongoing and leading work of ETC. The movement to come up with alternatives to TRIPS and modern capitalist systems of intellectual property and compensation

has been going on since the mid-1990s, under such rubrics as intellectual rights, collective rights, community rights, and, more recently, linked to questions of access to knowledge and open source (e.g., Gosh, ed. 2005).

24 There has been actually significant convergence in pointing at those three factors as the main threats to biodiversity since the publication of *Global Biodiversity Strategy* in the early 1990s (WRI/IUCN/UNEP 1992). Although I do not want to overlook differences in position among various actors in the established conservation network, it is fair to say that there is also an overall convergence on the analysis of causes and potential solutions, constituting what in chapter 6 I call resource management or globalocentric perspective. See also, for instance, the technical volumes of the Scientific Committee on Problems of the Environment, SCOPE (UNEP 1995); Swanson (1997); Wilson and Perlman (2000); the report of the National Research Council on the value of biodiversity (1999); and reports of the Global Biodiversity Forum (e.g., WRI/IUCN 1996), as well as the work of the Subsidiary Body on Scientific, Technical and Technological Advice (SBSTTA) of the CBD, which deal largely with specific scientific issues (e.g. forest, agricultural, or marine biodiversity; biosafety: genetic resources, etc.). These regularities of the biodiversity discourse at the level of causes and solutions in the dominant network can be easily identified by focusing on the hundreds of scientific and policy documents and reports of the key conservation organizations, whether NGOs (e.g., World Resources Institute, World Conservation Union, World Wildlife Fund) or international organizations (e.g., United Nations Environmental Program, World Bank). A doctoral dissertation by Mary King (University of Massachusetts, Amherst) analyzes these debates in depth (2006).

25 The "will to representation" linked to the biodiversity movement would itself deserve a book-length treatment. The concern with biodiversity seems to have awakened the imagination of many as they compete to convey the sense of loss and crisis through a combination of simplified statements, charts, figures, diagrams, interactive exhibits, documentaries, even commercial advertisements. To mention just two examples: the special section entitled "The Natural World" in the *New York Times*, June 2, 1998, to coincide with the opening of the Hall of Biodiversity at the American Museum of Natural History (itself a sophisticated experiment in interactive representation). Under the title "Lush Life: But as Species Vanish, What Will We Lose?," the twelve-page supplement includes, among other things, a front-page color composite of natural beings (which features fungi, plant, and animal species from California and Michigan to Australia, the Amazon, and "African elephants, from the San Francisco Zoo," all in one single representation, as if in a Garden of Eden before the Fall); two full-page ads by Monsanto, one of the sponsors of the museum exhibit, with appropriate scientific themes; a large evolutionary curve in a two-page display; the de rigueur articles on vanishing species, the need for classification, and economic utilization of biodiversity; and the full-page announcement of the museum exhibit, featuring Harrison Ford in a brown T-shirt and faded jeans against a background of

luscious rain forest and swerving tropical fish. This conflation of nature as science, business, and adventure unfortunately is at the root of many a popular representation of biodiversity. Here again we find strong deterministic views ("destroy the wrong genes, and the species might not be able to adapt and survive in the long run" [10], leading article). The second example, the *National Geographic* issue "Making Sense of the Millennium" (January 1998), certainly much more serious and accomplished, equally links biodiversity with science and population in a dazzling display of photographs and evolutionary schemas.

26 Compensation for ecosystems services to governments or communities by private corporations became salient in the late 1990s. The idea behind these programs is that the payment can be used for the conservation of particular ecosystems (e.g., a tract of forest). Ecotourism and green tourism also fit under this category. International agreements for the preservation of forests as carbon sinks have also become common, even if generating considerable controversy because of its implications, from new types of colonialism to "green whitewash" that may allow particularly northern polluters to go on with their activities without modification. These agreements are becoming central to what is called "the carbon economy." On the oppositional side is the movement for the calculation and payment of ecological debt, meaning by this the debt owed by rich countries to poor countries because of their disproportionate use of global resources, on the one hand, and of world pollution and destruction, on the other. In the late 1990s, some estimates assessed this debt as already much larger than the external financial debt owed currently by poor countries to banks and institutions in the rich ones. An Ecuadorian NGO, Acción Ecológica, has been among the pioneers in the ecological debt movement, although the idea has precursors in the 1980s (see Borrero 1994 for an initial study and summary) and well-known international proponents (e.g., Martínez Alier 2002).

27 See Escobar (1999a) for an extended exposition of the idea of nature regimes. This model follows an antiessentialist perspective; the three regimes are relationally defined, are not historical stages in the social history of nature, and they constantly overlap and are mutually hybridized. They are all marked by particular cultural constructions and should thus be seen as cultural models for the appropriation of nature.

28 The Hindu-Catalan theologian and philosopher Raimon Panikkar (1993) speaks of three kairological moments (not chronological, but qualitatively different), which he reads from many traditions and religions: an ecumenical moment, in which Man is fully integrated with nature, part of a sacred whole, the universe as living reality; an economic moment (roughly, that of modernity), ushered in by scientific humanism, with Man above Nature, in which Man rejects both his animality and his divinity and commits himself to understanding as Truth; this brings about an increasing estrangement from nature. This is followed by an "ecological interlude" that marks the end of modernity in which Man is conscious of his failure in the guise of a humanistic and ecological crisis. It is

an interlude because ecological consciousness, as in the case of the biodiversity discourse, yields only a rational response, that of a more reasonable exploitation of nature ("sustainable," as we would say), and ecology becomes "yet another tool for human mastery of the Earth" (43). This prepares the ground for a third, "catholic" moment (which has nothing to do with Catholicism as a religion), in which Man finally realizes that "the center is neither a merely transcendent Godhead, nor the cosmos, nor himself" (46), thus arriving at a new experience, what Panikkar calls *cosmotheandric*, in which the three dimensions of existence—cosmos, god, humanity; the biophysical, the divine, the human—are finally brought together again in a new synthesis. This remarkable conceptualization has yet to be explored by ecologists in the West.

29 This conceptualization comes from a lengthy report on the first phase of a project with which I have been involved. The project—to be discussed briefly in the conclusion—was funded mainly by a Danish NGO (Solstice Foundation) and supported the process of collective titling, territorial appropriation, and organizational strengthening in two areas, the Yurumanguí river south of Buenaventura and the Pílamo Hacienda, an Andean black community territory south of Cali. See PCN (2000, 2002, 2003a, 2003b).

30 See the text of the agreement, Movimiento Social de Comunidades Afrocolombianas (2002). A similar agreement was drawn up at the same time between the Park System and indigenous organizations of the Pacific.

31 The work of the NGO GROUP ETC is exemplary in channeling critical work in this area, and many of the pertinent documents and communiqués are found on their web site.

4. Development

1 As one of the top PLADEICOP planners told me candidly in an interview in 1993, they conceived of participation as a series of activities with local people geared toward getting the community to arrive at the same conclusions the planners had already envisioned. Participation in the 1980s was largely a functional requirement bureaucracies had to meet (through gathering of information, diagnoses exercises, and so forth), not a strategy to foster local empowerment.

2 The full title is *Plan de acción para la población afrocolombiana y raizal* (see DNP/UPRU/Plan Pacífico 1995). The term *raizal* refers to the Afro-Colombians of the archipelago of San Andrés y Providencia, off the coast of Nicaragua, a distinct black ethnic group in terms of language and culture that, while constituting about 30 percent of the population of the St. Andrew's archipelago, nevertheless has been systematically marginalized, especially since the start of government policies to encourage tourism development in the islands in the 1950s.

3 Wallerstein (2000) understands this transition in terms of the Kondratieff cycle started in 1945 and of the structural crisis of the world system since 1967–76. Using a Prigogine-type explanation (unpredictable chaotic behavior

leading to a bifurcation toward a new structure that is impossible to predetermine), Wallerstein (somewhat paradoxically) concludes that while the capitalist world economy will be recomposed after the crisis is over, the form of the new structure is open to political action.

4 I believe a Eurocentered view of modernity is present in most conceptualizations of modernity and globalization in philosophy, geography, sociology, anthropology, and communications, even if many of these works are important contributions to the understanding of modernity. Some of these works explicitly engage with Giddens's work and develop elegant conceptualizations of globalization (e.g., Tomlinson 1999); others follow an ethnographic orientation (e.g., Englund and Leach 2000, and Kahn 2001 for reviews; Appadurai 1996), or a cultural-historical orientation (e.g., Gaonkar, ed. 2001). Some assert the plurality of globalization yet go on to explain such plurality in political and economic terms, taking for granted a dominant cultural matrix (see the special issue of *International Sociology* entitled "Globalizations," vol. 15, no. 2 [June 2000]). The same assumptions are often found in sociological and political economy works on the Left. A Eurocentric notion of modernity is at play, for instance, in Hardt and Negri (2000). Their reinterpretation of the European history of sovereignty in light of current biopolitical structures of rule, as well as their elaboration of resistance in the Western philosophy of immanence, are novel elements for rethinking modernity. However, their Eurocentrism becomes particularly problematic in their identification of the potential sources of radical action and in their dictum that there is no outside to modernity (again, àla Giddens, even if coming from a view of the unification of world markets and biopolitical control).

In anthropology, the analysis of modernity was initially couched in terms of the ethnography of practices of reason and expert knowledge in the West and beyond (see the special issue of *Cultural Anthropology*, vol. 3, no. 4, edited by Paul Rabinow, and subsequent work by Rabinow and others in the group). Our own initial work in the Pacific (Escobar and Pedrosa, eds. 1996) was explicitly conceived as an anthropology of modernity, including ethnographic accounts of biologists, planners, capitalists, and NGOs and of social movement activists as agents of what we then called alternative modernities. More recently, the anthropology of modernity has focused on modernity abroad and on people's (largely nonexperts') engagement with it. As Kahn (2001) put it in a recent review, taken as a whole the works in this trend have pluralized modernity in various ways. Some posit the possibility of modernities that are not reducible to a north European matrix (Faubion 1993); others discuss alternative modernities as emerging in the encounter of dominant and nondominant (e.g., local, non-Western, regional) forms (e.g., Pred and Watts 1992; Gupta 1998; Sivaramakrishnan and Agrawal, eds. 2003; Arce and Long 2000). There is no unified conception in these works on what exactly constitutes modernity. Kahn is right in saying that stating that modernity is plural and then showing ethnographically the ways in which it is localized has limitations in terms of theory. As Ribeiro says in his commentary on

Kahn, "Modernity is subject to indigenization, but this does not amount to saying that it is a native category" (2001: 669). Englund and Leach (2000) make a related argument in their critique of the ethnographic accounts of multiple modernities; they argue, correctly in my mind, that these works reintroduce a metanarrative of modernity into the analysis, be it "the dialectic," a (European) core that remains invariant, or a self-serving appeal to "wider context" or "larger-scale perspective." The result is a pluralization of modernities that reflects the ethnographer's assumptions. What is lost in these debates, it seems to me, is the very notion of difference as both a primary object of anthropology and an anchoring point for theoretical construction and political action. Englund's and Leach's call is for renewed attention to ethnographic knowledge as a domain for ascertaining the very contexts that are relevant to investigation, before such a context is imputed to this or that version of modernity. From this perspective, a question remains: What other kinds of theoretical and political claims can one possibly make via the insights of the ethnographies of modernity? In short, it seems to me that in many recent anthropological works modernity is, first, redefined in a way that deprives it of historical coherence, let alone unitary, social, and cultural logic; and then, second, found ethnographically everywhere, always plural, changing, and contested. A new balance seems necessary. After all, why are we so ready still to ascribe to capitalism powerful and systematic effects, while denying modernity any coherent and dominant cultural logic? I shall return to this question in the conclusion.

5 I am referring here to what could be considered a sustained and coherent research program that has coalesced recently around the concepts of modernity/coloniality/decoloniality. This is an extremely inadequate presentation of this group's ideas (see Escobar 2003b for a more complete treatment, including earlier influences and world sources). This group is associated with the work of a few central figures, chiefly, the Argentinean/Mexican philosopher Enrique Dussel, the Peruvian sociologist Aníbal Quijano, and the Argentinean and U.S. literary theorist Walter Mignolo. There is a growing network of scholars associated with the group, particularly in the five Andean countries and a few places in the United States. The group already has a significant output in Spanish and a few collective works in English in preparation. A partial list of the works in Spanish includes Dussel (1976, 1992, 1993, 1996, 2000); Quijano (1988, 1993, 2000); Quijano and Wallerstein (1992); Mignolo (1995, 2000); Mignolo, ed. (2001); Lander, ed. (2000); Castro-Gómez and Mendieta, eds. (1998); Walsh, ed. (2003, 2005); Walsh, Schiwy, and Castro-Gómez, eds. (2002); Castro-Gómez (2005); Castro-Gómez and Grosfogel, eds. (2007); and the special issue of the journal *Nómadas* from Bogota, no. 26 (2007). In English, see Beverly and Oviedo, eds. (1993) for some of these authors' early works. A volume in English has been devoted to Dussel's work under the apposite title *Thinking from the Underside of History* (Alcoff and Mendieta, eds. 2000); the volume includes a feminist analysis of Dussel's work by the Finnish feminist theologian Elina Vuola (2000; see also Vuola 2002).

The journal *Nepantla: Views from South*, from Duke University, had a partial focus on the works of this group. See especially vol. 1, no. 3 (2000), with contributions by Dussel and Quijano among others. See also the special issue of *Cultural Studies* edited by Walter Mignolo, vol. 21, nos. 2–3 (2007). A forthcoming volume (Grosfogel, Maldonado-Torres, and Saldívar, eds.) includes chapters by a dozen of the main authors associated with the group.

6 The choice of origin point is not a simple matter of preference. The conquest and colonization of America is the formative moment in the creation of Europe's Other; the point of origin of the capitalist world system; the origin of Europe's own concept of modernity; the initiation point of Occidentalism as the overarching imaginary of the modern/colonial world system (which subalternized peripheral knowledge and created, in the eighteenth century, Orientalism as Other; see Coronil 1996). Whether the authors in the MCD program see modernity's origins as strictly European or as an always negotiated ensemble of forms of modernity/coloniality that shape the West as much as the rest remains a matter of debate.

7 Different authors emphasize different factors in the making and functioning of modernity/coloniality. For Quijano, for instance, the key process in its constitution is the colonial classification and domination in terms of race; a second key element is the structure of control of labor and resources. Dussel emphasizes the original violence created by modernity/coloniality (see also Rojas 2001), the importance of the first (Iberian) modernity for the structuring of coloniality, and the concealment of the non-European. Mignolo's project is that of conducting a genealogy of local histories leading to global designs, so as to enable other designs from other local histories to emerge from border thinking and the colonial difference.

8 Dussel derives his notion of exteriority from phenomenology and Marxism. For Mignolo, as for Quijano, "the modern world system looks different from its exteriority" (2000: 55); without the exteriority in which subaltern knowledges dwell, "the only alternative left is a constant reading of the great thinkers of the West in search of new ways to imagine the future" (Mignolo 2000: 302). Border thinking implies "thinking from another place, imagining another language, arguing from another logic" (313). Border thinking transcends the monotopic interpretation of modernity and moves toward a pluritopic hermeneutics that sees from various spaces at the same time (a concept Mignolo adapts from Panikkar's diatopic hermeneutics). The aim of border thinking is "to think otherwise, to move toward 'an other logic[,]' in sum to change the terms, not just the content[,] of the conversation" (70).

9 For a full response to the critiques of postdevelopment and the pertinent bibliography, see Escobar (2007). One additional point of clarification: Is the depiction of a monolithic development necessarily wrong? Here again one finds a misunderstanding of the poststructuralist project. Kamat (2002) is right in pointing out the fact that the depiction of the development discourse as having a certain relational coherence was necessary for understanding the links between

capital and the development discourse. A similar point has been made by Hardt and Negri when they argued that Marxist and poststructuralist analyses alike, while eliding differences among countries, were instrumental in highlighting "a tendential unity of political, social, and economic forms that emerge[d] in the long imperialist processes of formal subsumption" (2000: 334; see also 283). The need to recognize multiplicity, they add, "should not blind us to the fact that, from the point of view of capital in its march of global conquest, such a unitary and homogenizing conception did have a certain validity" (333).

10 Phenomenologically speaking, any feature entering a human group from anywhere is by necessity actively reabsorbed into the local background of understanding—shifting such a background and, of course, the incoming element.

11 In striving to explain this conceptualization, Arce and Long rely on metaphors that are also found in complexity theory: reassembly, dynamic changes, self-organization, recursivity. I find some resonance between their notion of counterwork as an endogenous process and Maturana's and Varela's notion of autopoiesis—change by drift with structural coupling to the environment and without loss of organization (1987).

12 These newer approaches, it seems to me, solve some problems but introduce new ones. The problems are found in much of contemporary theory that emphasizes a dispersed, networked logic of the production of the social. There are four interrelated problems originating in four claims: (1) Radical agentivity: everything and everybody has agency; questions: how can one differentiate among various kinds of agency? what happens to power? (2) Radical connectivity: everything is connected to everything; questions: how are things variously or differently connected? how do they fulfill different functions in an ensemble of connectedness? (3) Radical contextuality: everything is context-dependent; questions: what happens to difference? what is in there that is not fully produced by capitalism, development, and modernity, even if in touch with them? (4) Radical historicity: everything has a genealogy, a lineage; question: what happens to what is genuinely emergent, unpredictable? Final questions: in the newer approaches, how do authors construct their object of critique? what happens to praxis, i.e., the connection between theory and practice, knowledge and action? Sinha's framework of multiscale hegemony processes aims at working through some of these impasses.

13 Medeiros (2005) analyzes at length other idioms related to local development. Darkness, for instance, ever since the conquest, reflects not an inner inability to see but an externally imposed condition of "not knowing where to go." The notion that development is the light that enables one to know where to go makes sense in this context. The saying, "We [i.e., local people] don't know how to understand what is happening" means "We don't know how to grasp—how to read—the codes of the dominant society and hence to reveal how we have been systematically excluded from our rights as citizens of the nation and exploited as an ethnic group." The saying is thus a statement about the long history of asym-

metrical power relations, of which the locals are very much aware. Drunkenness is not a self-incurred state but a state of disorientation and distorted perception caused by a history of ever-changing laws and regimes, of messages of inclusion accompanied by practices of exclusion. Development as awakening—the way out of darkness—requires that locals get hold of those forces that exploit them and that at the same time would allow them to see where to go and to move forward and thus negotiate inclusion as citizens on their own terms. The problem development agencies must address is thus not that local people are outside of modernity, but that they have been produced by modernity in specific ways—that is, by the coloniality of power and the colonial difference. It is their historical experience of modernity that informs their vision, their suspicion, and their hopes.

14 The brief account of cooperatives in the Pacific and Coagropacífico that makes up this section is based on several sources: Interviews with Coagropacífico staff (Arismendi Aristizábal and Apolinar Granja) in Tumaco in 1993 (Escobar and Pedrosa), 1997 (Restrepo), 1998 (Escobar and Restrepo), and 2000 (Restrepo); interviews with Rubén Caicedo, a cooperative expert from Tumaco affiliated for a time with Coagropacífico (Escobar, Tumaco 1993); several field trips that included interviews on the Rosario river with participating *agricultores* (small farmers) in July and October 1993 (Escobar, Grueso, Lozano) and with former cooperative members in Guapi and Timbiquí. Interviews were also conducted with CVC staff (Proyecto CVC–Holanda) and PLADEICOP staff in Tumaco and Cali (1993, 1994). A more detailed treatment of cooperatives in the Pacífico Sur, including Coagropacífico, is found in Grueso and Escobar (1996).

15 This is from Arismendi Aristizábal, leader of the cooperative. Interview in Tumaco, August 1998. For him, Coagropacífico did a lot to give local people trust and pride in their own abilities to carry out ambitious projects.

16 In the Pacific, the *décima* (literally, a poem of ten verses whose origin is in the Spanish Golden Age period), is an oral poem made up of forty-four verses, most of them octosyllabic, distributed in five groups (four verses in the first one, and ten in each of the following four). It is one of the most important elements of the black oral literature of the Pacific, and legendary *decimeras* and *decimeros* are found throughout the region's history and geography.

17 A note on the sources and origins of the program. The roots of the program lay in the work on adult education by a group of faculty at the Universidad del Valle in Cali. This group started to work intensely with black communities in the Norte del Cauca region south of Cali in the early 1970s. Their approach was that of radical liberation and was thus critical of conventional literacy. By the time PLADEICOP came into the picture, the group had already been working with black peasants and indigenous communities for fifteen years. They had just begun working in some areas of the Pacífico Sur, with funding from Foster Parents International (Plan Padrinos). In 1986, my close friend and collaborator Alvaro Pedrosa, in conjunction with other faculty and students, proposed the creation

of a popular communications foundation, Fundación Habla/Scribe, with the aim of promoting graphic arts and popular communications among marginalized communities. The foundation was formally established in 1987 and soon became an important presence in the Pacific. A closely related program, Red de Radios (Radio Network) was also introduced and managed by people in the Communications Department at Universidad del Valle. Together, the printed media (the *entintados*) and the radio network constituted the Gente Entintada y Parlante and the Red de Editores Populares del Occidente Colombiano (Popular Editorial Network of Western Colombia). I was closely associated with Habla/Scribe since the late 1980s. This section is based on my collaboration of many years with Pedrosa (see also Pedrosa 1989a, 1989b), on my intimate knowledge of Habla/Scribe, and on my observation of some of the programs in Tumaco. Pedrosa was largely responsible for the creation of Habla/Scribe and for the development of the entire GEP approach, with the help of a young cadre of recent graduates of the Communications Department at Universidad del Valle, especially Jaime Ariza, Alberto Gaona, Jesús Alberto Valdés, Aurora Sabogal, and Jaime Rivas. In 2001, I commissioned a historical and analytical study of GEP by the Tumaco artist and communicator Jaime Rivas to complement my knowledge of the organization and of the communications movement in the Pacific (Rivas 2001). Some previous cultural activity in the Pacific served to support GEP on its arrival. In Tumaco, this included the promotion of traditional dances and musical forms since the late 1970s by groups funded by the Institute for Popular Culture and Plan Padrinos (e.g., the Fundación Cultural Escuela Danzas Ecos del Pacífico, founded by the Tumaco cultural activist Julio César Montaño).

18 The most important products were short, semiperiodical publications such as *El papel entintado*, *Voces del litoral*, *Hoja del agua*, and *Boletín de la red de radio*, which had various orientations—cultural or informational, with topics from local traditions to development and biodiversity—and circulated regionally in editions of about twenty-five hundred copies; collections (e.g., in the case of the anniversary in 1992 of the Conquest of America, under the rubric "Five Hundred Years of Future for the Pacific"; or about the new rights accorded by the Constitution of 1991); and a large variety of color printed material, often with texts from the oral traditions.

19 Fundación Habla/Scribe adapted the paper and ink supply and processing tools and services to the social, climatological, and ecological conditions of the Pacific. It developed simple technologies for binding, drawing, printing, recording, and filing that functioned well in the Pacific. Among the most notable popular communications foundations created in the Pacific were Esteros (Tumaco), El Chigualo (Barbacoas), Atarraya (Guapi), Sensemayá (Buenaventura), La Resaca and La Bahía (Bahía Solano), and Canalete (Istmina).

20 This is from the collection *Tus derechos son mis derechos*, from the print corresponding to "The Right to Express My Culture," referring to article 20 of the Constitution of 1991, designed under the direction of Alvaro Pedrosa and compiled

and printed by Habla/Scribe, 1993. Despite the fact that the collection was not designed by people from the region, it reproduced many of the linoleum prints made by locals and successfully captured the sentiment of the moment concerning the new rights. GEP lasted officially until 1991, but the popular communications movement endured much longer. Habla/Scribe went through a period of intense activity until about 1998, when the political and economic crisis of the country made it very difficult for popular NGOs to maintain their previous level of activities.

21 Information for this section comes from ethnographic research and interviews I carried out with PBP staff at various points between 1993 and 2000. I want to highlight particularly conversations and interviews with Libia Grueso (the regional PBP coordinator in Buenaventura), Alfredo Vanín (the regional coordinator in Tumaco), Enrique Sánchez (the coordinator of the *Valorar* program area), and Mary Lucía Hurtado (the coordinator of the *Movilizar* program area). Other valuable sources of information at the Bogotá office were Claudia Leal, José Manuel Navarrete, and Fernando Gast. For the PCN, Hernán Cortés (Tumaco) and Yellen Aguilar (Cali). My first formal involvement with PBP was in the spring of 1994, when I prepared a lengthy report for them on trends in intellectual property rights.

22 The gap between intentions and practice was heightened in the *Conocer* area. Scientific research in this area yielded a lot of valuable studies and information, including characterization of ecosystems through biogeographical, ecological, and mapping techniques, identification of particularly important species (e.g., vulnerable taxa), bioindicators and field assessment of particular species, and species inventories in various subregions. There was very little, however, of the "traditional knowledge" component. By 1998, PBP staff had developed an approach to the biodiversity that included a combination of approaches coming from hierarchy theory, biogeography, landscape and ecosystems ecology, and systematics, complemented with ethnobiology and the study of traditional knowledge. In terms of biogeography, the region was divided into seventeen biogeographical districts, identified through the study of a set of taxa and their corresponding species (plants [thirty-five families], butterflies, reptiles, birds, and mammals), previous information from the Ecological Zoning Project (IGAC 2000 for last published version), and field research. The seventeen districts clearly divided into two large units, those ecosystems closer to the Andes (900–2,600 meters above sea level) and those of the lowlands (less than 900 meters). This classification followed a combination of ecological and historical (chiefly geological) data. A second level was the study and classification of landscape units with similar structures and habitats. Scientific research expeditions were geared toward mapping ecological landscapes in a more thorough manner than before (e.g., for mangrove ecosystems or for a number of subregions). Attention was paid to the development of bioindicators on the status of particular ecosystems, local knowledge and use of the same ecosystems, anthropic processes that destroy biodiversity, and the viability of conserving biodiversity given the communi-

ties present in the area. Finally, species diversity was studied with reference to particular spatial scales, habitats, and degrees of endemism. PBP staff developed an interesting "ethno-biogeographic approach" that integrated three modules: flora, fauna, and ethnic group; the ethnic component related the scientific information to the local names and uses of specific flora and fauna. Almost one hundred thousand items were collected and systematized in the six taxa under consideration. Species threatened with extinction, including forty bird and thirty-three mammal species, and those with particular promise for a variety of uses—from medicinal and food to construction, ornamental, and cosmetic—were inventoried. The inventories were often carried out with participation by locals, who were first trained in Occidental taxonomy. Separate centers for ethnobiological research were established for black and indigenous communities with PBP support. Not all PBP studies, however, involved local people. Many of the ethnobiological and ethnobotanical projects were carried out from a standard scientific perspective. See PBP's summary report on biodiversity research and knowledge (PBP 1999) and the summary conclusions on conservation priorities (PBP 1998). The most thorough earlier study of plant species in the Pacific is that of Forero and Gentry (1989).

By 1998, most of the PBP staff were convinced of the validity of conservation and sustainability strategies anchored in the local communities. In a few instances, the projects with communities were quite comprehensive (e.g., projects in the Bajo Calima and the Farrallones de Cali-Naya-Munchique transect, for instance); they included aspects such as ethnoecology of local ecosystems, recuperation of native seeds and healing practices, identification of bioindicators (e.g., wasps, ants, birds of prey), projects of commercialization of particularly promising species (e.g., ornamental plants and fruit tree species), diagnoses of the state of vulnerable hunting fauna (e.g., *danta* [*Tapirus bairdii*], *nutria* [*Lutra longicaudis*], turtles, iguanas, parrots, *caimán* [*Crocodylus acutus*], local felines, and deer, etc.), possible steps for their protection, and projects for the strengthening of local organizations (PBP 1999). If it could be said that PBP arrived at general conclusions from the *Conocer* area, these were (a) there are still vast geographical areas and particular topics about which scientific knowledge is scant; (b) deforestation is currently the main source of destruction of diversity; (c) the ongoing erosion of diversity and ecosystems may have devastating consequences for the region, if not the country and the world; some of these are still poorly understood (e.g., potential impact of deforestation of forest wetlands on hydrological cycles); and (d) while many of the subregions are still salvageable and landscape restoration is possible, they will largely depend on strategies based on the strengthening of traditional production systems, on the one hand, and less damaging development policies, on the other.

23 Some degree of consultation with local organizations had taken place in 1993 and 1994 in order to appoint the coordinator for *Movilizar* as well as the regional coordinators (these posts went to representatives of ethnoterritorial organizations, including Mary Lucía Hurtado's appointment as Movilizar coordina-

tor) and for the first revision of the PBP's Operational Plan, which resulted in a new draft in September 1994 (see Ministerio del Medio Ambiente/PBP 1994). The external evaluation team was made up of representatives from the Ministry of the Environment, the Department of National Planning, UNDP, and the Swiss government; see Hernandez, Bidoux, Cortés, and Tresierra (1995) for the evaluation text. On the nature and impact of this evaluation, see Wilshusen (2001); Equipo Ampliado (1997a, 1997b).

24 The Equipo Ampliado was made up of the following: PBP's Technical Team (composed of heads of Thematic Areas and Regional Coordinators) plus seven representatives of ethnoterritorial organizations, including the Proceso de Comunidades Negras, PCN (for Valle, Cauca, and Nariño provinces), the Mesa de Trabajo de Chocó (for black organizations of the Chocó), the Indigenous Organization of Antioquia (OIA), the Regional Organization Embera Wounaan (OREWA), and the Regional Eperara Siapidara Organization (ORIES). Also important was the inclusion of one black and one indigenous representative (Hernán Cortés and Manuel Casamá, from PCN and OREWA, respectively) in the project's Comité Directivo, PBP's highest-level steering committee. Finally, the new structure also included *técnicos comunitarios* (community technicians, or experts), from the same organizations. By 1997, the Equipo Ampliado included eleven representatives from organizations. See Equipo Ampliado (1997a).

25 This and the following quotations are from the Piangua Grande document (Procesos Organizativos de Comunidades Negras e Indígenas 1995), also reprinted in PBP's 1995–1997 Operational Plan (Ministerio del Medio Ambiente/DNP/GEF 1996).

26 Like *concertación*, *socialización* was a political concept that was in vogue in the encounter between institutions and organizations and among the organizations themselves; it pointed at the continuous need to discuss, debate, and devolve to communities and organizations the discussions taking place at the interior of the project as well as project results. It was a way to ensure that knowledge was widely negotiated, disseminated, and appropriated. Ethnoterritorial organizations often referred to themselves as *procesos organizativos*, rather than as organizations, in order to signal the ever-changing character of social movement dynamics.

27 The transformation was the result of a long struggle that was initially resisted by some at the highest level of PBP and its governing bodies. The external evaluation mission in 1995 was instrumental in this regard, to the extent that it was very favorable for the organizing processes, particular in the southern Pacific. In this negotiation the point of view that prevailed—promoted chiefly by PCN—was that the aim should be to negotiate the project as a whole, rather than trying to get more positions, budget, or the like, as some other black organizations proposed. So the concertación was intense at many levels. Only part of the Technical Team supported the idea of the Equipo Ampliado at first, since they still wanted a list of projects and activities, which the political-conceptual document of Piangua Grande did not make particularly easy. Some PBP staff also

attempted to bolster black "civic" organizations to erode the basis for the ethno-territorial organizations, arguing that the Equipo Ampliado was biased in favor of the latter. Once the Expanded Team became a reality, however, most in the PBP supported it. A great battle was won in the fall of 1995 when the basic approach was changed from a (still largely scientific) focus on ecosystems to the territorial perspective of what became the Territorial Action Programs (TAPs). This framework considered (a) strategic ecosystems, but also (b) the kinds of organizing processes in the same areas, and (c) the presence or absence of institutions. The TAP approach took as a point of departure the belief that it is local people who make conservation happen. In 1996, TAPs were largely defined by the regional coordinators and their respective technical teams. By the beginning of 1997, the Expanded Team was facing opposition again. Some black and indigenous representatives withdrew from it, for different reasons; for some of the indigenous organizations, PBP simply lost priority. By the time of their last meeting, the Expanded Team had lost a great deal of clout, in part as a result of a strategy to erode the support for PCN on the part of a number of actors, including black liberal politicians (interview with Libia Grueso, PBP regional director and PCN member, Buenaventura, July 20, 1998).

28 This assessment was shared by activists such as Libia Grueso (regional coordinator, Buenaventura) and Hernán Cortés (Equipo Ampliado) of PCN, and by PBP staff such as Enrique Sánchez (coordinador, Valorar), Juan Manuel Navarrete (area Movilizar), and Alfredo Vanín (regional coordinator, Tumaco). Interviews conducted in Bogotá, Tumaco, and Buenaventura, Summer 1998.

29 The 1995–97 Operational Plan spelled out clearly the new planning strategy as it came out of the Equipo Ampliado. It is instructive to quote the view of the process adopted in the new document:

> When making the balance of the concertación process, the organizations state clearly the widening of a political space that had never been obtained before. They also highlight learning to negotiate, even without clearly identified goals. . . . The Project took on a more territorial dimension, becoming richer with the contributions of the organization representatives, even amidst the difficulties and conflicts. In the last instance, it was about coming up with a new way to look at ourselves, of mutual enrichment, without hiding the particular needs as organizations or as a Project with a short life and a specific mandate. One learned through doing. The path was constructed as one walked and the result was the convergence and fusion of three visions [black, indigenous, PBP] that, even if differences remained, enabled us to fulfill our mandate. What prevailed in the end was the fundamental concept that had been floating around in the Equipo Ampliado from the outset, namely, the conjunction of Territory-Culture-Biodiversity, a defining triad in terms of the wishes of the organizations and the essential vehicle for the accomplishment of PBP's goal as a societal project in the context of the emergence of ethnicity and the reappropriation of the territory in the wake

of the new legislation for the black and indigenous communities. (Ministerio del Medio Ambiente/DNP/GEF 1996: 7)

To sum up, "the Project had developed into a movement" (7). Nevertheless, the project objectives still talked about a new strategy of development, one based on the mobilization and autonomy of ethnic groups as well as on scientific knowledge and sustainability. New dimensions were introduced, however, including the idea of using the CBD's precautionary principle as a guide for development. Operationally, ten Territorial Action Programs were identified on biological, geographical, cultural, and political grounds, and most interventions (subcontracts) were directed toward them. Projects were implemented by organizations, NGOs, and individual researchers. Examples of projects by organizations included community management of biodiversity in the Golfo de Tribugá (including characterization of species), carried out by ethnocultural organizations (OBAPO and ACABA); a booklet on traditional production practices of communities in the Valle river, in the same subregion; evaluation of production systems on the Cajambre river (ODINCA); sustainable use of mangroves on the Saija river (ASOPRODESA); evaluation of agro-ecosystems on Tumaco (Coagropacífico), and others concerning medicinal plants, organizational strengthening, gender factors associated with biodiversity (e.g., a study of women's contribution to biodiversity by Fundación Chilanga in the Timbiquí-Tapaje area and studies of medicinal plants and ethnobotanical knowledge by women in several areas). Academic studies in various regions, conducted by academic researchers in collaboration with local coinvestigators, concerned ethnobotany, hunting fauna, biological corridors, social and satellite cartography, evaluation of productive systems, agroecology, environmental assessments, potential productive schemes, and so forth. Although many of these projects retained a conventional profile, many directly supported organizational processes by ethnic and women's groups.

5. Identity

1 This debate is particularly prominent in Melanesian anthropology. See Battaglia, ed. (1995) for a thoughtful perspective.

2 It is well known that in his later writings Foucault sought to develop a positive theory of the subject through his concept of the techniques of the self. See, for instance, Bernauer and Rasmussen, eds. (1988).

3 In other words, the subject is neither entirely free nor totally subjected to power. On the one hand, subjectivity implies subjection to the norms and forms of power that make identities possible; on the other, this subjection does not entirely limit the subject, since the power that the subject exercises overflows—and is of a different character from—the regulatory power that make him or her possible, so that the subject can always perform the norm in a different manner. These concepts can be applied to activists' identities in ways that show how norms are

destabilized and culture mobilized for the production of politics. I thank Juliana Flórez Flórez (Department of Social Psychology, Universitat Autonoma de Barcelona) for her discussion of these aspects of identity, chiefly based in Butler's work (personal communication, and see Flórez 2007).

4 Grossberg suggests a useful conception of individuality in terms of three planes: the subject as the source of experience; the agent as the basis for action; and the self as the site of social identity; in short, subjectivity, agency, self/identity. While subjectivity involves attachment to places, as a way of belonging agency entails a distribution of acts in space (or territory, in the parlance of the Pacific). If places are historical points of belonging and identification, agency is what defines the particular form places may take through the empowerment of particular populations. Agency creates places as strategic possibilities; it "maps how much room there is to move and where and how movement is possible. . . . Agency as a human problem is defined by the articulation of subject positions and identities into specific places and spaces . . . on socially constructed territories. Agency is the empowerment enabled at particular sites" (1996: 102). The consequences of this conceptualization for rethinking identity—leading to what Grossberg calls "a politics of singularity" and the possibility of belonging without identity—are significant.

5 The study of ethnic identities has become an important subject among *pacificólogos* and *pacificólogas*. The most detailed study of the ethnicization of black identities is that of Restrepo (2001, 2002, 2007a); a number of other anthropologists (Peter Wade, Mauricio Pardo, Manuela Alvarez, William Villa, Ann-Marie Losonczy, Juana Camacho, Bettina Ng'weno, and Michel Agier), sociologists (Carlos Agudelo, Christian Gros, Nelly Rivas, Fernando Urrea), historians (Oscar Almario), political scientists (Kiran Asher), and geographers (Ulrich Oslender and Odile Hoffmann) have contributed to this lively debate. Some of these works will be introduced in the pages that follow. Several volumes that have appeared since the mid-1990s include contributions to the study of identities and social movements in the Pacific, including Escobar and Pedrosa, eds. (1996), Restrepo and del Valle, eds. (1996), Uribe and Restrepo, eds. (1997), Camacho and Restrepo, eds. (1999), Pardo, ed. (2001), Mosquera, Pardo, and Hoffmann, eds. (2002), Mosquera and Barcelos, eds. (2007), Restrepo and Rojas, eds. (2004), and a special issue of the *Journal of Latin American Anthropology* edited by Peter Wade (Wade 2002).

6 The *chigualo*,or ritual performed upon the death of an infant or young child, is one of the most striking examples of this collective memory linked to African origin or slavery. The ritual is amply described in the literature. See, for instance, Whitten (1986 [1974]), Quiroga (1994), Restrepo (2002). The erasure of explicit memories of slavery does not mean that such memories are completely absent. Joseph Jordan maintains that it is possible these memories escape the field of categories of researchers (personal communication).

7 Some authors have attempted a long periodization of black identity. Almario (2001) distinguishes four phases: an ascending phase (from the end of the

eighteenth century to the middle of the nineteenth), when the end of the colonial period opened room for some expressions of alterity; a plateau period (up to the first decade of the twentieth century), when an identity linked to the territory but within the context of the nation-state became the norm; a phase of decline (1930–70), characterized by a weakening of the previous identity as the nation-state gained greater presence; and a new ascending phase that became consolidated with the Constitution of 1991 in the context of multiculturalism and pluriethnicity.

8 This process is explained in various terms: as the introduction of a "neo-Africanity as modern historicity" (Losonczy 1999); the imposition of a new cultural and institutional discourse (Villa 2001); a novel order of alterization (Wade 1997; Restrepo 2002; Almario 2001); a new script for identity (Pardo and Alvarez 2001); and a deliberate transition from race to ethnicity (Gros 1997, 2000). These evaluations can be said to be valid anthropological descriptions even though, as we shall see, they may not address some relevant aspects in the making of identity, particularly if one thinks from the perspective of collective action.

9 Restrepo's doctoral dissertation includes a thorough treatment of these processes in the Pacífico nariñense (2008); see also Alvarez (2002) for an insightful examination of state practices at the local level in the context of the encounter with activists and communities in Tumaco.

10 Among the most interesting trends in the cultural analysis of the state are those that combine Marxism and poststructuralism. Those inspired by the Foucaultian notion of governmentality have become prominent. Some journals have devoted special attention to these issues (e.g., *Economy and Society*, *Thesis Eleven*, plus now-defunct journals like *Ideology and Consciousness*). Some pioneering works in this regard were Corrigan and Sawyer (1985), Burchell, Gordon, Miller, eds. (1991), and Smith (1987). Other important theoretical works in this vein include Cruikshank (1999), Scott (1998), and Mitchell (1991). Among the more ethnographic works commonly cited are Gupta (1995), Joseph and Nugent, eds. (1994), Coronil (1997), and Hansen and Stepputat, eds. (2001). Some of these works are susceptible to the same critique as the ethnographies of modernity. The analysis proceeds in three stages: first, the state is theorized as being dispersed, contradictory, fragmented, and so forth. In a second moment, this new conceptualization is shown ethnographically. From this, third, the state is reintroduced as having a deessentialized, albeit still somewhat encompassing, logic; or, in other versions, the state appears so diluted as to have almost no coherent effects at all. This hampers attempts at recomposing a theory of the state, which would be the logical step in order to be able to make different claims after the fragmenting move. In some cases, the state is seen even as creating its own outside and then controlling it! In some ways, despite extremely insightful efforts, the old desire for the single enemy or cause creeps back into the analysis. I am in no way suggesting I have an answer to this predicament; I am pointing at the impasse. I find here again a useful parallel with Gibson-Graham's (2006)

project of recomposing a theory of the economy that suspends the desire for the single enemy.

11 Bakhtin's concepts of dialogism and self-fashioning and Vygotsky's views on human development are central to the work of these authors. For Bakhtin, people author the world through the making of meaning, although not as free-wheeling agents but under conditions of heteroglossia, or the simultaneity of languages and worldviews with their associated values and presuppositions. Similarly, according to Vygotsky, all individual inner speech is social speech. In other words, there is an authoring self, but authoring takes place through languages that are collective. This is why identity is dialogical through and through. In short, "dialogism makes clear that what we call identities remain dependent upon social relations and material conditions" (Holland et al. 1998: 189).

12 Throughout the 1990s and the early part of the present decade, the PCN comprised some 120 organizations in the southern Pacific and the Cauca Valley (Norte del Cauca and Norte del Valle) and a few elsewhere (Atlantic Coast). As I indicate below, these organizations are grouped into regional *palenques*. Some of these organizations have been more prominent or active than others. Most have only a few participants at any given moment. There is an enduring core of fifteen to twenty activists that includes those on the National Coordinating Committee and some of the main leaders of each palenque. In an informal conversation I had with them in June 2002 in Cali, Libia Grueso and Julia Cogollo estimated the number of active PCN members at that point at about one hundred. The core group remained together and fully active and committed for the better part of the 1990s and continues so today. A very small number left the core group during this period. Relatively few of the activists have college degrees, and most PCN activists in local organizations have no formal education beyond high school.

13 See also Grueso, Rosero, and Escobar (1998) for an extended analysis of the earlier phases of the movement.

14 This and the succeeding quotations in this section are from the proceedings of the Puerto Tejada conference of September 1993.

15 This declaration of principles has remained the same throughout the years, with minor adjustments. However, the principles have been progressively spelled out and operationalized, and the emphasis has shifted among them throughout time. In recent years, the fifth principle has taken on salience, as the PCN adopts more actively a diasporic position in light of the need for internation solidarity to face the conflict. I take up below the application of these principles to black women's struggles.

16 From these early years, conventional black politics started to capitalize on the newly open political and public spaces—getting public jobs for their constituencies, bureaucratic representation, the use of public funds to ensure reelection and political survival, etc. This affected the meaning of the demands raised by the ethnoterritorial organizations, which nevertheless have remained an important interlocutor ever since. It was the ethnoterritorial sector of the movement that

trained the majority of activists capable of carrying out a critical dialogue with the state and of endowing some river communities with a toolkit for the defense of their rights within the framework of Ley 70 and Ley 121 of 1991 (Ley 121 ratified agreement 169 of the International Labor Organization concerning indigenous and tribal communities). Today, the same organizations are organizing nationally and internationally against displacement and the free trade agreement with the United States, among other causes. These accomplishments became main ingredients of the political practice of many grassroots organizations. For greater detail, see Grueso, Rosero, and Escobar 1998.

17 This structure worked relatively well until 2000, when some of the palenques started to fall apart as a result of the armed conflict. The palenque of Tumaco was decimated by paramilitary action, causing most members to leave the region. By 2007, the Palenque el Congal from Buenaventura remained quite active, and there were attempts to reconstitute some of the other organizations in the southern Pacific that were hurt by displacement.

18 Conversation with Julia Cogollo, of the Palenque el Congal (Buenaventura) at the Cali PCN Office, June 2002.

19 The trends briefly referred to here that seek to problematize the binary between structure and agency are largely based on theories of complexity, self-organization, and nonlinearity. I shall have more to say about these approaches in the next chapter. Similar attempts from the perspective of more established sociological and anthropological theories are those by Tarrow (1994), Melucci (1989), and Alvarez, Dagnino, and Escobar eds. (1998). An early attempt at applying a framework of self-organization to social movements is found in Escobar (1992).

20 Conversation with Julia Cogollo, Alfonso Cassiani, and Libia Grueso, Cali, June 2002.

21 In the mid- to late 1990s these included, besides the PCN, the Working Group of Chocó Organizations; Afro-Colombian Social Movement; Cimarrón National Movement; National Afro-Colombian Home; Afro-Colombian Social Alliance; Afro-South; Afro-Antioquia; Malcom; Cali Black Community Council; Vanguard 21 of May; *Raizales*; and Federation of Organizations of the Cauca Coast. Some of the avatars of black movement organizing in Colombia, with attention to urban movements in particular, are discussed by Wade (1995).

22 There is much more to say regarding the relationship between the personal and the collective. These points were debated in a conversation I had with Libia Grueso, Julia Cogollo, and Alfonso Cassiani at the PCN office in Cali in June 2002. The occasion was a preliminary discussion on the need to approach identity at a personal level. If in the early years of the movement the pressure was for the *compañeros and compañeras* to "assume" their ethnicity (more explicitly, their militancy as part of a collective ethnic movement), today there seems to be a need to open up some room for personal (albeit not individualist) discussions and specificities. The main awareness continues to be, first, that it is collectivities that succeed in becoming important social actors in the long run; second, that

even if it is very important to create space for the personal (*reivindicar lo personal*), it is also important to point out that personal decisions—and even the costs of personal choices to particular individuals—have political and historical dimensions. Here the activists come close to the concept of history-in-person from a political vantage point.

The account of activists' personal experience that follows has three main sources: first, my own acquaintance (and in a handful of cases close collaboration) with about a dozen of the main activist leaders for over twelve years; second, a daylong discussion and sharing of personal narratives with a group of fourteen PCN activists in Cali (June 2002); and responses to a short interview sent over the Internet to a similar number of activists. The main questions I posed in both exercises were: (1) Where were you born and where did you grow up through the age of fifteen? (2) What was your first encounter with difference, and what memory do you have of it (first awareness of difference and of being black)? (3) What was your first instance of reaction or anger against injustice? How did the desire to be an activist emerge, and what were your first steps as an activist? (4) Could you express succinctly your vision of the future for the Pacific? How would you articulate your utopia in this regard? I am very much aware of the inadequacy of this methodology for dealing with such a complex issue as the personal narratives of activists. As will become clear, I am trying to make a few general points. An excellent recent doctoral dissertation based on the life histories of PCN activists deals with these issues in depth (Flórez 2007).

23 The last two quotations come from an interview-article by Mary Lucía Hurtado, who belonged to PCN in the early 1990s, and published in Escobar and Pedrosa (1996). See Hurtado (1996: 332–33). Hurtado's explanation is clearly influenced by black consciousness writings, especially Fanon.

24 This particular account is based on six hours of taped interviews conducted in 1994. Part of the interview was written up as a narrative and can be found in Escobar and Pedrosa, eds. (1996, 265–82; "Relato de Mercedes Balanta," a pseudonym). This courageous and creative activist remained an important actor in local activism until 2003, when she was forced to move to Bogotá, where she continues to work as a black cultural activist.

25 Response to interview with Libia Grueso, October 2002.

26 Microworlds and microidentities, however, do not add up to a centralized, unitary self, but result in a succession of shifting patterns—a selfless or "virtual" self. Contemporary cognitive science shows that there is no central system underlying the self; on the contrary, identity is an emergent property of a complex, distributed process mediated by social interactions and the construction of narratives and microworlds in everyday life (Varela 1999). "*The cognitive self is its own implementation: its history and its action are of one piece*," concludes Varela (1999: 54). There is much more to say about this synthetic effort, but I leave the discussion here for the sake of the overall argument. See also Varela, Thomson, and Rosch (1991). Martín Alcoff also resorts to hermeneutics and phenomenology in her

theory of identity (2006), and Gibson-Graham (2006) explicitly draw on Varela and on Spinosa, Flóres, and Dreyfus, in their exploration of economic activism.

27 Following these authors, it would be possible to characterize a similar ethics out of Catholic traditions (certainly among liberation theology practitioners, but also many other activists, including some of those in PCN—I would include myself here). One finds various expressions of this ethic—more or less explicitly stated—in practicing and nonpracticing Catholic intellectuals such as Luce Irigaray, Gianni Vattimo, Paimon Panikkar, Donna Haraway, and, of course, liberation theologists such as Leonardo Boff.

28 This perspective is presented as a critique of Cartesian modernity and advocates for the historical viability of an alternative modernity based on connections to place and relatively stable, albeit deeply historical, identities. Again, it is impossible to do justice to this complex argument in these few lines. The argument is based on and inspired largely by Heidegger's philosophy and the phenomenological biology of Maturana and Varela (discussed in chapter 6).

29 See Camacho (2004) for a comprehensive review of the literature. In anthropology, studies of black women started in the 1980s and focused largely on women's role in family and extended kinship (e.g., Friedemann and Espinosa 1995). The 1990s brought a wave of interest in women, particularly within women and development (WID) and gender and development (GAD) orientations. While this literature has been criticized as developmentalist and for failing to seriously address the issue of power between women and men (e.g., Lozano 1996, 2005; Escobar 1995), it did much to put the question of black women on the map. In this literature, gender is often equated with women. Studies of black male gender identities are scarce; the anthropologist Mara Viveros (2002a, 2002b; Viveros, Rivera, and Rodríguez, eds. 2006) has pioneered the analysis of masculinity in Colombia with some reference to black identities. Another important area of focus, also in the 1990s, was gender, biodiversity, and local knowledge (see chapter 3). Women's organizations have received some attention, particularly the Black Women's Network of the Pacific (e.g., Rojas 1996) and especially successful organizations such as Guapi's *Matanga y Guasá* (Asher 2004), and recently the overall intersection of gender and ethnicity has been examined (Asher 1998; Flórez Flórez 2004, 2007). The three most debated features related to gender dynamics of river communities in the Pacific have been the matrifocality of kinship, explained in terms of the dynamics of slavery (women remaining close to home, men traveling widely in the territory); polygyny (also explained in similar terms); and the gender complementary of tasks. There is a lot of active debate on the status of these concepts and their corresponding practices. See Camacho's review for an extended discussion.

30 E-mail interview with Libia Grueso, September 22, 2002.

31 Julia Cogollo in the daylong discussion and sharing of personal narratives I had with a group of fourteen PCN activists in Cali in June 2002. See note 22.

32 This is why, for instance, mothers "do not make babies" (311) in the sense that, as they do in commodity economies, women do not see themselves as "individual mothers" giving birth to a distinct entity called a baby; on the contrary, the whole process is entirely relational, according to which children are the outcome of relations involving multiple others, including clans and ancestors. Giving birth is not the act of an autonomous free individual, but one of yielding "what is already anticipated as a social object" (317). Similarly, in regard to men's imputed domination of society, as observed by an outsider: "It is not men's collective activity that 'creates society' or 'makes culture.' . . . The Melanesian material at least does not present us with an image of men promoting male values that also become the values of society at large, and thereby simply using female values in counter-point to their endeavors . . . however men are depicted it cannot be as authors of such an entity [society]" (318–19). In addition, "gender is not construed as a role 'imposed' on [an] individual by 'culture'" (324).

33 "The difference between man and woman is a difference in being, a difference between two worlds, which resists any quantitative assessment. Man and woman are irreducible one to the other, and the difference between them cannot be evaluated, calculated, and appropriated. It always remains insurmountable. . . . [The] dialogical relation between man and woman, men and women remains to be invented, almost entirely from scratch" (2000: 84, 85; see also 2001: 105). Conversely, "being able to identify with the other seems to me an ambiguous cultural improvement" (2000: 91). Again, let me clarify that Irigaray sees in equality before the law an important goal; however, it can only be a start. This strong position leads Irigaray to propose farsighted ideas. Intersubjectivity based on sexual difference can provide a good model for democracy, one based on respect for difference and on love rather than property. It would usher in not only a new way of relating between women and men—where one is not consumed by the other, as in the *I love to you* imaginary (1996: 109ff)—but a new politics of social relations as a whole, including interculturality, interreligiosity, and a novel practice of universals. Irigaray has been repeatedly accused of essentializing sexual difference. As Grosz (1989: 110–19) argues with conviction, however, this is a misreading of her position, according to which the body is clearly shaped by history and culture (hence her reference to a morphology and not an anatomy of the body, which entails a complex conception of the body grounded in experience, pleasure, and language). I believe Irigaray's arguments are valid—or have consequences—for thinking about relations among women and, of course, among men.

34 As we already know, the modernity/coloniality/decoloniality research has not dealt adequately with gender (Escobar 2003). I am suggesting an extension of coloniality (of being, knowledge, and power) through gender in two ways. First, coloniality needs to address what Irigaray might consider the "original

coloniality" of modernity, namely, the obliteration of gender difference; this operates through phallogocentric discourse, which also implies the mastery over nature through violence and the subjection of life to the individual (the male subject of philosophy) and the commodity. Second, the coloniality of racially marked women, such as the case of black women of the Pacific. Here, we have the vantage point of a double exteriority that places women in a particular position in terms of articulating both coloniality and the decolonizing act. For Irigaray, it is truth telling and language themselves that are at stake: "The device for thinking and transmitting truth must be reinvented and given a new foundation in sexual difference. In the absence of such a foundation, the word remains a form of power which is unreal and deceptive, a means of seduction which forgets reality and respect for the other" (2001: 105). Add "cultural" to "sexual" difference and one gets, I think, a more complete concept of coloniality. Reworking coloniality would thus involve both Irigaray's aim of representing women *otherwise* (than in phallogocentric terms; see Grosz 1989: 110) and Dussel's and Mignolo's thinking otherwise in relation to culturally Eurocentric modernity. The ability of women to speak as women—that is, not as "Universal Man," as Haraway (1989) would put it—and the ability of others to speak as others—not as Universal Occidental/White Man—necessitate a process of constructing structures of knowledge capable of authenticating different discourses. This process involves plural bodies, languages, knowledges, natures, truths.

35 The discussion about values was telling, and it was indicative of the lack of the same values in the conventional political culture of the country. These values included responsibility, solidarity, honesty, tolerance, understanding, transparency, respect, humility, and clarity and commitment in appropriating the political project of the movement.

36 Since the late 1990s I have been discussing with PCN activists the possibility of a collective volume on the PCN's political, ecological, and cultural thought. The project is currently under implementation within the scope of the *Otros Saberes* projects by the Latin American Studies Association, mentioned in the introduction.

6. Networks

1 The three representatives were Leyla Arroyo, Carlos Rosero, and Dionisio Miranda. SWISSAID was by then funding some projects in the Pacific, among them the publication of the journal *Esteros*, which became an activist and NGO forum for the discussion of things Pacific. The contacts were particularly fruitful with NGOs in Spain, Switzerland, and Germany; solidarity groups in Holland and Italy; and with some UN groups in Geneva. There is a report of the tour (PCN 1995a). This first tour was important in bringing to PCN's attention several key topics, such as the international status of the black communities in the United Nations system and the need to maintain a steady flow of information with European organizations, which underscored the importance of electronic mail. The

tour was also instrumental in enabling contacts that would result in funding for alternative development projects, particularly with Danish groups. These projects enabled an intense political and territorial organizing by PCN in two localities (see the conclusion).

2 IGGRI had included over the years a number of antidevelopment thinkers. By the time of the Hanasaari meeting, it included Smitu Kothari, Ponna Wignaraja from Sri Lanka, Luis Lopezllera and Gustavo Esteva from Mexico, Orlando Fals Borda from Colombia, Marja-Liisa Swantz, Thomas Wallgren, and Hillkka Pietala from Finland, Majid Rahnema (Iran/France), Wangaeo Surichai from Thailand, Yashpal Tandon from Zimbabwe, Judith Bizot and Franck Amalric from France, Claire Slatter from DAWN in Fiji, Siddhartha and D. L. Shet from India, Manfred Max Neff from Chile, Muto Ichiyo from Japan, Thierno Kane from Senegal, Ward Morehouse from the United States, and myself. IGGRI was firmly anchored in the alternative development movement in the global south; its emergence had been facilitated by meetings sponsored by the Society for International Development in Rome (SID) over the 1970s and 1980s. For a brief history of IGGRI, see the Hanasaari report (IGGRI 1999).

3 A year before, Fals Borda had convened a much larger gathering in Cartagena, Colombia, to celebrate the twentieth anniversary of the global launching of the PAR movement in 1977 in the same city. Attended by over one thousand people largely from Latin America but also other continents, the gathering brought together grassroots struggles against globalization. These more third world–oriented gatherings tend to go unreported in northern accounts of the global justice movement. I attended the Cartagena meeting and along with four PCN activists led a daylong workshop on the Pacific.

4 I will not deal here with the PGA but see the detailed analysis in the doctoral dissertations by Juris (2004) and King (2006); see also Routledge (2003). The people making up the PGA group at Hanasaari included a Maori, a well-known Indian activist from the Karnataka State Farmers Association (KRRS), a Salvadoran labor organizer, a Uruguayan environmentalist who was to become the main person of ETC Group in Mexico (Silvia Ribeiro), plus four or five Europeans who had been among the initiators of the PGA in Geneva. PCN activists played a salient role in PGA for a number of years, including through their participation in the organizing meetings in Geneva of January 1998, the Inter-Continental Caravan in Europe of June 1999, the PGA conferences in Bangalore (1999) and Cochabamba (2001), other events in various parts of the world in 1998–2002 (including preparation for and participation in demonstrations in Prague and Genoa), and as a member of the Committee of Conveners since 1998 and Latin American convener in the period 1999–2002. Despite the input from "the South," analyses of the PGA have shown the persistence of practices ascribing greater power to the European activists, what Routledge has called "the contested social relations within PGA" (2003: 343; see also Juris's chapter on the PGA [2004]). On the PGA Inter-continental Caravan, see Featherstone (2003).

5 I am combining notions from complexity, autopoiesis, and assemblage theory to be discussed later in the chapter. To anticipate a bit, a basic insight of these trends is the need to account for structure as the emergent property of a dynamic system, including (a) the bringing together of a set of heterogeneous elements in terms of functional complementarities; (b) the presence of catalysts; (c) a stable pattern of behavior, endogenously generated, that results from the interlocked heterogeneities (as, for example, in the case of self-organizing markets). Second, a characterization of meshworks as entities that (a) are self-organizing and grow in unplanned directions (such as Deleuze's and Guattari's rhizomes, 1987); (b) are made up of diverse human and nonhuman elements; (c) exist hybridized with other meshworks and hierarchies; (d) articulate together heterogeneous elements without imposing uniformity; (e) are determined by the degree of connectivity that enables them to become self-sustaining. There are other aspects of meshwork dynamics to be taken into account, such as (a) place-based strategies along with strategies of interweaving with other networks; (b) incorporation of new nodes with conservation of autopoiesis; (c) evolution by drift triggered by the environment (structural coupling between meshwork and environment with preservation of internal organization); (d) processes of destratification and reterritorialization of places, territories, regions, identities. See Escobar (2000) for a more detailed account of some of these concepts.

6 I characterized the knowledge production of PCN as having the following features (Escobar 2000): It is conjunctural without being punctual—it is cumulative and progressively refined; it is developed "on the run"; this means that there is little time to create the knowledge-production infrastructure that could make it more lasting; it is pragmatic without being just utilitarian or functional to the struggle; it is geared toward the articulation of demands but always with a sense of the long-term goal, namely, the defense of the historical life project of the communities. Other features are as follows: It is recursive to the extent that the same themes (territory, identity, cultural difference) are worked on in different ways at many scales—that is, knowledges have a fractal recursivity (self-similarity); it is "epistemologically dirty"—it grabs what it can and from whatever sources are at hand; it cares little or nothing about disciplines and proceeds more through bricolage than through systematic theory building, although theory is important to the process; it is profoundly interdisciplinary, although particularly disciplinary forms of knowledge have been important (anthropology, geography, ecology, gender studies).

7 Based primarily on interviews with Libia Grueso and Carlos Rosero in 2000 and 2001 and on conversations with other PCN participants in these events, including Hernán Cortés, Yellen Aguilar, and Leyla Arroyo; it is also based on my collaboration with the movement over the years on particular projects, my participation at some of the events mentioned (including partial overlap at events in 1995 and 1998 in Europe and many of the academic events), and assiduous

collaboration on a variety of funding proposals, international campaigns, and so forth.

8 Interview with Libia Grueso, Chapel Hill, October 2001.

9 On DESC, see PCN (2006). An initial statement on human rights and black communities is found in PCN (1995b). On the early debates on self-definition, see Grueso (1996), PCN (1999); this latter project, written by PCN, included Central American black communities such as the Garifuna. See also the *Declaración conjunta de las organizaciones de descendientes de africanos en el continente Americano*, issued by black organizations from the five Andean countries on April 28, 2000, as input to the Durban Conference. PCN digital archives contain materials on many of the meetings (e.g., binational meetings) but these are not systematized.

10 See, for instance, the "Cali Consensus" drafted by a coalition of black organizations for the Second Afro-Colombian Communal Council, led by President Uribe in Cali on June 7, 2007. Several members of the U.S. Congress Black Caucus participated in this event. For more information on Washington, D.C., actions, see the web pages of PCN and WOLA.

11 At least 60 percent of the speakers and 80 percent of the 120-plus people attending were Afro-Colombian intellectuals, activists, and academics. The event was organized by Claudia Mosquera, a professor of social work and researcher at the Grupo de Estudios Afrocolombianos, Universidad Nacional, Bogotá. I coordinated one of the sessions, "Reparations in the context of the armed conflict." See Mosquera and Barcelos, eds. (2007).

12 There was an acute sense that concepts that circulate globally need to be transformed in light of national conditions. To this extent, the debate on Afro-reparations resulted in convergences around a number of points, including (a) it has to link memory and justice, past and present; that is, while it should recover memories of slavery it also needs to contribute to redress racism and discrimination in the present; (b) it has to involve a sense of *justicia reparativa y transformadora* (reparative and transformative justice); (c) it needs to attend to the diversity of experiences, including issues of gender, age, and region; some participants suggested that it should embrace the project of healing the wounds caused by sexism and take the *terrritorio-cuerpo* as a focus of struggle; (d) it should aim at empowering subaltern subjects, going beyond the reconstruction of a collective memory and the decolonization of hierarchical geographies of race; (e) it should embrace equally action, reflection, and commitment. Conversely, some local activists' debates presaged international ones; what was called the principle of compensation—the first of six principles for development plans for the Pacific—anticipated international discussions on reparations. Enunciated in 1994 by PCN and other Afro-Colombian organizations, this principle identified the historical imbalance between the material and cultural contributions of the black communities to nation building and the meager retribution by the nation in terms of investment and conditions for development in areas where black communities predominate. The entire discussion on reparations connects with this book's

framework on economic, ecological, and cultural distribution conflicts—indeed, reparations involve the three dimensions of conflict. In the case of ecology, one could posit the existence of "ecological memories" linked to resistance and the study of "reparation ecologies" as a parallel project to the restoration ecology of biologists. The notion of ecological debt could be seen as an element in the framework of reparation ecologies.

13 Castells's characterization of networks is suggestive: "A network is a set of interconnected nodes. . . . Networks are open structures, able to expand without limits, integrating new nodes as long as they are able to communicate within the network, namely as long as they share the same communication codes. . . . Networks are appropriate instruments for a capitalist economy based on innovation, globalization, and decentralized concentration; for work, workers, and firms based on flexibility, and adaptability; for a culture of endless deconstruction and reconstruction. . . . Switches connecting the network . . . are the privileged instruments of power. . . . Since networks are multiple, the interoperating codes and switches between networks become the fundamental sources in shaping, guiding, and misguiding societies" (1996: 469–71). The consequences of this conceptualization assume a dystopian dimension: "Dominant functions are organized in networks pertaining to the space of flows that links them up around the world, while fragmenting subordinate functions, and people, in the multiple space of places, made of locales increasingly segregated and disconnected from each other. . . . Not that people, locales, or activities disappear. But their structural meaning does, subsumed in the unseen logic of the meta-network where value is produced, cultural codes are created, and power is decided" (476–77; 1997).

14 A well-known aspect of ANT is that this process greatly depends on materials that are not only human. For Latour (1993), moderns have been able to construct powerful networks to the extent that they have been able to enlist nonhuman elements—technologies, substances, scientific knowledge, etc.—in the creation of longer and more connected networks. A question of interest to social movements arises from this exposition: How does one compare networks? ANT's conclusion in this respect, it seems to me, is epistemologically weak: that one can talk only about longer and more powerful networks (invariably, those of the moderns) in terms of the methods and materials they utilize to generate themselves. Strathern (1996a) presents a corrective to this view in that "pre-moderns" (Latour's term) have a greater capacity for constructing (longer) hybrids and networks than one might suspect, by rolling into them equally unsuspected entities or materials—from clans and animals to the ancestors; premoderns might also be more adept at "cutting the network" than moderns, whose greed and sense of property (e.g., intellectual property) might compel them to arrive at premature closure (e.g., a patent, which forecloses the inventiveness of the previously existing network which produced it as a possibility). Castells's networks, like Latour's actor-networks, have a tendency for limitless expansion; this all-embracing logic is in keeping with a particular style of theorizing.

15 Cyberspace is seen by some as embodying a new model of life and world making; it is seen by enthusiasts as a space of collective intelligence or a "noosphere" (a sphere of collective thought, after Teilhard de Chardin) with the potential to constitute an internetworked society of intelligent communities centered on the democratic production of culture and subjectivity. Pierre Lévy (e.g., 1997) has most powerfully articulated this thesis in recent years. The recent work by the liberation theologian Leonardo Boff on *religación* (2000)—a "reconnecting" of humans with nature, each other, the earth, the cosmos, God—could also be interpreted in this light (he appeals explicitly to complexity). Discussions of the impact of ICTs on daily life abound, including those examining cybercultures (e.g., Harcourt, ed. 1999; Bell and Kennedy, eds. 2000; Burbano and Barragán, eds. 2002). As Terranova (2004: 75–97) warns, there are clear interfaces between capital and the digital economy. Cybercultural politics can thus be most effective if it fulfills two conditions: awareness of the dominant worlds that are being created by the same technologies on which the progressive networks rely; and an ongoing tacking between cyberpolitics and place-based politics, or political activism in the locations where networkers or netweavers live. This is the politics that some of today's movements are attempting to develop in creatively combining strategies for action at various scales (King 2006; Harcourt, ed. 1999; Escobar 2000).

16 As an advocate of this research said in a comprehensive introduction to the subject, "Networks will dominate the new century to a much greater degree that most people are yet ready to acknowledge" (Barabási 2002: 7, 222). The scientists' most striking claim is that there are some basic laws governing all networks. For instance, networks are highly interconnected, so that huge networks constitute "small worlds" in the sense that all elements in the network, owing to the presence of hubs and connectors, are only a few links away from all others. Networks follow certain rules that scientists refer to as "power laws" (e.g., Barabási 2002; King 2006; Duncan 2003).

17 Nobody has diagnosed the dangers of the worlds enabled by new information and communication technologies like Paul Virilio (see, e.g., 1997, 1999).

18 This section is largely based on Internet and library research over the period 1992–2000. See chapter 3, n. 19, for sources. I thank Henk Hobbelink and Nelson Alvarez of GRAIN in Barcelona for their time when I was there in 1999. Mary King's research with GRAIN in spring 1999 has also been important to my thinking in this section (e.g., King 2000 and 2006). The biodiversity area is changing so rapidly that is impossible for a single researcher to follow its multiple developments.

19 The SCOPE (Scientific Committee on Problems of the Environment) Program on Ecosystem Functioning of Biodiversity and the United Nations Environment Program's Global Biodiversity Assessment Program follow this approach. See SCOPE's technical volumes, particularly H. A. Mooney et al. (1995); and the useful review of the project in Baskin (1997). On the science of biodiversity,

see also the *Diversitas* newsletter launched in July 2002 by SCOPES and UNESCO among others.

20 Article 2 of the Convention on Biological Diversity provides the following definition: "'*Biological diversity*' means the variability among living organisms from all sources including, *inter alia*, terrestrial, marine and other aquatic ecosystems and the ecological complexes of which they are part; this includes diversity within species, between species and of ecosystems." This definition was further refined by the World Resources Institute (WRI) as comprising genetic diversity, the variation between individuals and populations within a species, and species and ecosystems diversity, to which some added functional diversity (WRI 1994: 147).

21 King's dissertation is based on field research and interviews conducted at a number of sites in the biodiversity network, including GRAIN, the CBD headquarters in Montreal, and CBD-related meetings elsewhere, among other sites.

22 For a more detailed presentation of this section, see Escobar (1998). There are differences and contestations within this dominant perspective, although they do not significantly alter the overall framework. At IUCN, for instance, a number of groups have pushed for benefit-sharing schemes, including comanagement and self-management, agroecology, and a range of cultural-political approaches for conservation; see, e.g., Borrini-Feyerabend, Pimbert, Farvar, Kothari, and Renard (2004), and the ongoing publications of IUCN's Commission on Environmental, Economic and Social Policy, particularly *Policy Matters*. I thank Grazia Borrini-Feyerabend and María Fernanda Espinosa of IUCN for discussion and materials on these approaches. Community-based resource management schemes are usefully discussed in Brosius, Tsing, and Zerner (1998).

23 This discussion reflects the state of the debate up to the late 1990s. Since then, there have been tremendously important developments in the field of intellectual property in relation to biodiversity and more broadly that would be impossible to summarize here; these challenge the mainstream framework with its resulting "harmonization model" of all intellectual property rights systems under a single cultural logic. These include conceptualizations and practices of creative commons, collective rights, communal frames, access to knowledge, ethical and spiritual guidelines, and so forth. Since 2000, the work of the Intergovernmental Committee (IGC) of the World Intellectual Property Organization (WIPO) on genetic resources, traditional knowledge, and folklore has been important in expanding the debate and possible outcomes in the field of genetic resources and intellectual property, including the protection of traditional knowledge. The debates became newly active at COP 7 in Kuala Lampur (2004) around traditional knowledge (article 8j of the CBD), the role of the WTO (TRIPS), and a push for greater participation by indigenous representatives. On these issues, see the papers sponsored by IUCN, including Ruiz (2004), de la Cruz (2004), Vivas-Eugui, Espinosa, and Winkler (2004). See also the recent volume by Gosh (2005), which addresses intellectual property issues in the natural and digital worlds; and

Hayden (2004) and Hirsch and Strathern, eds. (2004), which feature anthropological contributions to property debates.

24 There are at least four internationally known organizations in this area: GRAIN and RAFI/ETC; Vandana Shiva's Research Foundation for Science, Technology, and Natural Resource Policy in India; and the Malaysian-based Third World Network; and GRAIN. They are joined by a number of NGOs in Latin America, Asia, and Africa and a few others in North America and Europe. There are progressive NGOs in almost every country in Latin America with connections to this perspective (such as Acción Ecológica in Ecuador and ILSA and Grupo Semillas in Colombia). Together, these NGOs enact a loose assemblage. These groups are adamantly opposed to intellectual property rights as a mechanism for the protection of local knowledge and resources. On this perspective, see especially the works of Vandana Shiva (e.g., 1993, 1997); the Third World Network magazine, *Resurgence*; ETC's communiqués and Occasional Paper series; and GRAIN's newsletter, *Seedlings*. Two other groups that often issue sophisticated and timely material and analysis on related issues are Focus on the Global South, in Bangkok, and the Corner House in the United Kingdom. These organizations often coordinate action with social movement organizations, such as Vía Campesina, or anti-GMO campaigns. ETC has been particularly present at World Social Forum meetings in conjunction with social movement organizations working on similar issues.

25 On the differences between centralized, decentralized, and distributed networks, see Barabási (2002: 145).

26 The category flat, as used in this section, is completely different from the concept of flat files in mathematics and from the meaning given to it by Thomas Friedman in his book *The World Is Flat* (2005). I see the movement toward flat alternatives as taking place in a number of very different fields, including systems theory and informatics of the 1940s and 1950s; network theories in the physical, natural, and social sciences; some strands of thought in geography, cognitive science, and informatics/computing; complexity theories in biology; and Deleuze's and Guattari's neorealism. These are not altogether new, of course. I also see Foucault's work within this frame (e.g., Foucault's theory of the archaeology of knowledge may be seen as a theory of autopoiesis and self-organization of knowledge; his concept of "eventalization" resonates with recent proposals in assemblage theory; and his entire conception of power anticipates development in actor-network theory).

27 Deleuze deploys a difficult mathematical language that de Landa explains, particularly the concepts of multiplicity as a form of organization "which has no need whatsoever of unity in order to form a system" (2002: 13); manifolds, as the space of possible states of a system, regulated by the system's degrees of freedom; dynamical processes, in terms of trajectories in a space, recurrent behavior, and processes of differentiation; singularities that act as attractors around which a number of trajectories converge within the same sphere of influence (basin of attraction), possibly leading to a steady state (structural stability); and so forth.

De Landa summons complexity concepts to explain the Deleuzian world. Multiplicities are concrete universals; they are divergent and cannot be thought of in terms of three-dimensional Euclidean space but of nth dimensional (nonmetric) topological spaces, although the former is produced through differentiations in the latter. This happens through physical processes of differentiation of an undifferentiated continuous intensive space into extensive structures (i.e., discontinuous, divisible structures with metric properties) through processes that include phase transitions, symmetry breaking, etc. Multiplicities are thus immanent in material processes (see also the useful Primer on Complexity at the end of the volume by Haila and Dykes, eds. 2006).

Concrete realizations of a multiplicity are more accurately actualizations of a larger field of virtuality. This approach requires understanding the individuation of possible histories. This is complicated because the actualization of vector fields is rarely a linear process; on the contrary, it is shaped by nonlinear dynamics; trajectories may emerge out of an attractor even by accident or external shocks; they are always the result of a contingent history. Alternatives that are pursued at a given point (especially at bifurcation) may depend on chance fluctuations in the environment (a point underscored by complexity theorists, e.g., Prigogine and Nicolis 1989; Solé and Goodwin 2000), in a conjunction of chance and necessity. What matters in the investigation is to remain close to the specific individuation going on—that is, to the formation of spatiotemporal structures, boundaries, etc.

28 Hence "one of the tasks of a philosopher attempting to create a theory of virtuality is to locate those areas of the world where the virtual is still expressed, and *use the unactualized tendencies and capacities one discovers there as sources of insight into the nature of virtual multiplicities*" (2002: 67, as I suggested in the case of alternatives to modernity; emphasis added). This task could also be done by working backward from the concrete actualizations toward the virtual and by considering the population of multiplicities that exist in the plane of consistency. Alternative possibilities need to be shown as "historical results of actual causes possessing no causal power of their own" (75). Information can play a key role in these processes, for instance, in systems or networks poised at the edge of a threshold (see also, e.g., Kauffman 1995).

29 Some other aspects of assemblage theory are as follows: First, assemblage theory emphasizes the exteriority of relations. Second, it postulates two dimensions of analysis: (a) the role played by the components, from the purely material to the purely expressive; (b) processes of territorialization and deterritorialization that either stabilize or destabilize the identity of an assemblage. Third, it introduces several other mechanisms, particularly those of coding and decoding (by genes and language). Assemblage theory also seeks to account for the multiscaled character of social reality and provides adjustments to this end. First, it recognizes the need to explain the historical production of the assemblage but without emphasizing only the moment of birth (e.g., as in the origin of the col-

lectivity PCN) or original emergence of its identity at the expense of processes that maintain this identity through time. Second, assemblages (e.g., organizations) come into being in a world already populated by other assemblages; given a population of assemblages at one scale, they can generate larger-scale assemblages using members of existing population as components. Third, there are complex entities that cannot be treated as individuals. Here de Landa introduces other (nonmetric, topological) concepts from Deleuze, particularly those of possibility spaces or phase spaces (from physical chemistry) and attractors (universal singularities or topological invariants) that might be shared by many systems; and the concept of diagram as that which structures the space of possibilities of a particular assemblage.

Finally, there is the question of how assemblages operate at larger time scales—they often endure longer than their components and change at a slower rate. Does it take longer to effect change in organizations than in people, for example? At this level, one should identify (a) collective unintended consequences—slow, cumulative processes—that result from repeated interactions; (b) products of deliberate planning. The first item is more common in long-term historical change. In the second case, enduring change happens as a result of mobilization of internal resources (from material resources to solidarity). In general, the larger the social entity targeted for change, the larger the amount of resources that must be mobilized. *This implies that spatial scale does have temporal consequences since the necessary means may have to be accumulated over time. Said differently, the larger the spatial scale of the change desired, the more extensive the alliances among those involved need to be, and the more enduring their commitment to change.* There is no simple correlation, however, between larger spatial extension and long temporal duration. In the case of assemblages that do not have a well-defined identity, such as dispersed, low-density networks, this dynamic is a strength (the strength of weak links) and a weakness at the same time: On the one hand, "low density networks, with more numerous weak links, are for this reason capable of providing their component members with novel information about fleeting opportunities. On the other hand, dispersed networks are less capable of supplying other resources, like trust in a crisis, the resources that define the strength of strong links. They are also less capable of providing constraints, such as enforcement of local norms. The resulting low degree of solidarity, if not compensated for in other ways, implies that as a whole, dispersed communities are harder to mobilize politically and less likely to act as causal agents in their interaction with other communities" (2006: 35). De Landa applies this theory systematically to the worlds of persons, organizations, and governments.

30 Churchman (1968) and Laszlo (1972) were popular introductions to the systems approach. See also Emery, ed. (1969), Churchman (1971). On general systems theory, see Bertalanffy (1975), Laszlo, ed. (1972). Heinz von Foerster was among the first to formalize theories of self-organization (e.g., see the volume of papers assembled by Varela, von Foerster 1981); Jantsch (1980; Jantsch

and Waddington, eds. 1976) attempted a useful integration of theories of self-organization coming from physical chemistry (e.g., Prigogine and Stengers, 1984), cognitive science (Maturana and Varela 1980, 1987), mathematics, evolutionary biology, and philosophy. Margaret Mead and Gregory Bateson were important participants in these debates, always relating cybernetics to cultural mechanisms and back (see, for instance, Mead's Introduction to von Foerster et al., eds. 1968, entitled "The Cybernetics of Cybernetics").

31 To my knowledge, there is no integrated history of these movements, particularly from the perspective of how they have influenced today's trends. Two useful books that focus on the biographies and contributions of the main architects of these movements are Heims (1991), for the cybernetics group gathered around the famous Macy Conferences (1946–53), including Margaret Mead and Gregory Bateson, Norbert Wiener, John von Neumann and Warren McCulloch; and Hammond (2003), which features the main figures in general systems research, including Bertalanffy, Rappaport, Boulding, and others, paying attention to the interrelated contexts of cybernetics, systems, engineering, management, ecology, and social theory. Toward the end of his book, Heims discusses Heinz von Foerster's "second cybernetics" (or second-order cybernetics) group, which, from the 1960s on, worked on the conceptual (rather than technical) development of some of the insights of systems and cybernetics, particularly self-organization; Heims locates in this group researchers in France, Belgium, and England besides the United States, and the biologists Humberto Maturana and Francisco Varela in Chile (Heims 1991: 283–84), to which could be added the joint work of Terry Winograd and the Chilean Fernando Flores (1986). Contemporary complexity theories cannot be reduced to these earlier trends, yet they cannot be understood without them. Another important element in the development of complexity theories was the mathematical approaches in France of René Thom (catastrophe theory, 1975) and George Mandelbrot (fractals). Thom's mathematically derived theory of morphogenesis influenced biologists like Conrad Waddington and philosophers like Deleuze and Guattari. The development of dynamical systems and nonequilibrium thermodynamics are other factors commonly cited in this regard. Some additional references are found in Escobar (1994).

32 Maturana's and Varela's work can be seen as among the most profound of this entire group of thinkers in terms of its epistemological, biological, technological, and social implications. First, they bring back into the picture the centrality of body and experience; this has significant implications for science (particularly cognitive science), AI, epistemology (beyond representation and constructivism), and the ethics of society. In terms of AI, a main consequence is the need to abandon the idea that communication is the transfer of information; it is instead the modeling of a shared world through language. This implies bracketing the immediately utilitarian drive of much AI engineering: "If we emphasize the historical process that leads to emergent regularities without final fixed restrictions, we recover the biological condition of an open-ended finality; if, on the contrary, we

emphasize a network that acquires a very specific capacity in a well-demarcated domain [as in much AI or network talk], representation sneaks back in. . . . [it is thus important to] *remain closer to biological reality than to pragmatic considerations*" (Varela 1996 [1988]: 113). Although the enaction paradigm remains marginal in AI, there are small groups working from this perspective (Harry Halpin, personal communication). Winograd and Flores (1986) have elaborated a persuasive approach to computers and AI based on Heidegger's and Maturana's and Varela's ontology, introducing the fascinating notion of "ontological design"—the design of tools and technologies that reveal modes of being.

Maturana's and Varela's influence has been significant in theoretical biology, cognitive science, and social theory in Latin America and some European countries. Their work would require an entire separate treatment, yet I want to give at least some idea of their most fundamental insights. Beginning with Maturana's biology of cognition in the 1960s, continuing with his joint work with Varela in the 1970s and 1980s focused on the notion of autopoiesis, and Varela's own work on enaction until his death in 2001, and still going on today with the work of Maturana and colleagues in Chile, which foregrounds the biology of love, ethics, and social concerns, these two biologists have produced a compelling and complete theoretical system linking the biological and the social. The system is based on the fundamental insights that living beings are self-referential and self-creating entities, for which they coined the term *autopoiesis*, and that cognition is the most fundamental act of living. From their very first joint treatise, *Autopoiesis and Cognition* (1980; originally published in Spanish in 1972; see also Maturana and Varela 1987), they applied these basic insights to a broad range of biological and social phenomena—from simple living beings to ontogeny and phylogeny; from reproduction and heredity to evolution; from machines to society; from cognition to love—amounting to a veritable biologically based theory of systems and society that reconnects body and mind, person and world, biology and humanity.

Epistemologically, Maturana's and Varela's work is based on what they term the fundamental circularity of knowledge, action, and experience. Theirs can be said to be, in de Landa's terms, a neorealist epistemology but one that is also deeply constructivist in its own way, to the extent that although it might admit of a mind-independent reality, it argues that all knowledge is deeply embodied and thus never fully independent of the observer (their appeal is to "a *via media*: to understand the regularity of the world we are experiencing at every moment, but without any point of reference independent of ourselves" [1987: 241]; so the observer has no recourse other than to construct a metadomain of discourse in which she or he can couch her or his observations). Their work is derived from an established systems' view, but here again one finds their own particular, carefully developed set of notions. *Organization* refers to the set of relations among the components of a system that must be present for the system or living being to exist as such. Living beings are autonomous systems in that the only product of their organization is themselves. Throughout their history, systems go through structural changes without losing their basic organization. Changes in the environment trigger, but do not determine, structural changes, through what Maturana and Varela called "structural coupling" (loss of organization—that is, loss of the basic set of relations among components—leads to disintegration). In other words, structural change of living beings always implies conservation of autopoiesis; this is the meaning of adaptation. Evolution is the result of natural drift, meaning continuous structural coupling with the environment but with conservation of autopoiesis. This does not equate with the standard view of natural selection, in which living beings adapt progressively to an external world. In more complex social life, the key becomes the coordination of behavior between organisms constituting third-order unities through networks of "co-ontogenies"; this coordination takes place through communication in a linguistic domain.

33 A subtle but key distinction made in this paragraph is that between "characterization" and "knowledge" of the system. The first relates to the characterization of the autopoietic entity as such, centered on the domain in which the components establish the relations that result in the entity; this is the biophysical space of the living system. The second refers to a metadomain with respect to the biophysical domain, from which one describes it, which is part of the environment in which one beholds the entity in question. But this metadomain is not constitutive of the system since it does not participate in the making of the components or their relations. Representation and self-consciousness (the anthropologist's reflexivity) are one more domain of interaction in the structural coupling of living entities with their environment. However, I want to make clear that Maturana's and Varela's most basic point is that one is always immersed in a domain of description; indeed, reality is a domain of description, that is, of that with which one interacts. As describing system, one interacts with one's descriptions as if they were independent entities. Here and there Maturana and Varela seem to admit of a substratum in which the interactions occur, but it can never be characterized in terms of properties independent of the observer. It follows that reality, in the last instance, is the domain of descriptions of which one, as describing system, is part and that there is no such thing as a universe of independent entities (e.g., 1980: 22–56). Much of *Autopoiesis and Cognition* is devoted to working out this epistemological position (whether this framework provides a fully adequate solution to the problematic of the circularity of knowledge and the hermeneutic circle is a matter of debate).

34 This would be the beginning of another story altogether. These remarks come from a lecture by Goodwin at Schumacher College in Devon, England, February 20, 2006, and conversations with him throughout that week, and from Goodwin's most recent book (Goodwin 2007). Goodwin finds great hope in a holistic science that integrates mainstream science with the science of qualities, forms, and intensities that he sees present in the work of Goethe and also in many indigenous traditions. These sciences incorporate experience,

feelings, and intuition as modes of knowledge. In sum, Goodwin, among others, is trying to articulate anew the role of experience, feelings, intuition, and embodied knowledge as epistemological and ontological questions. For him, the issue is how to rethink humans' place within the flows of creative emergence on the planet on the basis of a deeper understanding of living process that moves back and forth between the life of form and forms of life. This has tremendous implications for ecological design. The analysis of biological life in terms of meaning has been developed by Markos (2002) by building bridges between hermeneutics and biology and as a complement to the more common genecentric and mechanistic biological thinking. Needless to say, proposals such as Goodwin's and Markos's are not accepted or even acknowledged in most of biology.

Conclusion

1 I participated in the discussions about the project throughout, and funding was secured largely through the mediation of the Danish anthropologist Søren Hvalkof. The project was implemented by the local ethnoterritorial organizations and community councils, with steady involvement by PCN activists from Buenaventura and Cali. The foundation's decision to fund directly a social movement (ethnoterritorial) organization was a rare occurrence in international development. The project was completely designed by PCN; I visited the river twice for a few days in conjunction with the project and also coincided with a five-person Danish team from Solstice in November 2002 in Cali for a day of discussions on the project with PCN. The project also received smaller amounts of support from the Global Greengrants Fund from Boulder. In recent years, it also arrived at an agreement with the regional development organization (CVC) and the Cali office of the World Wildlife Fund (WWF) to advance the *Plan de Manejo* for the Yurumanguí. See PCN/CVC/ WWF (2005).

2 Among the findings of the second evaluation of November 2000 was the need to support the political vision of the local organizations (the Asociación Popular de Negros Unidos del Río Yurumanguí [APONURY] and the community council and the equivalent ones for Pílamo) and the lack of continuing interaction (*mayor acompañamiento*) between the project in the locality, the PCN regional and national organization, and the Danish funding partner (see PCN 2000).

3 The reports contain detailed diagnoses of natural resource use as well as *planes de uso y manejo*, which would be impossible to summarize here. The plans were drawn by local organizations with the support of agronomists hired for the project. They are found in five lengthy project reports (PCN 2000, 2002, 2003a, 2003b; PCN/CVC/WWF 2005). The *temas organizativos* (aspects pertaining to the ethnoterritorial organizations) constitute an integral part of each of these reports, along with technical themes on specific "socioproductive" aspects or

projects. Most reports also have a section on the *enfoque político del* PCN (PCN's political perspective) that stresses the ethnoterritorial dynamics of the movement. According to Danish NGO evaluators, these are the best reports they have seen coming out of any such project in Latin America. The reports are extremely thorough in all of the social, political, ecological, economic (e.g., fishing, mining, handcrafts, and market-connected activities), and agronomical aspects, including detailed inventories of use spaces, species, hunting and gathering, and agricultural practices. In the case of the Yurumanguí, this was done for the entire length of the river, with its three distinct ecological zones (*zonas alta, media y baja*, or hillside, medium, and low-lying zones).

4 The *Territorios de Vida, Alegría y Libertad* is an important component of PCN's response to the armed conflict and displacement. It is related to the "peace communities" movement in other parts of the country but has its own conceptualization. It emphasizes the search for the communities' own perspective of the future based on cultural and environmental sustainability and the political recognition of cultural difference. The notion is linked to historical maroonism (*cimarronaje*) and intended as resistance to the homogenizing tendencies of development and the market. The concept takes as a point of departure the historical inequality of the black communities. This is linked to the notions of *proyecto de vida*, defined as the project based on those values and practices proper to the *cosmovisión* of the communities that are geared toward sustainability and self-subsistence; *vivir bien*, understood as an ideal state in which communities would have all of their basic needs satisfied, based on their own cultural practices; and *bienestar social colectivo*, which sees humans as an integral part of nature and nature as a guarantor of the *vivir bien*. These notions are amply elaborated in the reports and constitute part of the knowledge corpus developed by PCN. See also Grueso (2003). The reports systematically refer to the five fundamental principles (see, e.g., PCN 2003a: 31–45 for *lineamientos básicos*, or basic guidelines, following each of the five principles in the projects' context; PCN/CVC/WWF 2005). The principles were slightly redefined as follows: (1) identity: the reaffirmation of being (*la reafirmación del ser*); (2) territory: the space for being; (3) participation and autonomy: the exercise of being; (4) autonomous development (*desarrollo propio*) and collective well-being: the right to choose our own option of the future; and (5) alliances and solidarity: to be part of the black peoples of the world.

5 See, for instance, the action called for by American Friends Service Committee (AFSC) for the Yurumanguí of late December 2005, on its web site, but also others initiated from various sources, including academics with knowledge of the Pacific, in 2003 and 2006. One of the AFSC staff sent me the following message, for instance, on December 5, 2005: "La situación está tensa y las amenazas continúan. Los militares no dejan que entren o salgan las llamadas y están acampando debajo de las casas de la gente. Están amenazando a los líderes jóvenes que estaban tratando de tranquilizar a la gente" (The situation is tense

and the threats continue; the military does not allow calls in and out and they are camping under people's homes; they threaten the young leaders who are trying to calm people down). Strategies of intimidation such as these are common in the Pacific. These actions sometimes involved as many as several hundred navy or paramilitary elements.

6 See the critical evaluation of the putative market reforms on the continent over the past twenty-five years by an insider, CEPAL's former director, the Colombian economist José Antonio Ocampo (2004).

7 Grossberg's reading is much more complex than I can explain here; it includes an analysis of social, structural, and ontological discourses of modernity; a theory of modernity as virtual multiplicity, following Deleuze and Guattari; and a diagram of modernity in terms of the dialectic of space and time, structure and experience, change and presence, event and everyday life.

8 I hope my approach contributes to the discussion of the need to move anthropology in the United States away from the endless conversation with itself in which it seems to be mired at present.

9 These principles refer to the PCN-LASA project in question, *Otros Saberes*, mentioned in the introduction (see note 16).

10 To give an example: in *The End of Modernity*, Gianni Vattimo (1985) speaks about the philosophy of difference as a main vehicle for dismantling Western metaphysics, understood as a system of logocentrism and objectivizing thought—again, the ideas of progress and rational domination of the world; of logical truth as the foundation of all truth claims; and a humanism that places humans at the center of the universe. For Vattimo, the main goal of a postmodern philosophy is to pursue a "weakening" and "destructuration" of the structures of modernity from within. This is possible because of the generalized weakening of Being that has been the inevitable consequence of metaphysical thought. An important aspect of the decline of Being is that it involves "a condition of widespread contamination" (159) (a "Europeanization of earth and humanity" [151]). The world resembles a vast construction site of traces and residues, including those of the West itself, after alterity has been "entirely exhausted" (159). Postmodernity becomes a purposeful weakening of modernity that may open up new opportunities for thought and being. In this lies Vattimo's most radical proposal, one with which the MCD participants would agree up to a point. His alternative for a "weak thought" elicited by the "ontology of decline" of the West is indeed an opportunity for philosophical reconstruction; however, it denies the possibility of exteriority entertained by MCD perspective. In his more personal work (1996; entitled in English *Belief*), he again talks about the experience of the human condition at the end of modernity and about the "defeat of reason" as universal traits. Despite these features, Vattimo's attempt to build a radical historicized philosophy of events (eventalization, à la Foucault) and difference and its engagement with Catholic thought is illuminating.

11 The question about envisioning the kinds of decolonial societies one might wish to construct emerged at the Caracas meeting of the MCD group, May 14–20, 2007. The notion of "conditions of reexistence" (beyond resistance) was proposed by Adolfo Albán in the same context.

12 The idea that "war is the continuation of economics by other means" comes from the Salvadoran poet Roque Daltón.

13 "La dignidad, aspiración posible y necesaria, requiere para poder aportar a una construcción colectiva, de la decisión de mantenerse en el camino de los intereses propios, en el sentido en que estos representan en el terreno cultural, ecológico, y de construcción de sociedad, aspiraciones y alternativas que transcienden a los afrodescendientes. La guerra que se vive en Colombia no representa en principio ningún avance en el camino de la libertad que nuestros mayores delinearon para nosotros y que no se cerró hace 150 años cuando fue abolida legalmente la esclavitud. Es nuestro deber mantenernos en nuestros mandatos ancestrales que nos fueron dados y que aun en los tiempos más duros y adversos, han guiado nuestra resistencia. Hoy estos mandatos tienen en la defensa del Territorio, Identidad y Autonomía una razón de ser y una oportunidad que no es nueva en la medida que representa, aunque pueda expresarse de manera distinta en el tiempo y en distintos lugares, una búsqueda constante de los afrodescendientes. . . . Aunque la fuerza principal habrá que encontrarla y ganarla en la dura batalla con nosotros mismos, habrá que asumir también que en la crítica realidad del país, como lo dice un poema, 'uno solo no puede salvarse,' por lo que el futuro dependerá también de la capacidad que los afrodescendientes tengamos de juntar nuestras luchas y aspiraciones con las de otros sectores sociales excluidos y subordinados. . . . No asumir hoy la responsabilidad que tenemos con nuestro pasado y nuestro futuro, solo contribuirá a hacer más difícil y doloroso el camino para nuestras comunidades y sus renacientes. Con sus legados de Vida y Alegría, Esperanza y Libertad, los mayores nos dejaron también una senda; lo que hay que hacer hoy, no constituye un camino nuevo."

references cited

Acción Ecológica et al. 1997. "Choluteca Declaration." *World Aquaculture* 28(3): 38–39.

Agrawal, Arun. 2005. *Environmentality: Technologies of Government and the Making of Subjects*. Durham: Duke University Press.

Agudelo, Carlos. 2000. "El Pacífico colombiano: De 'remanso de paz' a escenario estratégico del conflicto armado." Presented at the international conference, "La societé prise en otage: Stratégies individuelles et collectives face à la violence–autour de cas colombien," Marseille.

Ahumada, Consuelo, et al. 2000. *¿Qué está pasando en Colombia?* Bogotá: El Ancora Editores.

Alberico, Michael. 1993. "La zoografía terrestre." In Pablo Leyva, ed., *Colombia Pacífico*, 1:232–38. Bogotá: Fondo FEN.

Alcoff, Linda, and Eduardo Mendieta, eds. 2000. *Thinking from the Underside of History: Enrique Dussel's Philosophy of Liberation*. Lanham: Rowman and Littlefield.

Allen, Timothy, and Thomas Hoekstra. 1992. *Toward a Unified Ecology: Complexity in Ecological Systems*. New York: Columbia University Press.

Almario, Oscar. 2005. *La invención del suroccidente colombiano*. 2 vols. Medellín: Universidad Pontificia Bolivariana.

———. 2004. "Dinámica y consecuencias del conflicto armado colombiano en el Pacífico: limpieza étnica y desterrorialización de afrocolombianos e indígenas y 'multiculturalismo' de Estado e indilencia nacional." In Eduardo Restrepo and Axel Rojas, eds., *Conflicto e (in)visibilidad: Retos en los estudios de la gente negra en Colombia*, 73–120. Popayán: Editorial Universidad del Cauca.

———. 2001. "Tras las huellas de los renacientes: Por el laberinto de la etnicidad e identidad de los grupos negros o 'afrocolombianos' del Pacífico sur." In M. Pardo, ed., *Acción colectiva, estado y etnicidad en el Pacífico colombiano*, 15–40. Bogotá: ICANH/Colciencias.

Almario, Oscar, and Ricardo Castillo. 1996. "Territorio, poblamiento y sociedades negras en el Pacífico Sur colombiano." In E. Restrepo and I. del Valle, eds., *Renacientes del guandal*, 57–120. Bogotá: Universidad Nacional/Biopacífico.

Altieri, Miguel. 1995 [1987]. *Agroecology: The Science of Sustainable Agriculture*. Boulder: Westview Press.

Álvarez, Manuela. 2002. "Altered States: Culture and Politics in the Colombian Pacific." MA thesis, Department of Anthropology, University of Massachusetts, Amherst.

Alvarez, Sonia E. Forthcoming. *Contentious Feminisms: Critical Readings of Social Movements, NGOs, and Transnational Organizing in Latin America*. Durham: Duke University Press.

———. 1998. "Latin American Feminisms 'Go Global': Trends in the 1990s and Challenges for the New Millennium." In Sonia E. Alvarez, Evelina Dagnino, and Arturo Escobar, eds., *Cultures of Politics/Politics of Cultures: Revisioning Latin American Social Movements*, 293–324. Boulder: Westview Press.

Alvarez, Sonia, Evelina Dagnino, and Arturo Escobar, eds. 1998. *Cultures of Politics/Politics of Cultures: Revisioning Latin American Social Movements*. Boulder: Westview Press.

Amin, Samir. 2003. "For Struggles, Global and National: Interview." *Frontline* 20(2): 1–10.

———. 1976. *Unequal Development*. New York: Monthly Review Press.

Anderson, David, and Eeva Berglund, eds. 2003. *Ethnographies of Conservation*. Oxford: Berghahn Books.

Angel, Augusto. 2001. *El retorno de Icaro*. Cali: Corporación Regional Autónoma de Occidente.

Angulo, Nianza del Carmen. 1996. *Los impactos socioculturales causados por las industrias palmicultoras y camaroneras en el municipio de Tumaco*. Bogotá: Proyecto Biopacífico.

Anzaldúa, Gloria, and Analouise Keatin, eds. 2002. *The Bridge We Call Home*. New York: Routledge.

Aparicio, Juan Ricardo. 2007. "Mattresses, Folders, and Internally Displaced Persons: Towards an Anthropology of Failure." MA thesis, Department of Anthropology, University of North Carolina, Chapel Hill.

Apffel-Marglin, Frèdèrique, ed. 1998. *The Spirit of Regeneration: Andean Culture Confronting Western Notions of Development*. London: Zed Books.

Apffel-Marglin, Frèdèrique, and Loyda Sánchez. 2002. "Developmentalist Feminism and Neocolonialism in Andean Communities." In Kriemild Saunders, ed., *Feminist Post-Development Thought: Rethinking Modernity, Post-Colonialism, and Representation*, 159–79. London: Zed Books.

Appadurai, Arjun. 1996. *Modernity at Large*. Minneapolis: University of Minnesota Press.

Aprile-Gniset, Jacques. 1993. *Poblamiento, hábitats y pueblos del Pacífico*. Cali: Universidad del Valle.

Arce, Alberto, and Norman Long, eds. 2000. *Anthropology, Development, and Modernities*. London: Routledge.

Ariza, Jaime. 1997. "Diez años imprimiendo huellas." *Gente del Agua*, no. 3: 3–5.

Arocha, Jaime. 1991. "La ensenada de Tumaco: invisibilidad, incertidumbre e innovación." *América Negra* 1: 87–112.

Arroyo, Jesús Eduardo, Juana Camacho, Mireya Leyton, and Marbell González, eds. 2001. *Zoteas: Biodiversidad y relaciones culturales en el Chocó biogeográfico colombiano*. Bogotá: Fundación Natura/Fundación SWISSAID.

Asher, Kiran. 2004. "Texts in Context: Afro-Colombian Women's Activism in the Pacific Lowlands of Colombia." *Feminist Review* 78: 38–55.

———. 2001. "The 'Global Environment' Discourse and Biodiversity Research in Colombia." Presented at the XXIII International Congress of the Latin American Studies Association, Washington, D.C., September 6–8.

———. 2000. "Mobilizing the Discourse of Sustainable Economic Development and Biodiversity Conservation in the Pacific Lowlands of Colombia." *Strategies* 13(1): 111–25.

———. 1998. "Constructing Afro-Colombia: Ethnicity and Territory in the Pacific Lowlands." PHD diss., Department of Political Science, University of Florida, Gainesville.

ASOCARLET. 1999a. *Restauración de áreas de manglar en el sector comprendido entre el Pindo y el estero de Aguaclara en el municipio de San Andrés de Tumaco, Nariño*. Unpublished project proposal. Tumaco: ASOCARLET.

———. 1999b. *Fortalecimiento operativo del camaricultivo artesanal de la Asociación de Carboneros y Leñeteros de Tumaco, ASOCARLET, en el municipio de Tumaco (Nariño), Colombia*. Unpublished project proposal. Tumaco: ASOCARLET.

Balick, Michael, Elaine Elisabetsky, and Sarah Laird, eds. 1996. *Medicinal Resources of the Tropical Forests*. New York: Columbia University Press.

Barabási, Albert-László. 2002. *Linked: The New Science of Networks*. Cambridge, Mass.: Perseus Publishing.

Barkin, David. 1998. *Riqueza, pobreza y desarrollo sustentable*. Mexico City: Editoral Jus.

Baskin, Yvonne. 1997. *The Work of Nature: How the Diversity of Life Sustains Us*. Washington: Island Press.

Battaglia, Debbora, ed. 1995. *Rhetorics of Self-Making*. Berkeley: University of California Press.

Bauman, Zygmunt. 1996. "From Pilgrim to Tourist—or a Short History of Identity." In S. Hall and P. du Gay, eds., *Questions of Cultural Identity*, 18–36. London: Sage.

Bebbington, Anthony. 2005. "Donor–NGO Relations and Representations of Livelihood in Nongovernmental Aid Chains." *World Development* 33(6): 937–50.

———. 2004. "NGOs and Uneven Development: Geographies of Development Intervention." *Progress in Human Geography* 28(6): 725–45.

———. 2000. "Re-encountering Development: Livelihood Transitions and Place Transformations in the Andes." *Annals of the Association of American Geographers* (90) 3: 495–520.

Bebbington, Anthony, and S. P. J. Batterbury. 2001. "Transnational Livelihoods and Landscapes: Political Ecologies of Globalization." *Ecumene* 8(4): 369–80.

Belausteguigoitia, Marisa. 1998. "Visualizing Places: She Looks, Therefore, Who Is?" *Development* 41(2): 44–52.

Bell, David, and Barbara Kennedy, eds. 2000. *The Cybercultures Reader*. London: Routledge.

Bender, Barbara. 1998. *Stonehenge: Making Space*. Oxford: Berg.

Berlin, Brent. 1992. *Ethnobiological Classification*. Princeton: Princeton University Press.

Bernal, Rodrigo, and Gloria Galeano. 1993. "Las palmas del andén Pacífico." In Pablo Leyva, ed., *Colombia Pacífico*, 1:220–30. Bogotá: Fondo FEN.

Bernauer, James, and David Rasmussen, eds. 1988. *The Final Foucault*. Cambridge, Mass.: MIT Press.

Berry, Thomas. 1999. *The Great Work: Our Way into the Future*. New York: Bell Tower.

Bertalanffy, Ludwig von. 1975. *Perspectives on General Systems Theory*. New York: Braziller.

Beverly, John, and José Oviedo, eds. 1993. *The Postmodernism Debate in Latin America*. Durham: Duke University Press.

BID/Plan Pacífico/ ACABA. n.d. *Proyecto titulación colectiva en el río Baudó: Sistematización de la información y guías de recolección*. Bogotá: BID/PLAN PACÍFICO/ACABA.

Biersack, Aletta. 2006. Introduction. In Aletta Biersack and James Greenberg, eds. *Re-imagining Political Ecology*. Durham: Duke University Press.

———. 1999. "Introduction: From the 'New Ecology' to the New Ecologies." *American Anthropologist* 101(1): 5–18.

Biersack, Aletta, and James Greenberg, eds. 2006. *Re-Imagining Political Ecology*. Durham: Duke University Press.

Blaser, Mario. Forthcoming. "Storytelling Globality: A Border Dialogue Ethnography of the Paraguayan Chaco." Durham: Duke University Press.

Blaser, Mario, Harvey Feit, and Glenn McRae, eds. 2004. *In the Way of Development: Indigenous Peoples, Life Projects, and Globalization*. London: ZED Books

Boff, Leonardo. 2004. *Ecologia: Grito da Terra, Grito dos Pobres*. Rio de Janeiro: Editorial Sextante.

———. 2002. *El Cuidado Esencial*. Madrid: Editorial Ttrotta.

———. 2000. *La Oración de San Francisco*. Santander: Sal Terrae.

Bogues, Anthony. 2003. *Black Heretics, Black Prophets*. New York: Routledge.

Bookchin, Murray. 1990. *The Philosophy of Social Ecology*. Montreal: Black Rose.

———. 1986. *Post-scarcity Anarchism*. 2d ed. Montreal: Black Rose.

Borrero, José María. 1994. *The Ecological Debt*. Cali: FIPMA.

Borrini-Feyerabend, Grazia, Michael Pimbert, M. Taghi Farvar, Ashis Kothari, and Yves Renard. 2004. *Sharing Power: Learning-by-Doing in Co-Management of Natural Resources throughout the World*. Teheran: IIED and IUCN/CEESP/CMWG.

Braudel, Fernand. 1977. *Afterthoughts on Material Civilization and Capitalism*. Baltimore: Johns Hopkins University Press.

Bravo, Elizabeth. 2006. *Encenciendo el debate sobre biocombustibles*. Buenos Aires: Capital Intelectual/Le Monde Diplomatique.

Bravo, Hernando. 1998. *Diversidad cultural y manglares en el Pacífico Colombiano*. Bogotá: Ministerio del Medio Ambiente.

Brockway, Lucille. 1979. *Science and Colonial Expansion: The Role of the British Royal Botanical Gardens*. New York: Academic Press.

Brosius, Peter. 1999. "Analyses and Interventions: Anthropological Engagements with Environmentalism." *Current Anthropology* 40 (3): 277–309.

Brosius, Peter, Anna Tsing, and Charles Zerner. 1998. "Representing Communities: Histories and Politics of Community-Based Natural Resource Management." *Society and Natural Resources* 11: 157–68.

Browdy, Craig, and Stephen Hopkins, eds. 1995. *Swimming through Troubled Water: Proceedings of the Special Session on Shrimp Farming*. Baton Rouge: World Aquaculture Society.

Brush, Stephen. 1998. "Prospecting the Public Domain." Presented at the Globalization Project, Center for Latin American Studies, University of Chicago, February 12.

Brush, Stephen, and Doreen Stabinski, eds. 1996. *Valuing Local Knowledge*. Washington: Island Press.

Bryant, Raymond. 2000. "Politicized Moral Geographies: Debating Biodiversity Conservation and Ancestral Domain in the Philippines." *Political Geography* 19: 673–705.

Bryant, Raymond, and Sinéad Bailey. 1997. *Third World Political Ecology*. London: Routledge.

Burbano, Andrés, and Hernando Barragán, eds. 2002. *hipercubo/ok/. arte, ciencia y tecnología en contextos próximos*. Bogotá: Universidad de los Andes / Goethe Institut.

Burchell, Graham, Colin Gordon, and Peter Miller, eds. 1991. *The Foucault Effect: Studies in Governmentality*. Chicago: University of Chicago Press.

Butler, Judith. 1993. *Bodies that Matter: On the Discursive Limits of "Sex."* New York: Routledge.

———. 1990. *Gender Trouble*. London: Routledge.

Calvino, Italy. 1974. *Invisible Cities*. New York: Harcourt.

Camacho, Juana. 2004. "Silencios elocuentes, voces emergentes: Reseña bibliográfica de los estudios sobre la mujer afrocolombiana." In Claudia Mosquera, M. Clemencia Ramírez, and Mauricio Pardo, eds., *Panorámica afrocolombiana: Estudios sociales en el Pacífico*, 167–210. Bogotá: ICANH/Universidad Nacional.

———. 1998. "Huertos de la Costa Pacífica chocoana: Prácticas de manejo de plantas cultivadas por parte de mujeres negras." MA thesis, Desarrollo Sostenible de Sistemas Agrarios, Universidad Javeriana, Bogotá.

Camacho, Juana, and Carlos Tapia. 1997. *Conocimiento y manejo del territorio y los recursos naturales entre las poblaciones negras de Nuquí y Tribugá, Costa Pacífica chocoana*. Final report presented to the Dutch Government. Bogotá: Fundación Natura.

Camacho, Juana, and Eduardo Restrepo, eds. 1999. *De montes, ríos y ciudades: Territorios e identidades de la gente negra en Colombia*. Bogotá: ICANH/Natura/Ecofondo.

Camazine, Scott, Jean-Louis Deneubourg, Nigel Franks, James Sneyd, Guy Theraulaz, and Eric Bonabeau. 2001. *Self-Organization in Biological Systems*. Princeton: Princeton University Press.

Campbell, Ben, ed. Forthcoming. Replacing Nature: Connecting Ethnographies and Environmental Protection.

Cantera, Jaime, and Rafael Contreras. 1993. "Ecosistemas costeros." In P. Leyva, ed., *Colombia Pacífico*, 1:64–79. Bogotá: Fondo FEN.

Carrere, Ricardo, and Larry Lohmann. 1996. *Pulping the South*. London: ZED Books.

Casas Cortés, María Isabel. 2006a. "Precarious Europe and Its Insurgent Knowledges:The Movement of Activist Research in Spain." PHD diss. prospectus, Department of Anthropology, University of North Carolina, Chapel Hill.

———. 2006b. "WAN and Activist Research: Towards Building Decolonial and Feminist Projects." *Journal of the World Anthropologies Network* 1(2): 75–89.

Casey, Edward. 1993. *Getting Back into Place: Toward a Renewed Understanding of the Place-World*. Bloomington: Indiana University Press.

Casimir, Jean. 2004. *Pa Bliye 1804 : Souviens-Toi de 1804*. Port-au-Prince, Haiti: Imprimerie Lakay Delmas.

Cassiani, Alfonso, Catalina Achipiz, and Angela Umaña. 2002. *Cultural y Derecho*. Cali: Universidad Libre.

Castaño, Sandra. 1996. *Informe final del impacto socioeconómico y ambiental de las industrias de palma africana y el camarón de cautiverio en el municipio de Tumaco, Departamento de Nariño*. Bogotá: Proyecto Biopacífico.

Castells, Manuel. 1997. *The Power of Identity*. Oxford: Blackwell.

———. 1996. *The Rise of the Network Society*. Oxford: Blackwell.

Castro-Gómez, Santiago. 2005. *La hybris del punto cero*. Bogotá: Universidad Javeriana.

Castro-Gómez, Santiago, and Ramón Grosfogel, eds. 2007. *El giro decolonial*. Bogotá: Universidad Javeriana/IESCO/Siglo del Hombre.

Castro-Gómez, Santiago, and Eduardo Mendieta, eds. 1998. *Teorías sin disciplina, latinoamericanismo, poscolonialidad y globalización en debate*. Mexico City: Miguel Angel Porrúa Editorial/University of San Francisco.

CEGA (Centro de Estudios Ganaderos y Agrícolas). 1999. *Análisis de factibilidad y diseño institucional para el desarrollo de cinco núcleos de cultivo de palma de aceite en Tumaco, Nariño*. Bogotá: CEGA.

CENSAT. 2005. *Los nuevos mercaderes: La vida como mercancía*. Bogotá: CENSAT and Global Forest Coalition.

Chakrabarty, Dipesh. 2000. *Provincializing Europe*. Princeton: Princeton University Press.

Chesters, Graeme. 2003. "Shape Shifting: Civil Society, Complexity and Social Movements." Manuscript, Centre for Local Policy Studies, Edge Hill University College, Lancashire, UK.

Chesters, Graeme, and Ian Welsh. 2006. *Complexity and Social Movements: Multitudes at the Edge of Chaos*. London: Routledge.
Churchman, C. West. 1971. *The Design of Inquiring Systems*. New York: Basic Books.
———. 1968. *The Systems Approach*. New York: Laurel.
Cogollo, Julia. 2005. "Fortalecimiento de la identidad en Comunidades Negras como estrategia de resistencia en el territorio: construcción de un plan de desarrollo a través de Intervención participativa." Thesis for degree in Social Pyschology, Universidad Nacional a Distancia, Palmira, Colombia.
Colmenares, Germán. 1979. *Historia económica y social de Colombia*. Volume 2: *Popayán: Una Sociedad Esclavista*. Bogotá: Tercer Mundo Editores.
———. 1973. *Historia económica y social de Colombia, 1537–1719*. Volume 1. Cali: Universidad del Valle.
Comaroff, Jean, and John Comaroff. 1991. *Of Revelation and Revolution*. Chicago: University of Chicago Press.
Comaroff, John. 1996. "Ethnicity, Nationalism, and the Politics of Difference." In E. Wilmsen and P. McAllister, eds., *The Politics of Difference: Ethnic Premises in a World of Power*, 163–205. Chicago: University of Chicago Press.
Consalvo, Mia, and Susanna Passonen, eds. 2002. *Women and Everyday Uses of the Internet*. New York: Peter Lang.
Conway, Janet. 2006. *Praxis and Politics: Knowledge Production in Social Movements*. New York: Routledge.
Cordeagropaz. 2000. *Memorias taller de evaluación experiencia Convenio CVC–Holanda y Coagropacífico, en cinco ríos de la Ensenada de Tumaco*. Tumaco, unpublished report.
Coronil, Fernando. 1997. *The Magical State*. Chicago: University of Chicago Press.
———. 1996. "Beyond Occidentalism: Toward Non-Imperial Geohistorical Categories." *Cultural Anthropology* 11(1): 52–87.
Corrigan, Philip, and Derek Sayer. 1985. *The Great Arch:English State Formation as Cultural Revolution*. Oxford: Blackwell.
Cortés, Hernán. 1999. "Titulación colectiva en comunidades negras del Pacífico nariñense." In Juana Camacho and Eduardo Restrepo, eds., *De montes, ríos y ciudades: Territorios e identidades de la gente negra en Colombia*, 131–42. Bogotá: ICANH/Natura/Ecofondo.
Cortés Severino, Catalina. 2007. "Landscapes of Terror: In Between Hope and Memory." MA thesis, Department of Communications, University of North Carolina, Chapel Hill.
Crosby, Alfred. 1986. *Ecological Imperialism: The Overseas Migration of Western Europeans as a Biological Phenomenon*. Cambridge: Cambridge University Press.
Cruikshank, Barbara. 1999. *The Will to Empower*. Ithaca: Cornell University Press.
Crumley, Carole, ed. 1994. Historical Ecology: Culture, Knowledge, and Changing Landscapes. Santa Fe: SAR Press.
Cuomo, Chris. 1998. *Feminism and Ecological Communities*. New York: Routledge.
de Chardin, Pierre Teilhard. 1963. *El fenómeno humano*. Madrid: Taurus.

de la Cruz, Rodrigo. 2004. "Vision of Indigenous Peoples in the Context of Decisions Pertaining to Access to Genetic Resources and Benefit Sharing (ABS) and Article 8j: An Analysis of the Impacts of CBD/COP Decisions with respect to WIPO's IGC Mandate." Gland, Switzerland: IUCN (Policy, Biodiversity and International Agreements Unit).

de Landa, Manuel. 2006. *A New Philosophy of Society. Assemblage Theory and Social Complexity*. New York: Continuum Press.

———. 2003. "1000 Years of War: CTHEORY Interview with Manuel de Landa." From http://www.ctheory.net/

———. 2002. *Intensive Science and Virtual Philosophy*. New York: Continuum Press.

———. 1997. *A Thousand Years of Non-Linear History*. New York: Zone Books.

———. n.d. "Meshworks, Hierarchies and Interfaces" (in http://www.to.or.at/delanda/).

Deleuze, Gilles, and Félix Guattari. 1987. *A Thousand Plateaus*. Minneapolis: University of Minnesota Press.

Descola, Phillippe. 1996. "Constructing Natures: Symbolic Ecology and Social Practice." In P. Descola and G. Pálsson, eds., *Nature and Society: Anthropological Perspectives*, 82–102. London: Routledge.

Descola, Philippe, and Gísli Pálsson, eds. 1996. *Nature and Society: Anthropological Perspectives*. London: Routledge.

de Vries, Pieter. 2007. "Don't Compromise Your Desire for Development: A Lacanian/Deleuzian Rethinking of the Anti-Politics Machine." *Third World Quarterly* 28(1): 25–43.

De Walt, Billie, Philippe Vergne, and Mark Hadin. 1996. "Shrimp Aquaculture Development and the Environment: People, Mangroves and Fisheries on the Gulf of Fonseca, Hunduras." *World Development* 24(7): 1193–1208.

Diani, Mario, and Doug McAdam, eds. 2003. *Social Movements and Networks*. Oxford: Oxford University Press.

Diawara, M. 2000. "Globalization, Development Politics and Local Knowledge." *International Sociology* 15(2): 361–71.

Dirlik, Arif. 2001. "Place-based Imagination: Globalism and the Politics of Place." In R. Prazniak and A. Dirlik, eds., *Places and Politics in the Age of Globalization*, 15–52. New York: Rowman and Littlefield.

DNP (Departamento Nacional de Planeación de Colombia). 1997. *Programa BID–Plan Pacífico. Plan Operativo*. Bogotá: DNP.

DNP/CORPES. 1992. *Plan Pacífico: Una estrategia de desarrollo sostenible para la Costa Pacífica Colombiana*. Bogotá: DNP.

DNP/CVC/UNICEF. 1983. *Plan de desarrollo integral para la Costa Pacífica*, PLADEICOP. Cali: CVC.

DNP/UPRU/Plan Pacífico. 1995. *Plan de acción para la población afrocolombiana y raizal*. Bogotá: DNP.

Dumont. Louis. 1977. *From Mendeville to Marx: The Genesis and Triumph of Economic Ideology*. Chicago: University of Chicago Press.

Duncan, Watts. 2003. *Six Degrees: The Science of a Connected Age*. New York: W. W. Norton.
Duque-Caro, Hermann. 1993. "Los Foraminíferos." In P. Leyva. ed., *Colombia Pacífico*, 1:96–109. Bogotá: Fondo FEN.
Dussel, Enrique. 2006. *20 Tesis de Política*. Mexico City: Siglo XXI.
———. 2000. "Europe, Modernity, and Eurocentrism." *Nepantla* 1(3): 465–78.
———. 1996. *The Underside of Modernity*. Atlantic Highlands, N.J.: Humanities Press.
———. 1993. "Eurocentrism and Modernity." In J. Beverly and J. Oviedo, eds., *The Postmodernism Debate in Latin America*, 65–76. Durham: Duke University Press.
———. 1992. *1492: El encubrimiento del Otro*. Bogotá: Antropos.
———. 1976. *Filosofía de la Liberación*. Mexico City: Editorial Edicol.
Emery, F. E. 1969. *Systems Thinking: Selected Readings*. New York: Penguin Books.
Englund, Harri, and James Leach. 2000. "Ethnography and the Meta-Narratives of Modernity." *Current Anthropology* 41(2): 225–48.
Equipo Ampliado. 1997a. *El Proyecto Biopacífico*. La Esperanza: PBP.
———. 1997b. *Síntesis Proyecto Biopacífico: A. Proceso de conocimiento de la biodiversidad*. Perico Negro: PBP.
Errázuriz, Jaime. 1980. *Tumaco—La Tolita*. Bogotá: Carlos Valencia Editores.
Escalante, Aquiles. 1971. *La minería del hambre: Condoto y la Chocó Pacífico*. Medellín: Ediciones Universidades Medellín, Córdoba y Simón Bolívar.
Escobar, Arturo. 2007. "Post-development as Concept and Social Practice." In Aram Ziai, ed., *Exploring Post-development*, 18–32. London: Zed Books.
———. 2006 [1999]. "An Ecology of Difference: Equality and Conflict in a Glocalized World." *Focaal* 47 (originally prepared for UNESCO's *World Culture Report*, 1999).
———. 2004a. "Beyond the Third World: Imperial Globality, Global Coloniality, and Anti-Globalisation Social Movements." *Third World Quarterly* 25(1): 207–30.
———. 2004b. "Other Worlds Are (Already) Possible: Self-Organisation, Complexity, and Post-Capitalist Cultures." In Jai Sen, Anita Anad, Arturo Escobar, and Peter Waterman, eds., *The World Social Forum: Challenging Empires*, 349–58. Delhi: Viveka.
———. 2003a. "Displacement and Development in the Colombian Pacific." *International Social Science Journal* 175: 157–67.
———. 2003b. "'*World and Knowledges Otherwise*': The Latin American Modernity/Coloniality Research Program." *Cuadernos del* CEDLA 16: 31–67.
———. 2001. "Culture Sits in Places: Reflections on Globalism and Subaltern Strategies of Localization." *Political Geography* 20: 139–74.
———. 2000. "Beyond the Search for a Paradigm? Post-development and Beyond." *Development* 43(4): 11–14.
———. 1999a. "After Nature: Steps to an Anti-essentialist Political Ecology." *Current Anthropology* 40(1): 1–30.

———. 1999b. "Gender, Place and Networks: A Political Ecology of Cyberculture." In Wendy Harcourt, ed., *Women@Internet: Creating New Cultures in Cyberspace*, 31–54. London: Zed Books.

———. 1998. "Whose Knowlege, Whose Nature? Biodiversity Conservation and the Political Ecology of Social Movements." *Journal of Political Ecology* 5: 53–82.

———. 1997. *Biodiversidad, Naturaleza y Cultura: Localidad y Globalidad en las Estrategias de Conservación*. Colección El Mundo Actual. Mexico City: UNAM/CIICH.

———. 1996. "Constructing Nature: Elements for a Poststructuralist Political Ecology." In R. Peet and M. Watts, eds., *Liberation Ecologies*, 46–68. London: Routledge.

———. 1995. *Encountering Development*. Princeton: Princeton University Press.

———. 1994. "Welcome to Cyberia: Notes on the Anthropology of Cyberculture." *Current Anthropology* 35(3): 211–31.

———. 1992. "Imagining a Postdevelopment Era? Critical Thought, Development, and Social Movements." *Social Text* 31/32: 20–56.

Escobar, Arturo, and Alvaro Pedrosa, eds. 1996. *Pacífico: ¿Desarrollo o diversidad? Estado, Capital y Movimientos Sociales en el Pacífico Colombiano*. Bogotá: CEREC.

Escobar, Marta. 1990. *La frontera imprecisa: Lo natural y lo sagrado en la cultura negra del norte de Esmeraldas*. Quito: Centro Cultural Afro-Ecuatoriano.

Espinosa, Mónica. 2004. "Of Visions and Sorrows: Manuel Quintín Lame's Indianist Thought and the Violences of Modern Colombia." PHD diss., Department of Anthropology, University of Massachusetts, Amherst.

Fagan, G. H. 1999. "Cultural Politics and (post) Development Paradigms." In Ronald Munck and Denis O'Hearn, eds., *Critical Development Theory: Contributions to a New Paradigm*, 179–95. London: Zed Books.

Fajardo, Darío, ed. 1991. *Campesinos y desarrollo en América Latina*. Bogotá: Tercer Mundo.

Fals Borda, Orlando. 1970. *Ciencia Propia y Colonialismo Intelectual*. Mexico City: Editorial Nuestro Tiempo.

Fast, Arlo, and James Lester, eds. 1992. *Marine Shrimp Culture: Principles and Practices*. Amsterdam: Elsevier.

Faubion. James. 1993. *Modern Greek Lessons: A Primer in Historical Constructivism*. Princeton: Princeton University Press.

Featherstone, David. 2003. "Spatialities of Transnational Resistance to Globalization: The Maps of Grievance of the Inter-Continental Caravan." *Transactions of the Institute of British Geography* NS 28: 404–21.

FEDEPALMA. 1999. *Censo Nacional de Palma de Aceite, Colombia 1997–1998*. Bogotá: FEDEPALMA.

———. 1993. *Anexo Estadístico 1992–93*. Bogotá: FEDEPALMA.

Feld, Stephen, and Keith Baso, eds. 1996. *Senses of Place*. Santa Fe: School of American Research.

Ferguson, James. 1990. *The Anti-Politics Machine*. Cambridge: Cambridge University Press.

Fernandez, Maria, Faith Wilding, and Michelle Wright, eds. 2002. *Domain Errors! Cyberfeminist Practices*. Brooklyn: Autonomedia.
Ferrand, Maurice. 1959. *Informe sobre posibilidades de las oleaginosas en Colombia*. Bogotá: IFA.
Filose, John. 1995. "Factors Affecting the Processing and Marketing of Farm-Raised Shrimp." In Craig Browdy and J. Stephen Hopkins, eds., *Swimming through Troubled Waters: Proceedings of the Special Session on Shrimp Farming*, 227–34. Baton Rouge: World Aquaculture Society.
Fischer, Michael. 2003. *Emergent Forms of Life and the Anthropological Voice*. Durham: Duke University Press.
Flórez Flórez, Juliana. 2007. "Tácticas de des-sujeción: disensos, subjetividades y deseo em los movimientos sociales. Relaciones de género en la red 'Proceso de Comunidades Negras' del Pacífico colombiano." PHD diss., Programa de Doctorat en Psicología Social, Universitat Autónoma de Barcelona.
———. 2004. "Implosión identitaria y movimientos sociales: desafíos y logros del Proceso de Comunidades Negras ante las relaciones de género." In E. Restrepo and A. Rojas, eds., *Conflicto e (in)visibilidad: retos de los estudios de la gente negra en Colombia*, 219–46. Popayán: Editorial de la Universidad del Cauca.
Foerster, Heinz von. 1981. *Observing Systems*. Seaside, Calif.: Intersystems Publications.
Foerster, Heinz von, John White, Larry Peterson, and John Russel, eds. 1968. *Purposive Systems*. New York: Spartan Books.
Forero, Enrique, and Alwyn Gentry. 1989. *Lista anotada de las plantas del departamento del Chocó*. Bogotá: Museo de Historia Natural, Universidad Nacional.
Fortun, Kim. 2003. "Ethnography In/Of/As Open Systems." *Reviews in Anthropology* 32: 171–90.
Foucault, Michel. 1991. "Governmentality." In G. Burchell, C. Gordon, and P. Miller, eds., *The Foucault Effect*, 87–104. Chicago: University of Chicago Press.
———. 1979. *Discipline and Punish*. New York: Vintage Books.
———. 1973. *The Order of Things*. New York: Vintage Books.
Friedemann, Nina S. de. 1989. *Críele críele son: Del Pacífico negro*. Bogotá: Planeta.
———. 1984. "Estudios de negros en la antropología colombiana: presencia e invisibilidad." In Jaime Arocha and Nina S. de Friedemann, eds., *Un siglo de investigación social: antropología en Colombia*, 507–72. Bogotá: Etno.
———. 1974. *Minería, descendencia y orfebrería artesanal: Litoral Pacífico, Colombia*. Bogotá: Imprenta Universidad Nacional.
Friedemann, Nina S. de, and Alfredo Vanín. 1991. *El Chocó: Magia y leyenda*. Bogotá: Arco.
Friedemann, Nina S. de, and Jaime Arocha. 1984. *Un siglo de investigación social: Antropología en Colombia*. Bogotá: Etno.
Friedemann, Nina S. de, and Mónica Espinosa. 1995. "La mujer negra en la historia de Colombia." In M. Velásquez, ed. *Las mujeres en la historia de Colombia*. Volume I. Bogotá: Editorial Norma.

Fundación Habla/Scribe. 1995. *Territorio, etnia, cultura e investigación en el Pacífico Colombiano.* Cali: Fundación Habla/Scribe.

Galvis, Jaime, and Jairo Mojica. 1993. "Geología." In Pablo Leyva, ed., *Colombia Pacífico*, 1:80–95. Bogotá: Fondo FEN.

Gaonkar, Dilip, ed. 2001. *Alternative Modernities.* Durham: Duke University Press.

Garay, Luis, ed. 2002. *Repensar a Colombia: Hacia un nuevo contrato social.* Bogotá: PNUD.

García Hierro, Pedro. 1998. *Apropiación, manejo y control social de territorios de comunidades negras a través de procesos de titulación colectiva. Informe de consultoría.* Copenhagen/Cali: IWGIA/PCN.

Gardner, Katty, and David Lewis. 1996. *Anthropology, Development, and the Postmodern Challenge.* London: Pluto Press.

Gari, Josep. 2001. "Biodiversity and Indigenous Agroecology in Amazonia: The Indigenous Peoples of Pastaza." *Etnoecológica* 5(7): 21–37.

GEF/PNUD. 1993. *Conservación de la biodiversidad del Chocó Biogeográfico. Proyecto Bipacífico.* Bogotá: GEF-PNUD COL/92/G31.

Gentry, Alwin. 1993. "Riqueza de especies y composición florística." In Pablo Leyva, ed., *Colombia Pacífico*, 1:200–19. Bogotá: Fondo FEN.

Gibson, Katherine. 2002. "Women, Identity and Activism in Asian and Pacific Community Economies." *Development* 45(1): 74–80.

———. 2000. "Improving Productivity of the Small Holder Oil Palm Sector in Papua New Guinea" Unpublished research proposal, Department of Human Geography, Australian National University.

Gibson-Graham, J. K. 2006. *A Postcapitalist Politics.* Minneapolis: University of Minnesota Press.

———. 2005. "Building Community Economies: Women and the Politics of Place." In W. Harcourt and A. Escobar, eds., *Women and the Politics of Place*, 130–57. Bloomfield, Conn.: Kumarian Press.

———. 2003. "Politics of Empire, Politics of Place." Manuscript, Department of Geography, University of Massachusetts, Amherst, and Department of Geography, Australian National University, Canberra.

———. 1996. *The End of Capitalism (As We Knew It).* Oxford: Blackwell.

Giddens, Anthony. 1990. *The Consequences of Modernity.* Stanford: Stanford University Press.

González, Pedro H. 2002. *Marginalidad y exclusión en el Pacífico Colombiano.* Cali: Universidad Santiago de Cali.

Gonzales, Tirso. 2004. "Coloniality and Place: A Historical and Contemporary Overview of Cultural and Biological Diversity in the Americas." Manuscript, University of California, Berkeley, July.

Goodman, Alan, and Thomas Leathermann, eds. 1998. *Building a New Biocultural Synthesis.* Ann Arbor: University of Michigan Press.

Goodwin, Brian. 2007. *Nature's Due: Healing Our Fragmented Culture.* Edinburgh: Floris Books.

Gosh, Rishab Aiyer, ed. 2005. CODE: *Collaborative Ownership and the Digital Economy*. Cambridge, Mass.: MIT Press.

Graeber, David. 2002. "The New Anarchists." *New Left Review* 13: 61–73.

GRAIN. 2000. "Growing Diversity Project." Project description. Barcelona: GRAIN.

GRAIN (Genetic Resources Actions International). 1998. "Patenting Life: Progress or Piracy?" *Global Biodiversity* 7(4): 2–6.

GRAIN. 1995. "Towards a Biodiversity Community Rights Regime." *Seedling* 12(3): 1–24.

GRAIN/GAIA. 1998a. "TRIPS versus CBD." *Global Trade and Biodiversity in Conflict*, no. 1. Barcelona: GRAIN.

GRAIN/GAIA. 1998b. "Intellectual Property Rights and Biodiversity: The Economic Myths." *Global Trade and Biodiversity in Conflict*, No. 3. Barcelona: GRAIN

Grillo, R. D., and R. L. Stirrat, eds. 1997. *Discourses of Development. Anthropological Perspectives*. Oxford: Berg.

Gros, Christian. 2000. *Políticas de la etnicidad: Identidad, estado y modernidad*. Bogotá: ICANH.

———. 1997. "Indigenismo y etnicidad: el desafío neo-liberal." In M. V. Uribe and E. Restrepo, eds., *Antropología en la modernidad*, 15–60. Bogotá: ICANH.

Grosfogel, Ramón, Nelson Maldonado-Torres, and José Saldívar, eds. Forthcoming. *Unsettling Postcolonial Studies*. Durham: Duke University Press.

Grossberg, Lawrence. 2007. "Critical Studies in Search of Modernities." Manuscript, Chapel Hill.

———. 2006. "Does Cultural Studies Have Futures? Should It? (Or What's the Matter with New York?): Cultural Studies, Contexts, and Conjunctures. *Cultural Studies* 20(1): 1–32.

———. 1996. "Identity and Cultural Studies—Is That All There Is?" In S. Hall and P. du Gay, eds., *Questions of Cultural Identity*, 87–107. London: Sage.

Grosz, Elizabeth. 1989. *Sexual Subversions: Three French Feminists*. Sydney: Allen and Unwin.

Grubacic, Andrej. 2004. "Towards Another Anarchism." In J. Sen, A. Anand, A. Escobar, and P. Waterman, eds., *The World Social Forum: Challenging Empires*, 35–43. Delhi: Viveka.

Grueso, Jesús Alberto, and Arturo Escobar. 1996. "Las cooperativas agrarias y la modernización de los agricultures." In A. Escobar and A. Pedrosa, eds., *Pacífico: ¿Desarrollo o diversidad?* 90–108. Bogotá: CEREC/Ecofondo.

Grueso, Libia. 2005. "Producción de conocimiento y movimiento social: Un análisis desde la experiencia del Proceso de Comunidades Negras (PCN) en Colombia." *LASA Forum* 36(1): 21–23.

———. 2003. "Territorios de Vida, Alegría y Libertad: una opción frente al conflicto en el territorio-región del Pacífico colombiano." Buenaventura: PCN.

———. 1996. "Hacia una autodefinición de los pueblos negros en América Latina." Manuscript. Cali: PCN.

Grueso, Libia, and Leyla Arroyo. 2005. "Women and the Defense of Place in Colombia's Black Movement Struggles." In W. Harcourt and A. Escobar, eds., *Women and the Politics of Place*, 100–114. Bloomfield, Conn.: Kumarian Press.

Grueso, Libia, Carlos Rosero, and Arturo Escobar. 1998. "The Process of Black Community Organizing in the Southern Pacific Coast Region of Colombia." In S. Alvarez, E. Dagnino, and A. Escobar, eds., *Cultures of Politics/Politics of Cultures: Re-Visioning Latin American Social Movements*, 196–219. Oxford: Westview Press.

Guattari, Félix. 1995. *Chaosophy*. New York: Semiotext(e).

———. 1990. *Las tres ecologías*. Valencia: Pre-Textos.

Gudeman, Stephen. 2001. *The Anthropology of Economy*. Oxford: Blackwell.

———. 1986. *Economics as Culture*. New York: Routledge.

Gudeman, Stephen, and Alberto Rivera. 1990. *Conversations in Colombia*. Cambridge: Cambridge University Press.

Guerra, Jairo Miguel. 1997. *Sistematización de la experiencia PBP*. Bogotá: PBP.

Guha, Ramachandra, and Joan Martínez Alier. 1997. *Varieties of Environmentalism*. London: Earthscan.

Guha, Ranajit. 1988. "The Prose of Counter-Insurgency." In R. Guha and G. Spivak, eds., *Selected Subaltern Studies*, 37–44. Delhi: Oxford University Press.

Gupta, Akhil. 1998. *Postcolonial Developments*. Durham: Duke University Press.

———. 1995. "Blurred Boundaries: The Discourse of Corruption, the Culture of Politics, and the Imagined State." *American Ethnologist* 22 (2): 375–402.

Habermas, Jürgen. 1987. *The Philosophical Discourse of Modernity*. Cambridge, Mass.: MIT Press.

———. 1973. *Legitimation Crisis*. Boston: Beacon Press.

Haenn, Nora. 2005. *Fields of Power, Forests of Discontent: Culture, Conservation, and the State in Mexico*. Tucson: University of Arizona Press.

Haenn, Nora, and Richard Wilk, eds. 2005. *The Environment in Anthropology*, New York: New York University Press.

Haila, Yrjö, and Chuck Dyke, eds. 2006. *How Nature Speaks. The Dynamic of the Human Ecological Condition*. Durham: Duke University Press.

Hakken, David. 2003. *The Knowledge Landscapes of Cyberspace*. London: Routledge

Hale, Charles. 2006. "Activist Research vs. Cultural Critique: Indigenous Land Rights and the Contradictions of Politically Engaged Anthropology." *Cultural Anthropology* 21(1):96–120.

Hall, Stuart. 1996. "Introduction: Who Needs 'Identity'"? In S. Hall and P. du Gay, eds., *Questions of Cultural Identity*, 1–17. London: Sage.

———. 1990. "Cultural Identity and Diaspora." In J. Rutherford ed., *Identity, Community, Culture, Difference*, 392–403. London: Lawrence and Wishart.

Hammond, Debora. 2003. *The Science of Synthesis*. Boulder: University Press of Colorado.

Hansen, Thomas, and Finn Stepputat, eds. 2001. *States of Imagination: Ethnographic Explorations of the Postcolonial State*. Durham: Duke University Press.

Haraway, Donna. 1997. *Modest_Witness@Second_Millennium: FemaleMan_Meets_OncoMouse*. New York: Routledge.

———. 1991. *Simians, Cyborgs and Women: The Reinvention of Nature*. New York: Routledge.

———. 1989. *Primate Visions*. New York: Routledge.

———. 1988. "Situated Knowledges: The Science Question in Feminism and the Privilege of Partial Perspective." *Feminist Studies* 14(3): 575–99.

Harcourt, Wendy, and Arturo Escobar, eds. 2005. *Women and the Politics of Place*. Bloomfield, Conn.: Kumarian Press.

Harcourt, Wendy, ed. 1999. *Women@Internet: Creating New Cultures in Cyberspace*. London: Zed Books.

Hardt, Michael, and Antonio Negri. 2000. *Empire*. Cambridge, Mass.: Harvard University Press.

Hayden, Cory. 2004. *When Nature Goes Public*. Princeton: Princeton University Press.

Heidegger, Martin. 1977a. "The Age of the World Picture." In M. Heidegger, *The Question Concerning Technology*, 115–54. New York: Harper and Row.

———. 1977b. *The Question Concerning Technology*. New York: Harper and Row.

Heims, Steve. 1991. *The Cybernetics Group*. Cambridge, Mass.: MIT Press.

Heller, Chaia. 2005. "From Scientific Risk to Paysan Savoir-Faire: Divergent Rationalities of Science and Society in the French Debate over GM Crops." PHD diss., Department of Anthropology, University of Massachusetts, Amherst.

———. 2002. "From Scientific Risk to *Paysan* Savoir-Faire: Peasant Expertise in the French and Global Debate over GM Crops." *Science as Culture* 11(1): 5–37.

———. 2000. *Ecology of Everyday Life*. Montreal: Black Rose.

Heller, Chaia, and Arturo Escobar. 2003. "From Pure Genes to GMOs: Transnational Gene Landscapes in the Biodiversity and Transgenic Food Networks." In A. Goodman, D. Heath, and S. Lindee, eds., *Genetic Nature/Culture*, 155–75. Berkeley: University of California Press.

Hernandez, Jorge, Alain Bidoux, Edgar Cortés, and Julio Tresierra. 1995. *Proyecto Biopacífico: Primera Evaluación Interna. Informe Final*. Bogotá: PBP.

Hess, David. 2001. "Ethnography and the Development of Science and Technology Studies." In P. Atkinson et al., eds., *Handbook of Ethnography*, 234–45. London: Sage.

Hirsch, Eric, and Marilyn Strathern, eds. 2004. *Transactions and Creations: Property Debates and the Stimulus of Melanesia*. Oxford: Berghanh.

Hoffmann, Odile. 1999. "Territorialidades y alianzas: construcción y activación de espacios locales en el Pacífico." In J. Camacho and E. Restrepo, eds., *De montes, ríos, y ciudades*, 75–94. Bogotá: ICANH/Natura/Ecofondo.

Holland, Dorothy, and Jean Lave. 2001. "History in Person: An Introduction." In D. Holland and J. Lave, eds., *History in Person: Enduring Struggles, Contentious Practice, Intimate Identities*, 3–36. Santa Fe: School of American Research.

Holland, Dorothy, William Lachicotte, Debra Skinner, and Carole Cain. 1998. *Identity and Agency in Cultural Worlds*. Cambridge, Mass.: Harvard University Press.

Hurtado, Mary Lucía. 1996. "La construcción de una nación multiétnica y pluricultural." In A. Escobar and A. Pedrosa, eds., *Pacífico: ¿Desarrollo o diversidad?* 329–52. Bogotá: ICANH/CEREC.

Hvalkof, Søren, and Arturo Escobar. 1998. "Political Ecology and Social Practice: Notes Towards an Academic and Political Agenda." In A. Goodman and T. Leathermann, eds., *Building a New Biocultural Synthesis*, 425–50. Ann Arbor: University of Michigan Press.

IGAC (Instituto Geográfico Agustín Codazzi). 2000. *Zonificación ecológica de la región Pacífica colombiana*. Bogotá: Ministerio del Ambiente.

IGGRI. 1999. *Expanding People's Spaces in the Globalising Economy: Report of a Gathering*. Hanasaari, Finland: IGGRI.

Illich, Ivan. 1982. *Gender*. New York: Pantheon Books.

INCODER. 2005. *Los cultivos de palma de aceite en los territorios colectivos de las comunidades negras de los ríos Curvaradó y Jiguamiandó, en el departamento del Chocó*. Bogotá: INCODER, Subgerencia de Ordenamiento Social de la Propiedad, Grupo de Asuntos Etnicos.

INCORA (Instituto Colombiano de la Reforma Agraria). 1998. *Tierras de las comunidades Negras: Guía para la constitución de consejos comunitarios y formulación de solicitudes de titulación colectiva de las tierras de comunidades negras*. Bogotá: INCORA.

Ingold, Tim. 2000a. *The Perception of the Environment*. London: Routledge.

———. 2000b. "Three in One: On Dissolving the Distinctions Between Body, Mind and Culture." Manuscript, Department of Anthropology, University of Manchester.

———. 2000c. "Ancestry, Generation, Substance, Memory, Land." In T. Ingold, *The Perception of the Environment*, 132–52. London: Routledge.

———. 1993. "The Temporality of the Landscape." *World Archaeology* 25(2): 152–73.

———. 1992. "Culture and the Perception of the Environment." In E. Croll and D. Parkin, eds., *Bush Base: Forest Farm*, 39–56. London: Routledge.

———. 1990. "An Anthropologist Looks at Biology." *Man* 25(2): 208–29.

Instituto de Investigaciones Ambientales del Pacífico. 2000. *Agenda Pacífico XXI*. Quibdó: IIAP.

Irigaray, Luce. 2001. *To Be Two*. New York: Routledge.

———. 2000. *Why Different?* New York: Semiotext(e).

———. 1996. *I Love To You*. New York: Routledge.

Jackson, Michael, ed. 1996. *Things as They Are: New Directions in Phenomenological Anthropology*. Bloomington: Indiana University Press.

Jantsch, Erich. 1980. *The Self-Organizing Universe*. New York: Pergamon Press.

Jantsch, Erich, and Conrad Waddington, eds. 1976. *Evolution and Human Consciousness*. London: Addison-Wesley.

Janzen, Daniel. 1992. "A South–North Perspective on Science in the Management, Use and Economic Development of Biodiversity." In O. T. Sanlund, K. Hindar and A. H. D. Brown, eds., *Conservation of Biodiversity for Sustainable Development*, 27–52. Oslo: Scandinavian University Press.

Jordan, Joseph. 2006. "March 6, 2006, Manifesto. AFAM Diaspora Lecture." Manuscript, Sonja Haynes Stone Center for Black Culture and History, University of North Carolina, Chapel Hill.

Joseph, Gilbert, and Daniel Nugent. 1994. *Everyday Forms of State Formation: Revolution and the Negotiation of Rule in Modern Mexico*. Durham: Duke University Press.

Joxe, Alain. 2002. *Empire of Disorder*. New York: Semiotext(e).

Juris, Jeff. 2005. "The New Digital Media and Activist Networking within Anti-Corporate Globalization Movements." *Annal*, AAPSS 597: 189–208.

———. 2004. "Digital Age Activism: Anti-Corporate Globalization and the Cultural Politics of Transnational Networking." PHD diss., Department of Anthropology, University of California, Berkeley.

Kahn, Joel. 2001. "Anthropology and Modernity." *Current Anthropology* 42(5): 651–80.

Kamat, Sangeeta. 2002. *Development Hegemony: NGOs and the State in India*. Delhi: Oxford University Press.

Kaufman, Stuart. 1995. *At Home in the Universe: The Search for Laws of Self-Organization and Complexity*. Oxford: Oxford University Press.

Keck, Margaret, and Kathryn Sikkink. 1998. *Activists Beyond Borders: Advocacy Networks in International Politics*. Ithaca: Cornell University Press.

Keller, Evelyn Fox. 1995. *Refiguring Life*. New York: Columbia University Press.

King, Mary. 2006. "Emergent Socialities: Networks of Biodiversity and Anti-Globalization." PHD diss., Department of Anthropology, University of Massachusetts, Amherst.

———. 2005. "Looking out from Within, Confronting the Dilemmas of Activist Research." Presented at the AAA Annual Meeting, Washington, D.C. November 30–December 4.

———. 2000. "Of Unknown Quantity: NGO Network Organizing and Global Environmental Politics." Presented at the Session on Actors, Networks, Meanings: Environmental Social Movements and the Anthropology of Activism, AAA Annual Meeting, San Francisco, November 15–19.

Kirsch, Stuart. 2001. "Lost Worlds: Environmental Disaster, 'Culture Loss,' and the Law." *Current Anthropology* 42(2): 167–98.

Klandermans, Bert. 1992. "La Construcción Social de la Protesta y los Campos Pluriorganizativos." In A. Morris and C. Mueller, eds., *The Frontiers in Social Movement Theory*. New Haven: Yale University Press.

Koczberski, Gina. 2002. "Plots, Plates and Tinpis: New Income Flows and the Strengthening of Women's Gendered Identities in Papua New Guinea." *Development* 45(1): 88–93.

Kuletz, Valerie. 1998. *The Tainted Desert*. New York: Routledge.

Laclau, Ernesto. 1996. *Emancipation(s)*. London: Verso.

Laclau, Ernesto, and Chantal Mouffe. 1985. *Hegemony and Socialist Strategy*. London: Verso.

Lander, Edgardo. 2002. "Los derechos de propiedad intelectual en la geopolítica del saber de la sociedad global." In Catherine Walsh, F. Schiwi, and S. Castro-Gómez, eds., *Interdiscplinar las ciencias sociales*, 73–102. Quito: Abya-Yala.

———, ed. 2000. *La colonialidad del saber: eurocentrismo y ciencias sociales*. Buenos Aires: CLASCO.

Laszlo, Ervin. 1972. *The Systems View of the World*. New York: Braziller.

———, ed. 1972. *The Relevance of General Systems Theory*. New York: Braziller.

Latour, Bruno. 1997. "The Trouble with Actor-Network Theory." In http://www.ensmp.fr/ffilatour/poparticles/poparticle/po67.html.

———. 1993. *We Have Never Been Modern*. Cambridge, Mass.: Harvard University Press.

———. 1988. *The Pasteurization of France*. Cambridge, Mass.: Harvard University Press.

Lave, Jean, and Etienne Wenger. 1991. *Situated Learning: Legitimate Peripheral Participation*. Cambridge: Cambridge University Press.

Law, John. 2000 [1992]. "Notes on the Theory of the Actor Network." Downloaded from http://tina.lancs.ac.uk/sociology/soc054jl.html.

Leal, Claudia. 2003. "Natural Treasures and Racial Tensions: The Pacific Lowlands of Colombia at the Turn of the 19th Century, 1880–1930." Presented at the Congress of the Latin American Studies Association, LASA, Dallas, March 27–29.

———. 2004. "Black Forests: The Pacific Lowlands of Colombia, 1850–1930." PHD diss., Department of Geography, University of California, Berkeley.

———, ed. 1995. *Economías de las comunidades rurales en el Pacífico colombiano*. Bogotá: Proyecto Biopacífico.

Leal, Claudia, and Eduardo Restrepo. 2003. *Unos bosques sembrados de aserríos: Historia de la extracción maderera en el Pacífico colombiano*. Medellín: Editorial de la Universidad de Antioquia.

Leal, Francisco, ed. 1999. *Los laberintos de la guerra: Utopías e incertidumbres sobre la paz*. Bogotá: Tercer Mundo.

Leathermann, Thomas, and Brooke Thomas. 2001. "Political Ecology and Constructions of Environment in Biological Anthropology." In C. Crumley, ed., *New Directions in Anthropology and Environment: Intersections*, 113–31. Walnut Cove, Calif.: Altamira Press.

Leesberg, July, and Emperatriz Valencia. 1987. *Los sistemas de producción en el medio Atrato (Chocó)*. Quibdó: Proyecto DIAR.

Leff, Enrique. 1998a. *Saber Ambiental*. Mexico City: Siglo XXI.

———. 1998b. "Murray Bookchin and the End of Dialectical Materialism." *Capitalism, Nature, Socialism* 9(4): 67–93.

———. 1995a. *Green Production: Toward an Environmental Rationality*. New York: Guilford Press.

———. 1995b. "De quién es la naturaleza? Sobre la reapropriación social de los recursos naturales." *Gaceta Ecológica* 37: 58–64.

———. 1993. "Marxism and the Environmental Question." *Capitalism, Nature, Socialism* 4(1): 44–66.

———, ed. 1986. *Los problemas del conocimiento y la perspectiva ambiental del desarrollo*. Mexico City: Siglo XXI.

Levins, Richard, and Richard Lewontin. 1985. *The Dialectical Biologist*. Cambridge, Mass.: Harvard University Press.

Lévy, Pierre. 1997. *Collective Intelligence: Mankind's Emerging World in Cyberspace*. New York: Plenum Trade.

Leyva, Pablo, ed. 1993. *Colombia Pacífico*. 2 vols. Bogotá: Fondo Fen.

Leyva Solano, Xochitl. 2003. "Concerning the Hows and Whys in the Ethnography of Social Movement Networks." Presented at the Department of Anthropology, University of North Carolina, Chapel Hill, March 17.

———. 2002. "Neo-Zapatismo: Networks of Power and War." PHD diss., Department of Anthropology, University of Manchester.

Llano, María Clara, ed. 1998. *La gente de los ríos: Junta Patía*. Bogotá: ICANH/PNR.

Lobo-Guerrero, Alberto. 1993. "Hidrología e hidrogeología." In Pablo Leyva, ed., *Colombia Pacífico*, 1:120–34. Bogotá: Fondo FEN.

Losonczy, Anne-Marie. 1999. "Memorias e identidad: los negro-colombianos del Chocó." In J. Camacho and E. Restrepo, eds., *De montes, ríos y ciudades: Territorios e identidades de la gente negra en Colombia*, 13–23. Bogotá: ICANH/Natura/Ecofondo.

———. 1997a. *Les Saints et la Foret*. Paris: L'Harmattan.

———. 1997b. "Hacia una antropología de lo inter-étnico: una perspectiva negro-americana e indígena." In M. V. Uribe and E. Restrepo, eds., *Antropología en la modernidad*, 253–78. Bogotá: ICANH.

———. 1993. "De lo vegetal a lo humano: Un modelo cognitivo afro-colombiano del Pacífico." *Revista Colombiana de Antropología* 30: 39–57.

———. 1989. "Del ombligo a la comunidad: Ritos de nacimiento en la cultura negra del Litoral Pacífico colombiano." *Revindi* 1: 49–54.

Lozano, Betty Ruth. 2005. "Las afro-reparaciones: el punto de vista de las mujeres afrocolombianas." Presented at the Conference "Afro-reparaciones: Memorias de la esclavitud y justicia social contemporánea," Cartagena, Octubre.

———. 1996. "Mujer y desarrollo." In A. Escobar and A. Pedrosa, eds., *Pacífico: ¿Desarrollo o diversidad?* 176–204. Bogotá: ICANH/CEREC.

Lu, Flora. 2007. "Integration to Market among Indigenous Peoples: A Cross-Cultural Perspective from the Ecuadorian Amazon." *Current Anthropology* 48(4): 593–602.

Luke, Timothy. 1999. *Capitalism, Democracy, and Ecology*. Chicago: University of Chicago Press.

Mander, Jerry, and Victoria Tauli-Corpuz, eds. 2006. *Paradigm Wars: Indigenous People's Resistance to Globalization*. San Francisco: Sierra Club.

Marcus, George, ed. 1999. *Critical Anthropology Now*. Santa Fe: School of American Research.

Markos, Anton. 2002. *Readers of the Book of Life: Contextualizing Developmental Evolutionary Biology*. Oxford: Oxford University Press.

Marston, Sally, John Paul Jones III, and Keith Woorward. 2005. "Human Geography without Scale." *Transactions of the Institute of British Geography* NS 30: 416–32.

Martín Alcoff, Linda. 2006. *Visible Identities: Race, Gender, and the Self*. Oxford: Oxford University Press.

Martín Alcoff, Linda, Michael Hames-García, Satya Mohanty, and Paula Moya, eds. 2006. *Identity Politics Reconsidered*. New York: Palgrave/Macmillan.

Martínez, José. 2004. "Critical Anthropo-Geographies: Contested Place, Nature, and Development in the 'Zona Maya' of Quintana Roo, Mexico." PHD diss., Department of Anthropology, University of Massachustts, Amherst.

Martínez Alier, Joan. 2002. *The Environmentalism of the Poor: A Study of Ecological Conflicts and Valuation*. London: Elgar.

———. 1996. "Merchandising Biodiversity." *Capitalism, Nature, Socialism* 7(1): 37–54.

Martínez M., Jaime Orlando. 1993. "Geomorfología." In P. Leyva, ed., *Colombia Pacífico*, 1:110–19. Bogotá: Fondo Fen.

Massey, Doreen. 1997. "A Global Sense of Place." In A. Gray and J. McGuigan, eds., *Studying Culture*, 232–40. London: Edward Arnold.

———. 1994. *Space, Place and Gender*. Minneapolis: University of Minnesota Press.

Mato, Daniel. 2005. "Estudios intelectuales latinoamericanos en cultura y poder." In Daniel Mato, ed., *Cultura, polîtica y sociedad: Perspectivas latinoamericanas*, 471–97. Buenos Aires: CLASCO.

Maturana, Humberto, and Francisco Varela. 1987. *The Tree of Knowledge*. Berkeley: Shambhala.

———. 1980. *Autopoiesis and Cognition*. Boston: Reidel.

Mayr, Ernst. 1982. *The Growth of Biological Thought*. Cambridge, Mass.: Harvard University Press.

McAfee, Katherine. 1999. "Setting Nature to Save It? Biodiversity and Green Developmentalism." *Environmental Planning D: Society and Space* 17:133–54.

Medeiros, Carmen. 2005. "The Right "To Know How to Understand": Coloniality and Contesting Visions of Development and Citizenship in the Times of Neo-Liberal Civility." PHD diss., Department of Anthropology, CUNY Graduate Center.

Meertens, Doreen. 2000. "El futuro nostálgico: desplazamiento, terror y género." *Revista colombiana de antropología* 36: 112–35.

Melucci, Alberto. 1989. *Nomads of the Present*. Philadelphia: Temple University Press.

Merchant, Carolyn. 1980. *The Death of Nature*. New York: Harper and Row.

Merizalde, Bernardo. 1921. *Estudio de la costa colombiana del Pacífico*. Bogotá: Imprenta del Estado Mayor General.

Mesa Dishington, Jans. 1993. "La Palma debería ser un propósito nacional." *El Palmicultor* 260: 32–39.

Mies, Maria. 1986. *Patriarchy and Accumulation on a World Scale*. London: Zed Books.

Mignolo, Walter. 2001. "Local Histories and Global Designs: An Interview with Walter Mignolo." *Discourse* 22(3): 7–33.

———. 2000. *Local Histories/Global Designs*. Princeton: Princeton University Press.

———. 1995. *The Darker Side of the Renaissance*. Ann Arbor: University of Michigan Press.

———, ed. 2001. *Capitalismo y geopolítica del conocimiento*. Buenos Aires: Ediciones del Signo.

Miller, David, and Peter Reill, eds. 1996. *Vision of Empire: Voyages, Botany, and Representatives of Nature*. Cambridge: Cambridge University Press.

Milton, Kay. 1996. *Environmentalism and Cultural Theory*. London: Routledge.

———, ed. 1993. *Environmentalism: The View from Anthropology*. London: Routledge.

Ministerio del Medio Ambiente de Colombia. 2000. *Zonificación ecológica de la región pacífica colombiana*. Bogotá: IGAC/PMNR

Ministerio del Medio Ambiente/DNP/GEF. 1996. *Hacia una estrategia concertada de conservación y uso sostenible de la biodiversidad en el Pacífico colombiano. Plan Operativo 1995–1997*. Bogotá: Ministerio del Ambiente-DNP-GEF COL/92/G31.

Ministerio del Medio Ambiente/PBP. 1994. *Política de conservación de la biodiversidad para el desarrollo sostenible de la región biogeográfica del Pacífico (propuesta)*. Bogotá: Ministerio del Ambiente/PBP COL 92/G31.

Mitchell, Timothy. 1991. "The Limits of the State: Beyond the Statist Approach and Their Critics." *American Political Science Review* 85(1): 78–96.

Mooney, H. A., et al. 1995. "Biodiversity and Ecosystem Functioning." In United Nations Environment Program, ed., *Global Biodiversity Assessment*, 275–452. Cambridge: Cambridge University Press.

Mooney, Pat. 1979. *Seeds of the Earth*. London: ICDA.

Moran, Emilio. 1991. "Ecosystems Ecology in Biology and Anthropology: A Critical Assessment." In E. Moran, ed., *The Ecosystems Approach in Anthropology*, 3–40. Michigan: University of Michigan Press.

Mosquera, Claudia, and Luiz C. Barcelos, eds. 2007. *Afro-reparaciones: Memorias de la esclavitud y justicia reparativa para negros, afrocolombianos y raizales*. Bogotá: Universidad Nacional.

Mosquera, Claudia, Mauricio Pardo, and Odile Hoffmann, eds. 2002. *Afrodescendientes en las Américas: Trayectorias sociales e identitarias. (150 años de la abolición de*

la esclavitud en Colombia). Bogotá: Universidad Nacional de Colombia-ICANH-IRD-ILAS.
Mosquera, Gilma. 1999. "Hábitats y espacio productivo y residencial en las aldeas parentales del Pacífico." In J. Camacho and E. Restrepo, eds., *De montes, ríos y ciudades*, 49–74. Bogotá: ICANH/Natura/Ecofondo.
Mosse, David. 2005. *Cultivating Development: An Ethnography of Aid Policy and Practice*. London: Pluto Press.
Mouffe, Chantal. 1993. *The Return of the Political*. London: Verso.
Movimiento Social de Comunidades Afrocolombianas. 2002. "Principios de relacionamiento y agenda común." Cali: Movimiento Social de Comunidades Negras, Consejos Comunitarios del Pacífico, y Unidad Administrativa Especial del Sistema de Parques Nacionales Naturales.
Mueller, Adele. 1987. "Peasants and Professionals: The Social Organization of Women in Development Knowledge." PHD diss., Ontario Institute for Studies in Education.
Munda, Giuseppe. 1995. *Multicriteria Evaluation in a Fuzzy Environment: Theory and Applications in Ecological Economics*. Heidelberg: Physica Verlag.
Múnera, Alfonso. 2005. *Fronteras imaginadas: La construcción de las razas y de la geografía en el siglo XIX colombiano*. Bogot·: Planeta.
Nazarea, Virginia, ed. 1999. *Ethnoecology: Situated Knowledge/Located Lives*. Tucson: University of Arizona Press.
———. 1998. *Cultural Memory and Biodiversity*. Tucson: University of Arizona Press.
Ng'weno, Bettina. 2007. *Tuif Weis: Territory and Citizenship in the Contemporary State*. Stanford: Stanford University, Press.
Nonini, Donald, Charles Price, and Erich Fox-Tree. Forthcoming. "Grounded Utopian Movements: Subjects of Neglect." In press, *Anthropological Quarterly*.
Norgaard, Richard. 1995. *Development Betrayed*. New York: Routledge.
Norval, Aletta. 1996. "Thinking Identities: Against a Theory of Ethnicity." In E. Wilmsem and P. McAllister, eds. *The Politics of Difference: Ethnic Premises in a World of Power*. 59–70. Chicago: University of Chicago Press.
Notes from Nowhere. 2003. *We Are Everywhere: The Irresistible Rise of Global Anti-Capitalism*. London: Verso.
NRC (National Research Council). 1999. *Perspectives on Biodiversity*. Washington, D.C.: National Academy Press.
Ocampo, José Antonio. 2004. *Reconstruir el futuro: Globalización, desarrollo y democracia en América Latina*. Bogotá: Editorial Norma.
OCN (Organización de Comunidades Negras de Buenaventura). 1996. "Movimiento Negro, Identidad y Territorio: Entrevista con OCN." In A. Escobar and A. Pedrosa, eds., *Pacífico: ¿Desarrollo o diversidad?* 245–82. Bogotá: ICANH/CEREC.
O'Connor, James. 1998. *Natural Causes*. New York: Guilford Press.
O'Connor, Martin. 1993. "On the Misadventures of Capitalist Nature." *Capitalism, Nature, Socialism* 4(3): 7–40.

Offen, Karl. 2003. "The Territorial Turn: Making Black Territories in Pacific Colombia." *Journal of Latin American Geography* 2(1): 43–73.
Olesen, Thomas. 2005. "Long Distance Zapatismo: Globalization and the Construction of Solidarity." PHD diss., Department of Political Science, University of Aarhus.
Olinto Rueda, José. 1993. "Población y poblamiento." In Pablo Leyva, ed., *Colombia Pacífico*, 1:464–86. Bogotá: Fondo FEN.
Ong, Aihwa. 1987. *Spirits of Resistance and Capitalist Discipline*. Albany: SUNY Press.
Oslender, Ulrich. 2004. "Geografía de terror y desplazamiento forzado en el Pacífico colombiano: conceptualizando el problema y buscando respuestas." In Eduardo Restrepo and Axel Rojas, eds., *Conflicto e (in)visibilidad: retos en los estudios de la gente negra en Colombia*, 35–52. Popayán: Editorial Universidad del Cauca.
———. 2002. "'The Logic of the River': A Spatial Approach to Ethnic-Territorial Mobilization in the Colombian Pacific Region." *Journal of Latin American Anthropology* 7(2): 86–117.
———. 2001. "Black Communities on the Colombian Pacific Coast and the 'Aquatic Space': A Spatial Approach to Social Movement Theory." PHD diss., Department of Geography, University of Glasgow.
———. 1999. "Espacio e identidad en el Pacìfico colombiano." In J. Camacho and E. Restrepo, eds., *De montes, ríos y ciudades: Territorios e identidades de la gente negra en Colombia*, 25–48. Bogotá: ICANH/Natura/Ecofondo.
Osterweil, Michal. 2006. "Theoretical-Practice: Il Movimento dei Movimenti and (Re)Inventing the Political?" Presented at Fondazione Giangiacamo Feltrinelli, Annual Colloquium, "Cultural Conflicts, Social Movements and New Rights: A European Challenge," October, Cortona, Italy. See http://www.fondazionefeltrinelli.it/it/risorse_digitali/papers/colloquio-di-cortona-2006
———. 2005a. "Place-Based Globalism: Locating Women in the Alternative Globalization Movement." In W. Harcourt and A. Escobar, eds., *Women and the Politics of Place*, 174–89. Bloomfield, Conn.: Kumarian Press.
———. 2005b. "Social Movements as Knowledge Practice Formations: Towards an Ethno-Cartography of Italy's Movimento dei Movimenti." PHD diss. prospectus, Department of Anthropology, University of North Carolina, Chapel Hill.
———. 2004. "A Cultural-Political Approach to Reinventing the Political." *International Social Science Journal* 182:495–506.
Otoya, Jaime. 1993. "La industria del camarón en Tumaco." Presented at the Primer Seminario sobre Constitución, Desarrollo, y Autonomía, Tumaco, October 15–16.
Oyama, Susan. 2006. "Speaking of Nature." In Yrjö Haila and Chuck Dyke, eds., *How Nature Speaks: The Dynamic of the Human Ecological Condition*, 49–66. Durham: Duke University Press.
———. 2000. *Evolution's Eye: A Systems View of the Biology–Culture Divide*. Durham: Duke University Press.

Palacios Santamaría, Aída. 1993. "Cultura material indígena o artesanías?" In Pablo Leyva, ed., *Colombia Pacífico*, 1:362–67. Bogotá: Fondo FEN.

Palenque el Congal. 2003a. "Primer taller de evaluación y planificación estratégica: Palenque regional el Congal—Proceso de Comunidades Negras. Taller 'Análisis Situacional.' July 5–8." Unpublished report. Buenaventura: PCN.

———. 2003b. "Segundo taller de evaluación y planificación estratégica: Palenque regional el Congal—Proceso de Comunidades Negras. Taller 'Construcción del Sueño Estratégico.' July 5–8." Unpublished report. Buenaventura: PCN.

Panikkar, Raimon. 1993. *The Cosmotheandric Experience*. New York: Orbis Books.

Pardo, Mauricio, ed. 2001. *Acción colectiva, estado y etnicidad en el Pacífico colombiano*. Bogotá: ICANH/Colciencias.

Pardo, Mauricio, and Manuela Alvarez. 2001. "Estado y movimiento negro en el Pacìfico colombiano." In M. Pardo, ed., *Acción colectiva, estado y etnicidad en el Pacífico colombiano*, 229–58. Bogotá: ICANH/Colciencias.

Parpart, J. L., and M. H. Marchand, eds. 1995. *Feminism/Postmodernism/Development*. London: Routledge.

Patiño, Víctor Manuel. 1948. "Información preliminar sobre la Palma de Aceite Africana (*elaeis guineensis*) en Colombia." Cali: Imprenta Departamental, Serie Botánica Aplicada, no. 2.

Paulson, Susan, and Lisa Gezon, eds. 2005. *Political Ecology across Spaces, Scales, and Social Groups*. New Brunswick: Rutgers University Press.

PBP (Proyecto Biopacífico). 1999. *El estudio de la biodiversidad regional: Aportes al conocimiento y a la práctica investigativa. Proyecto Biopacífico, Informe final general*. Volume 1. Bogotá: PBP.

———. 1998. *Diversidad amenazada: Prioridades de manejo y conservación. Proyecto Biopacfico, Informe final general*. Volume 2. Bogotá: PBP.

PCN (Proceso de Comunidades Negras). 2007. "Territorio y conflicto desde la perspectiva del Proceso de Comunidades Negras de Colombia. Reporte Proyecto PCN-LASA *Otros Saberes*." Cali: PCN.

———. 2006. "Propuesta para la apropriación y posicionamiento de las Comunidades Negras como Pueblo en torno a los derechos económicos, sociales, y culturales." Project proposal. Buenaventura: PCN.

———. 2003a. *Fortalecimiento organizativo y apropiación territorial en los territorios colectivos de las comunidades negras del* PCN *en el rio Yurumanguí (Valle) y el Territorio Pílamo (Cauca). Proyecto* PCN*—Solsticio. Quinto informe técnico. Octubre 2002—Junio 2003*.

———. 2003b. *Proyecto Contribución al fortalecimiento de la autonomía alimentaria y dinámica organizativa con jóvenes en el río Yurumanguí y sus familias. Reporte presentado a Global Greengrants Funds* (Boulder).

———. 2002. *Fortalecimiento de las dinámicas organizativas del Proceso de Comunidades Negras del Pacífico sur colombiano, en torno al ejercicio de los derechos étnicos, culturales y territoriales. Proyecto* PCN*-Solsticio. Cuarto informe técnico Febrero—Septiembre 2002*.

———. 2000. *Fortalecimiento de las dinámicas organizativas del Proceso de Comunidades*

Negras del Pacífico sur colombiano, en torno al ejercicio de los derechos étnicos, culturales y territoriales. Proyecto PCN-Solsticio. Segundo informe técnico trimestral, Septiembre—Noviembre 2000.
———. 1999. "Seminario/Conferencia sobre la situación y la auto-identificación de las Comunidades Negras de Centroamérica, el Caribe, y Suramérica." Presented to IGWIA for Funding. Buenaventura: PCN.
———. 1995a. "Informe viaje a Europa." Unpublished report. Buenaventura, Colombia.
———. 1995b. "Comunidades Negras y Derechos Humanos en Colombia." Unpublished report. Buenaventura: PCN.
———. 1994. "Principios para el Plan de Desarrollo de Comunidades Negras." Presentado a la Comisíon Técnica Formuladora del Plan de Desarrollo para las Comunidades Negras, Bogotá.
Peck, Jamie. 2000. "Political Economies of Scale." Presented at the workshop Producing Place(s), Miami University, Oxford, Ohio, May 12, 13.
Pedrosa, Alvaro. 1996a. "Paisaje y cultura." In A. Escobar and A. Pedrosa, eds., *Pacífico: ¿Desarrollo o diversidad?* 29–40. Bogotá: CEREC/EcoFondo.
———. 1996b. "La institucionalización del desarrollo." In A. Escobar y A. Pedrosa, eds., *¿Pacífico: ¿Desarrollo o diversidad?* 41–65. Bogotá: CEREC/EcoFondo.
———. 1989a. "Desarrollo sostenible del alfabetismo y la literalidad en el Pacífico colombiano." Informe final presentado a Colciencias, Universidad del Valle, Cali.
———. 1989b. "Sistema integrado de servicios editoriales y Red de Editores del occidente colombiano." Manuscript, Cali, Fundación Habla/Scribe.
Pedrosa, Alvaro, and Alfredo Vanín. 1994. *La vertiente afropacífica de la tradición oral.* Cali: Universidad del Valle.
Peet, Richard, and Michael Watts, eds. 1996. *Liberation Ecologies: Environment, Development, Social Movements.* London: Routledge.
Peltonen, Lasse. 2006. "Fluids on the Move: An Analogical Account of Environmental Mobilization." In Yrjö Haila and Chuck Dyke, eds., *How Nature Speaks. The Dynamics of the Human Ecological Condition,* 150–76. Durham: Duke University Press.
Pinto-Escobar, Polidoro. 1993. "José Cuatrasecas y la flora y la vegetación." In Pablo Leyva, ed., *Colombia Pacífico,* 1:168–79. Bogotá: Fondo FEN.
Plant, Roger, and Søren Hvalkof. 2001. *Land Titling and Indigenous Peoples.* Washington, D.C.: Inter-American Development Bank.
Polanyi, Karl. 1957. *The Great Transformation.* Boston: Beacon Press.
Polanyi, Karl, Conrad Arensberg, and Harry Pearson, eds. 1944. *Trade and Market in the Early Empires.* Glencoe, Ill: Free Press.
Prahl, H. von, Jaime Cantera, and Rafael Contreras. 1990. *Manglares y hombres del Pacífico Colombiano.* Bogotá: Fondo FEN.
Pratt, Mary. 1992. *Imperial Eyes: Travel Writing and Transculturation.* New York: Routledge.

Pred, Alan, and Michael Watts. 1992. *Reworking Modernity*. New Brunswick: Rutgers University Press.

Prigonine, Ilya, and Grégoire Nicolis. 1989. *Exploring Complexity*. New York: W. H. Freeman.

Prigonine, Ilya, and Isabelle Stengers. 1984. *Order Out of Chaos*. New York: Bantam Books.

Procesos Organizativos de Comunidades Negras e Indígenas. 1995. *Propuesta de Reformulación del PBP. Documento Marco*. Buenaventura: PCN.

Quijano, Aníbal. 2002. "El Nuevo imaginario anti-capitalista." In http://www.forumsocialmundial.org.br/dinamic/es/tbib_Anibal_Quijano.asp

———. 2000. "Coloniality of Power, Ethnocentrism, and Latin America." *Nepantla* 1(3): 533–80.

———. 1993. "Modernity, Identity, and Utopia in Latin America." In J. Beverly and J. Oviedo, *The Postmodernism Debate in Latin America*, 140–55. Durham: Duke University Press.

———. 1988. *Modernidad, identidad y utopía en América Latina*. Lima: Sociedad y Política Ediciones.

Quijano, Aníbal, and Immanuel Wallerstein. 1992. "Americanity as a Concept, or the Americas in the Modern World-System." *International Social Science Journal* 134: 459–559.

Quiroga, Diego. 1994. "Saints, Virgins, and the Devil: Witchcraft, Magic, and Healing in the Northern Coast of Ecuador." PHD diss., Department of Anthropology, University of Illinois.

Rabinow, Paul. 2003. *Anthropos Today. Reflections on Modern Equipment*. Princeton: Princeton University Press.

Rahnema, Majid, and Victoria Bawtree, eds. 1997. *The Post-Development Reader*. London: Zed Books.

Ramírez, Sergio. 1991. "Prólogo. Visibilidades refractadas." In N. S. de Friedemann and A. Vanín, eds., *El Chocó: magia y leyenda*. Bogotá: Litografía Arco.

Rangel, J. Orlando, and Peter Lowy. 1993. "Tipos de vegetación y rasgos fitogeográficos." In Pablo Leyva, ed., *Colombia Pacífico*, 1:182–98. Bogotá: Fondo FEN.

Rappaport, Joanne. 2005. *Intercultural Utopia: Public Intellectuals, Cultural Experimentation, and Ethnic Pluralism in Colombia*. Durham: Duke University Press.

Rappaport, Roy. 1991. "Ecosystems, Population, and People." In Emilio Moran, ed., *The Ecosystems Approach in Anthropology*, 41–73. Ann Arbor: University of Michigan Press.

Redclift, Michael. 1987. *Sustainable Development: Exploring the Contradictions*. London: Routledge.

Restrepo, Eduardo. 2008. "Eventalizing Blackness in Colombia." PHD diss., Department of Anthropology, University of North Carolina, Chapel Hill.

———. 2007. "Antropologìa y descolonialidad." In Santiago Castro-Gómez and Ramón Grosfogel, eds., *El giro descolonial: Reflexiones para una diversidad*

epistèmica más allá del capitalismo global, 289–304. Bogotá: Universidad Central/ Universidad Javeriana.
———. 2005. *Políticas de la teoría y dilemas de los estudios de las colombias negras*. Popayán: Editorial Universidad del Cauca.
———. 2004. "Un océano verde para extraer aceite: hacia una etnografía del cultivo de la palma africana en Tumaco." *Universitas Humanística* 38(58): 72–81.
———. 2002. "Memories, Identities, and Ethnicity: Making the Black Community in Colombia." MA thesis, Department of Anthropology, University of North Carolina, Chapel Hill.
———. 2001. "Imaginando comunidad negra: Etnografía de la etnización de las poblaciones negras en el Pacífico sur colombiano." In M. Pardo, ed. *Acción colectiva, estado y etnicidad en el Pacífico colombiano*, 41–70. Bogotá: ICANH/ Colciencias.
———. 1996a. "Economía y simbolismo del Pacífico negro." B.A. thesis, Universidad de Antioquia, Medellín.
———. 1996b. "Los tuqueros negros del Pacífico sur colombiano." In E. Restrepo and Jorge I. del Valle, eds., *Renacientes del Guandal*, 243–350. Bogotá: Universidad Nacional/Biopacífico.
———. 1996c. "Cultura y biodiversidad." In A. Escobar and A. Pedrosa, eds., *Pacífico: ¿Desarrollo o diversidad?* 220–44. Bogotá: CEREC/Ecofondo.
Restrepo, Eduardo, and Arturo Escobar. 2005. "'Other Anthropologies and Anthropology Otherwise': Steps to a World Anthropologies Framework." *Critique of Anthropology* 25(2): 99–128.
Restrepo, Eduardo, and Alejandro Rojas, eds. 2004. *Conflicto e (in)visibilidad: retos de los estudios de la gente negra en Colombia*. Popayán: Editorial de la Universidad del Cauca.
Restrepo, Eduardo, and Jorge I. del Valle, eds. 1996. *Renacientes del Guandal*. Bogotá: Proyecto Biopacífico/Universidad Nacional.
Rhoades, Robert, and Virginia Nazarea. 1999. "Local Management of Biodiversity in Traditional Agroecosystems." In W. Collins and C. Qualset, eds., *Biodiversity in Agroecosystems*, 215–36. Boca Raton: CRC Press.
Ribeiro, Gustavo Lins. 2000. *Cultura e política no mundo contemporaneo*. Brasilia: Editora UNB.
———. 1998. "Cybercultural Politics: Political Activism at a Distance in a Transnational World." In Sonia Álvarez, Evelina Dagnino, and Arturo Escobar eds., *Cultures of Politics/Politics of Cultures: Re-visioning Latin American Social Movements*, 325–52. Boulder: Westview Press.
Ribeiro, Gustavo, and Arturo Escobar, eds. 2006. *World Anthropologies: Disciplinary Transformations within Systems of Power*. Oxford: Berg.
Riles, Annelise. 2000. *The Network Inside Out*. Ann Arbor: University of Michigan Press.
Rivas, Jaime. 2001. "Gente Entintada y Parlante del Pacífico colombiano." Investigative report, Fundación Renacientes, Cali.

Riveros, Daniel C. 1996. *Evaluación de sistemas productivos en el río Cajambre, Pacífico vallecaucano.* Report prepared for the Organization for the Defense of the Cajambre River, CDINCA.

Robledo, Jorge. 2000. *Balance y perspectivas.* Bogotá: El Ancora Editores.

Rocheleau, Dianne. 2000. "Complex Communities and Relational Webs: Stories of Surprise and Transformation in Machakos." Presented at workshop "Communities, Uncertainty and Resources Management," Institute of Development Studies, Sussex, November 6–8.

———. 1995a. "Environment, Development, Crisis and Crusade: Ukambani, Kenya, 1890–1990." *World Development* 23(6): 1037–51.

———. 1995b. "Maps, Numbers, Text, and Context: Mixing Methods in Feminist Political Ecology." *Professional Geographer* 47(4): 458–66.

Rocheleau, Dianne, and L. Ross. 1995. "Trees as Tools, Trees as Text: Struggles over Resources in Zambrana-Chacuey, Dominican Republic." *Antipode* 27(4): 407–28.

Rocheleau, Dianne, Barbara Thomas-Slater, and Esther Wangari, eds. 1996. *Feminist Political Ecology.* New York: Routledge.

Rojas, Cristina. 2002. *Civilization and Violence.* Minneapolis: University of Minnesota Press.

Rojas, Jeannette. 1996. "Las mujeres en movimiento: Crónicas de otras miradas." In A. Escobar and A. Pedrosa, eds., *Pacífico: ¿Desarrollo o diversidad?* 205–19. Bogotá: ICANH/CEREC

Romero, Mario Diego. 1997. *Historia y etnohistoria de las comunidades Afrocolombianas del Río Naya.* Cali: Gobernación del Valle del Cauca.

———. 1995. *Poblamiento y sociedad en el Pacífico Colombiano, siglos XVI al XVIII.* Cali: Universidad del Valle.

Rosenberry, Robert. 1997. *World Shrimp Farming 1997.* San Diego: Shrimp News International.

Rosero, Carlos. 2002. "Los afrodescendientes y el conflicto armado en Colombia: La insistencia en lo propio como alternativa." In Claudia Mosquera, Mauricio Pardo, and Odile Hoffmann, eds., *Afrodescendientes en las Américas: Trayectorias sociales e identitarias,* 547–60. Bogotá: Universidad Nacional/ICANH.

Routledge, Paul. 2003. "Convergence Space: Process Geographies of Grassroots Mobilization Networks." *Transactions of the Institute of British Geography* NS 28: 333–49.

Ruiz, Manuel. 2004. "Access to Genetic Resources, Intellectual Property Rights and Biodiversity: Processes and Synergies." Gland, Switzerland: IUCN (Policy, Biodiversity and International Agreements Unit).

Ruiz-Palma, Javier. 1998. *Aprovechamiento de la fauna silvestre en las comunidades negras del transecto Naya (ríos Cajambre, Yurumanguí y Naya), Fase II.* Buenaventura: PBP.

Ruiz-Palma, Javier, Aníbal Moya, and Orlando Moya. 1997. *Encuentro-taller de indígenas Wounaan sobre cacería y alternativas de manejo de fauna silvestre.* Cali: PBP/Comunidades Wounaan del Bajo San Juan.

Sachs, Wolfgang, ed. 1992. *The Development Dictionary: A Guide to Knowledge as Power*. London: Zed Books.

Salgado, Sebastião. 2000. *Éxodos*. Madrid: Fundación Retevisión.

Sánchez, Enrique. 1998. *Los sistemas productivos tradicionales: Una opción propia de desarrollo sostenible. Proyecto Biopacífico. Informe Final General*. Volume 4. Bogotá: PBP.

———. 1996. "La conservación de la biodiversidad y gestión territorial de las comunidades negras." In W. Villa, ed., *Comunidades negras: territorio y desarrollo*, 189–212. Bogotá: SWISSAID.

Sánchez, Enrique, and Claudia Leal. 1995. "Elementos para una Evaluación de Sistemas Productivos Adaptativos en el Pacífico Colombiano" In Claudia Leal, ed., *Economías de las Comunidades Rurales en el Pacfico Colombiano*, 73–88. Bogotá: Proyecto Biopacífico.

Sánchez, Enrique, and Roque Roldán. 2001. *Titulación de los territorios comunales afrocolombianos e indígenas en la Costa Pacífica de Colombia*. Dirección sectorial para el desarrollo social y ecológicamente sostenible, Préstamo No. 3692-CO. Washington: World Bank.

Santos, Boaventura de Sousa. 2004. "The World Social Forum: Towards a Counter-Hegemonic Globalization (Part I)." In J. Sen, A. Anand, A. Escobar, and P. Waterman, eds., *The World Social Forum: Challenging Empires*, 235–45. Delhi: Viveka.

———. 2002. *Towards a New Legal Common Sense*. London: Butterworth.

———. 1992. "A Discourse on the Sciences." *Review* 15(1): 9–47.

Schech, Susanne, and Jane Haggis. 2000. *Culture and Development: A Critical Introduction*. Oxford: Blackwell.

Scheller, Mimi. 2001. *The Mechanisms of Mobility and Liquidity: Re-thinking the Movement in Social Movements*. Department of Sociology, Lancaster University. See www.comp.lancs.ac.uk/sociology/soc076ms.html

Schild, Verónica. 1998. "New Subjects of Rights? Women's Movements and the Construction of the Citizenship in the 'New Democracies.'" In Sonia Álvarez, Evelina Dagnino, and Arturo Escobar, eds., *Cultures of Politics/Politics of Cultures: Re-visioning Latin American Social Movements*, 93–117. Boulder: Westview Press.

Schiwy, Freya. 2005a. "La Otra Mirada: Video Indígena y Descolonización." *Revista del audiovisual* 8 (on-line journal). http://www.miradas.eictv.co.cu/index.php (23-page typescript).

———. 2005b. "Entre Multiculturalidad e Interculturalidad: Video indígena y la descolonización del pensar." In Josefa Salmón, ed., *Construcción y poética del imaginario boliviano*, 127–47. La Paz: Plural.

———. 2003. "Decolonizing the Frame: Indigenous Video in the Andes." *Framework* 44(1): 116–32.

———. 2002. "Intelectuales Subalternos? Notas sobre las dificultades de pensar un diálogo intercultural." In Catherine Walsh, Freya Schiwy, and Santiago Castro-Gómez, eds., *(In)disciplinar las ciencias sociales*, 101–34. Quito: Abya-Yala.

Schmink, Marianne, and Charles Wood. 1987. "The 'Political Ecology' of Amazonia." In P. Little and M. Horowitz, eds., *Lands at Risk in the Third World*, 38–57. Boulder: Westview Press.

Scott, David. 1999. *Refashioning Futures: Criticism After Postcoloniality*. Princeton: Princeton University Press.

Scott, James C. 1998. *Seeing Like a State*. New Haven: Yale University Press.

Sharp, William. 1970. *Forsaken But for Gold: An Economic Study of Slavery and Mining in the Colombian Choco, 1680–1810*. Ann Arbor: University Microfilms International.

Shiva, Vandana. 2000. *Stolen Harvest: The Hijacking of the Global Food Supply*. Boston: South End Press.

———. 1997. *Biopiracy: The Plunder of Nature and Knowledge*. Boston: South End Press.

———. 1993. *Monocultures of the Mind*. London: Zed Books.

Sinha, Subir. 2006a. "Trans-national Development Regimes: Towards a Conceptual Framework." Unpublished paper, SOAS, University of London.

———. 2006b. "Lineages of the Developmentalist State: Transnationality and Village India, 1900–1965." Unpublished paper, SOAS, University of London.

Sivaramakrishnan, K., and Arun Agrawal, eds. 2003. *Regional Modernities: The Cultural Politics of Development in India*. Stanford: Stanford University Press.

Slater, Candace, ed. 2003. *In Search of the Rainforest*. Durham: Duke University Press.

Slater, David. 2004. *Geopolitics and the Postcolonial*. Oxford: Blackwell.

Smith, Dorothy. 1987. *The Everyday World as Problematic: A Feminist Sociology*. Boston: Northeastern University Press.

Smith, Jackie, Charles Chatfield, and Ron Pagnucco, eds. 1997. *Transnational Social Movements and Global Politics*. Syracuse: Syracuse University Press.

Sodikoff, Genese. 2005. "Reverse Labor: A Moral Economy of Conservation in Madagascar." PHD diss., Department of Anthropology, University of Michigan.

Solé, Ricard, and Brian Goodwin. 2000. *Signs of Life: How Complexity Pervades Biology*. New York: Basic Books.

Soulé, Michael, and Gary Lease, eds. 1995. *Reinventing Nature? Responses to Postmodern Deconstruction*. Washington, D.C.: Island Press.

Spinosa, Charles, Fernando Flores, and Hubert Dreyfus. 1997. *Disclosing New Worlds*. Cambridge, Mass.: MIT Press.

Stemper, David, and Héctor Salgado. 1995. "Local Histories and Global Theories in Colombian Pacific Coast Archaeology." *Antiquity* 69(263): 248–69.

Steward, Julian. 1968. "Cultural Ecology." In *International Encyclopedia of the Social Sciences*, 4:337–44. New York: Macmillan.

Strathern, Marilyn. 1998. "Cultural Property and the Anthropologist." Paper presented at Mt. Holyoke College, December 8.

———. 1996a. "Cutting the Network." *Journal of the Royal Anthropological Institute* (n.s.) 2: 517–35.

———. 1996b. "Potential Property: Intellectual Rights in Property and Persons." *Social Anthropology* 4(1): 17–32.
———. 1992. *Reproducing the Future: Essays on Anthropology, Kinship, and the New Reproductive Technologies*. New York: Routledge.
———. 1991. *Partial Connections*. New York: Rowman and Littlefield.
———. 1988. *The Gender of the Gift*. Berkeley: University of California Press.
———. 1980. "No Nature, No Culture: The Hagen Case." In C. MacCormack and M. Strathern, eds., *Nature, Culture, and Gender*, 174–222. Cambridge: Cambridge University Press.
Sturgeon, Noel. 1997. *Ecofeminist Natures*. New York: Routledge.
Summer, Kay, and Harry Halpin. 2005. "The End of the World as We Know It." In D. Harvie, K. Milburn, B. Trott, and D. Watts, eds., *Shut Them Down! The G8, Gleneagles 2005 and the Movement of Movements*, 351–60. Brooklyn: Autonomedia.
Swanson, Timothy. 1997. *Global Action for Biodiversity*. London: Earthscan/IUCN.
Swyngedouw, Erik. 1998. "Homing In and Spacing Out: Re-configuring Scale." In H. Gebhart, ed., *Europa im Globalisieringsporzess von Wirtschaft und Gesellschaft*, 81–100. Stuttgart: Franz Steiner Verlag.
———. 1997. "Neither Global nor Local: Glocalisation and the Politics of Scale." In K. Cox, ed., *Spaces of Globalization: Reasserting the Power of the Local*, 137–66. New York: Guilford Press.
Sylvester, Christine. 1999. "Development Studies and Postcolonial Studies: Disparate Tales of the Third World." *Third World Quarterly* 20(4): 703–21.
Takacs, David. 1996. *The Idea of Biodiversity*. Baltimore: Johns Hopkins University Press.
Tapia, Carlos, Rocío Polanco, and Claudia Leal. 1997. *Los sistemas productivos de la communidad negra del río Valle, Bahía Solano, Chocó*. Bogotá: PBP/Natura.
Tarrow, Sydney. 1994. *Power in Movement: Social Movements, Collective Action, and Politics*. Cambridge: Cambridge University Press.
Taussig, Michael. 2000. "Special Effects: Heat (In Age of Global Warming)." Presented at Conference "Special Effects," Stanford University, February.
———. 1987. *Shamanism, Colonialism, and the Wild Man*. Chicago: University of Chicago Press.
———. 1980. *The Devil and Commodity Fetishism in South America*. Chapel Hill: University of North Carolina Press.
Taylor, Charles. 2004. *Modern Social Imaginaries*. Durham: Duke University Press.
Terranova, Tiziana. 2004. *Network Culture*. London: Pluto Press.
Thom, René. 1975. *Structural Stability and Morphogenesis*. London: Benjamin/Cummins.
Tilley, Christopher. 1994. *A Phenomenology of Landscape*. Oxford: Berg.
Tirado, Cristina. 1998. *Langostino tropical: Impactos medioambientales y socioeconómicos de la acuicultura del langostino tropical*. Madrid: Greenpeace.
Toledo, Víctor. 2005. "Repensar la conservación: áreas naturales protegidas o estrategia bioregional?" *Gaceta Ecológica* 77: 67–83.

———. 2000a. "Indigenous Peoples and Biodiversity." In S. Levin et al., eds., *Encyclopedia of Biodiversity*. Volume 5, 451–64. New York: Academic Press.

———. 2000b. "Ethnoecology: A Conceptual Framework for the Study of Indigenous Knowledge of Nature." Presented at the Seventh International Congress of Ethnobiology, Athens, Georgia, October 2000.

Tomlinson, John. 1999. *Globalization and Culture*. Chicago: University of Chicago Press.

Trinh, Minh-ha T. 1989. *Woman, Native, Other*. Bloomington: Indiana University Press.

Ulloa, Astrid, Heidi Rubio and Claudia Campos. 1996. *Trua Wandra*. Bogotá: OREWA/Fundacion Natura.

UNEP (United Nations Environment Program). 1995. *Global Biodiversity Assessment*. Cambridge: Cambridge University Press.

Uribe, María Victoria. 2004. *Antropología de la inhumanidad: Un ensayo interpretativo sobre el terror en Colombia*. Bogotá: Editorial Norma.

Uribe, María Victoria, and Eduardo Restrepo, eds. 1997. *Antropología en la modernidad*. Bogotá: ICANH.

Urrea, Fernando, and Alfredo Vanín. 1995. "Religiosidad popular no oficial alrededor de la lectura del tabaco: Instituciones sociales y procesos de modernidad en poblaciones negras de la Costa Pacífica colombiana." *Boletín Socioeconómico* CISDE, no. 28.

Vaneigen, Raoul. *The Movement of the Free Spirit*. New York: Zone Books.

Vanín, Alfredo. 1998. *Islario*. Cali: Ediciones Pájaro del Agua.

———. 1996. "Lenguaje y modernidad." In A. Escobar and A. Pedrosa, eds., *Pacífico: ¿Desarrollo o diversidad?* 41–65. Bogotá: CEREC/Ecofondo.

———. 1995. "Naturaleza: abundancia y lejanía en la literatura popular del Pacífico." *El Hilero* 2: 21–22.

———. 1986. *El príncipe Tulicio: Cinco relatos orales del Litoral Pacífico*. Buenaventura: Centro de Publicaciones del Pacífico.

Varela, Francisco. 1999. *Ethical Know-How: Action, Wisdom, and Cognition*. Stanford: Stanford University Press.

———. 1996 [1988]. *Conocer: Las ciencias cognitivas: tendencias y perspectivas*. Barcelona: Gedisa.

Varela, Francisco, Evan Thompson, and Eleanor Rosch. 1991. *The Embodied Mind*. Cambridge, Mass.: MIT Press.

Vargas, Patricia. 1999. "Propuesta metodológica para la investigación participativa de la percepción territorial en el Pacífico." In Juana Camacho and Eduardo Restrepo, eds., *De montes, ríos y ciudades: territorios e identidades de gente negra en Colombia*, 143–76. Bogotá: Ecofondo-Natura-Instituto Colombiano de Antropología.

———. 1993. "Los Embera, los Waunana, y los Cuna." In P. Leyva. ed., *Colombia Pacífico*, 1:292–309. Bogotá: Fondo FEN.

Vattimo, Gianni. 1996. *Creer que se cree*. Barcelona: Paidós.

———. 1991. *The End of Modernity*. Baltimore: Johns Hopkins University Press.
Vayda, Andrew, and Roy Rappaport. 1968. "Ecology, Cultural and Noncultural." In J. A. Clifton, ed., *Introduction to Cultural Anthropology*, 476–97. New York: Houghton Mifflin.
Velásquez, Julie. 2005. "'And the Creator Began to Carve Us from Cocobolo': Culture, History, Forest Ecology, and Conservation among the Wounaan of Eastern Panama." PHD diss., Department of Anthropology and School of Forestry and Environmental Studies, Yale University.
Velásquez, Rogelio. 2000. *Fragmentos de historia, etnografía y narraciones del Pacífico colombiano negro*. Bogotá: ICANH.
———. 1957. "La medicina popular de la costa colombiana del Pacífico." *Revista Colombiana de Antropología* 6: 195–258.
Villa, William. 2001. "La sociedad negra del Chocó: Identidad y movimientos sociales." In M. Pardo, ed., *Acción colectiva, estado y etnicidad en el Pacífico colombiano*, 207–28. Bogotá: ICANH/Colciencias.
———. 1998. "Movimiento social de comunidades negras en el Pacífico colombiano: La construcción de una noción de territorio y región." In Adriana Maya, ed., *Los afrocolombianos*. Geografía Humana de Colombia, 4:431–48. Bogotá: Instituto Colombiano de Cultura Hispánica.
———. 1996. "Ecosistema, territorio y desarrollo." In W. Villa, ed., *Comunidades Negras: Territorio y Desarrollo*, 17–28. Medellín: Editorial Endymión/SWISSAID.
———, ed. 1996. *Comunidades Negras: Territorio y Desarollo*. Medellín: Editorial Endymión/SWISSAID.
Virilio, Paul. 2000. *Politics of the Very Worst*. New York: Semiotext(e).
———. 1990. *Speed and Politics*. New York: Semiotext(e).
———. 1997. *The Open Sky*. New York: Verso.
Vivas-Eugui, David, María Fernanda Espinosa, and Sebastián Winkler. 2004. "Internacional Negotiations on Biodiversity, Genetic Resources, and Intellectual Property: Implications of the WIPO Intergovernmental Committee's New Mandate." Gland, Switzerland: IUCN (Policy, Biodiversity and International Agreements Unit).
Viveros, Mara. 2002a "La imbricación de los estereotipos racistas y sexistas: El caso de Quibdó." In *150 años de la abolición de la esclavización en Colombia: Desde la marginalidad a la construcción de la nación*, 508–29. Bogotá: Aguilar.
Viveros, Mara. 2002b. *De quebradores y cumplidores: Sobre hombres, masculinidades y relaciones de género en Colombia*. Bogotá: CES/Universidad Nacional.
Viveros, Mara, Claudia Rivera, and Manuel Rodríguez, eds. 2006. *De mujeres, hombres y otras ficciones: Género y sexualidad en América Latina*. Bogotá: Tercer Mundo/Universidad Nacional.
von Foerster, Heinz. 1981. *Observing Systems*. Seaside, Calif.: Intersystems Publications.
Vuola, Elina. 2002. *Ethics of Liberation: Feminist Theology and the Limits of Poverty and Reproduction*. London: Sheffield Academic Press.

———. 2000. "Thinking Otherwise: Dussel, Liberation Theology, and Feminism." In L. Alcoff and E. Mendieta, eds., *Thinking from the Underside of History*, 149–80. Lanham: Rowman and Littlefield.

Wade, Peter. 2002. "Introduction. The Colombian Pacific in Perspective." *Journal of Latin American Anthropology* 7(2): 2–33.

———. 1995. "The Cultural Politics of Blackness in Colombia." *American Ethnologist* 22(2): 341–57.

———. 1997. *Race and Ethnicity in Latin America*. London: Pluto Press.

———. 1993. *Blackness and Race Mixture in Colombia*. Baltimore: Johns Hopkins University Press.

Wakker, E. 1998. *"Lipstick Traces in the Rainforest": Palm Oil, Crisis and Forest Loss in Indonesia*. Report to the World Wildlife Fund, Germany, http://forests.org/archive/indomalay/oilpalm.txt.

Wallerstein, Immanuel. 2000. "Globalization, or the Age of Transition? A Long-Term View of the Trajectory of the World System." *International Sociology* 15(2): 249–65.

Walsh, Catherine. 2007. "Shifting the Geopolitics of Critical Knowledge: Decolonial Thought and Cultural Studies 'Others' in the Andes." *Cultural Studies* 21(2–3): 224–39.

———, ed. 2005. *Pensamiento critico y matriz (de)colonial*. Quito: Universidad Andina Simón Bolívar/Abya-Yala.

———, ed. 2003. *Estudios culturales latinoamericanos: Restos desde y sobre la región andina*. Quito: Universidad Andina Simón Bolívar/Abya-Yala.

Walsh, Catherine, Freya Schiwy, and Santiago Castro-Gómez, eds. 2002. *Indisciplinar las ciencias sociales*. Quito: Universidad Andina Simón Bolívar/Abya-Yala.

WAN Collective. 2003. "A Conversation about a World Anthropologies Network." *Social Anthropology* 11(2): 265–69.

Waterman, Peter. 2004. "The Global Justice and Solidarity Movement and the World Social Forum: A Backgrounder." In J. Sen, A. Anand, A. Escobar and P. Waterman, eds., *The World Social Forum: Challenging Empires*, 55–63. Delhi: Viveka.

West, Paige. 2006. *Conservation Is Our Government Now: The Politics of Ecology in Papua New Guinea*. Durham: Duke University Press.

West, Robert. 1957. *The Pacific Lowlands of Colombia: A Negroid Area of the American Tropics*. Baton Rouge: Louisiana State University Press.

Whitten, Norman. 1986 [1974]. *Black Frontiersmen: Afro-Hispanic Culture of Ecuador and Colombia*. Prospect Heights, Ill.: Waveland Press.

Wilshusen, Peter. 2006. "Territory, Nature, and Culture: Negotiating the Boundaries of Biodiversity Conservation in the Colombian Pacific Coast Region." In Steve Brechin, Peter Wilshusen, Crystal Fortwrangler, and Patrick West, eds. *Contested Nature: Promoting International Biodiversity with Social Justice in the Twenty-First Century*, 73–88. Albany: SUNY Press.

Wilson, Edward O. 1995. *Naturalist*. New York: Warner Books.

———. 1993. *The Diversity of Life*. New York: W. W. Norton.

Wilson, Edward O., and Dan Perlman. 2000. *Conserving Earth's Biodiversity*. Washington, D.C.: Island Press (CD-ROM).

Winograd, Terry, and Fernando Flores. 1986. *Understanding Computers and Cognition*. Norwood, N.J.: Ablex.

Woodhouse, Edward, David Hess, Steve Breyman, and Brian Martin. 2002. "Science Studies and Activism: Possibilities and Problems for Reconstructivist Agendas." *Social Studies of Science* 32(2): 297–319.

Wouters, Mieke. 2001. "Derechos étnicos bajo el fuego: el movimiento campesino negro frente a la presión de grupos armados en el Chocó." In M. Pardo, ed., *Acción colectiva, estado y etnicidad en el Pacífico colombiano*, 259–85. Bogotá: ICANH/Colciencias.

WRI (World Resources Institute). 1993. *Biodiversity Prospecting*. Oxford: Oxford University Press.

———. 1994. *World Resources, 1994–95*. Washington: WRI.

WRI/IUCN. 1996. *Global Biodiversity Forum 1996*. Gland, Switzerland: IUCN.

WRI/IUCN/UNEP. 1992. *Global Biodiversity Strategy*. Washington: WRI.

Yacup, Sofonías. 1934. *Litoral Recóndito*. Bogotá: Editorial Renacimiento.

Yang, Mayfair. 2000. "Putting Global Capitalism in Its Place." *Current Anthropology* 41(4): 477–509.

Yehia, Elena. 2006. "Decolonizing Knowledge and Practice: An Encounter between the Latin American Modernity/Coloniality/Decoloniality Research Program and Actor Network Theory." *Journal of the World Anthropologies Network* 1(2): 91–108.

Zamocs, León. 1986. *The Agrarian Question and the Peasant Movement in Colombia*. Cambridge: Cambridge University Press.

index

Pages in italics indicate figures and tables.

Arturo Escobar is Kenan Distinguished Teaching Professor of Anthropology at the University of North Carolina at Chapel Hill. His books include *Más allá del Tercer Mundo: Globalización y Diferencia* (2005), *El Final del Salvaje* (1999), and *Encountering Development: The Making and Unmaking of the Third World* (1995). He has coedited (with Sonia Alvarez) *The Making of Social Movements in Latin America: Identity, Strategy and Democracy* (1992); (with Sonia Alvarez and Evelina Dagnino) *Cultures of Politics/ Politics of Cultures: Revisioning Latin American Social Movements*; (with Wendy Harcourt) *Women and the Politics of Place* (2005); and (with Gustavo Lins Ribeiro) *World Anthropologies: Disciplinary Transformations within Systems of Power* (2006).

Library of Congress Cataloging-in-Publication Data
Escobar, Arturo, 1951–
Territories of difference : place, movements, life, *redes* / Arturo Escobar.
p. cm — (New ecologies for the twenty-first century)
Includes bibliographical references (p.) and index.
ISBN 978-0-8223-4344-8 (cloth : alk. paper)
ISBN 978-0-8223-4327-1 (pbk : alk. paper)
1. Social movements—Colombia—Pacific Coast—Case studies.
2. Regionalism—Colombia—Pacific Coast.
3. Proceso Nacional de Comunidades Negras.
4. Blacks—Colombia—Pacific Coast—Politics and government.
5. Pacific Coast (Colombia)—Social conditions.
6. Pacific Coast (Colombia)—Politics and government. I. Title.
HN310.P33E83 2008
306.09861'5—dc22 2008028958